# GLOBALIZING POLAR SCIENCE

PALGRAVE STUDIES IN THE HISTORY OF
SCIENCE AND TECHNOLOGY

James Rodger Fleming (Colby College) and Roger D. Launius (National Air and Space Museum), Series Editors

This series presents original, high-quality, and accessible works at the cutting edge of scholarship within the history of science and technology. Books in the series aim to disseminate new knowledge and new perspectives about the history of science and technology, enhance and extend education, foster public understanding, and enrich cultural life. Collectively, these books will break down conventional lines of demarcation by incorporating historical perspectives into issues of current and ongoing concern, offering international and global perspectives on a variety of issues, and bridging the gap between historians and practicing scientists. In this way they advance scholarly conversation within and across traditional disciplines but also to help define new areas of intellectual endeavor.

Published by Palgrave Macmillan:

*Continental Defense in the Eisenhower Era: Nuclear Antiaircraft Arms and the Cold War*
By Christopher J. Bright

*Confronting the Climate: British Airs and the Making of Environmental Medicine*
By Vladimir Jankovic

*Globalizing Polar Science: Reconsidering the International Polar and Geophysical Years*
Edited by Roger D. Launius, James Rodger Fleming, and David H. DeVorkin

# Globalizing Polar Science

## Reconsidering the International Polar and Geophysical Years

Edited by
Roger D. Launius,
James Rodger Fleming,
and
David H. DeVorkin

First published in 2010 by
PALGRAVE MACMILLAN®
in the United States—a division of St. Martin's Press LLC,
175 Fifth Avenue, New York, NY 10010.

Where this book is distributed in the UK, Europe and the rest of the world,
this is by Palgrave Macmillan, a division of Macmillan Publishers Limited,
registered in England, company number 785998, of Houndmills,
Basingstoke, Hampshire RG21 6XS.

Palgrave Macmillan is the global academic imprint of the above companies
and has companies and representatives throughout the world.

Palgrave® and Macmillan® are registered trademarks in the United States,
the United Kingdom, Europe and other countries.

ISBN: 978–0–230–10532–4 Hardcover
ISBN: 978–0–230–10533–1 Paperback

Library of Congress Cataloging-in-Publication Data

Globalizing polar science : reconsidering the International Polar and
Geophysical years / edited by Roger D. Launius, James R. Fleming, and
David H. DeVorkin.
     p. cm.—(Palgrave studies in the history of science and technology)
     ISBN 978–0–230–10532–4
     1. Research—Polar Regions—International cooperation—History.
2. Geophysics—International cooperation. I. Launius, Roger D. II. Fleming,
James Rodger. III. DeVorkin, David H., 1944–

Q180.A3G56 2010
509.11—dc22                                                    2010013627

A catalogue record of the book is available from the British Library.

Design by Newgen Imaging Systems (P) Ltd., Chennai, India.

First edition: December 2010

10 9 8 7 6 5 4 3 2 1

Printed in the United States of America.

# Contents

# Part Two   National Case Studies

# Part Three   Networked Personalities and Programs

# Part Four   National Roles for International Science:
# Quests and Questions in the International Geophysical Year

**Part Five   Legacies of Global Science: Space Science,
Anthropology, and Earth Science**

# Illustrations

# Contributors

**Dian Olson Belanger**, independent scholar, is the author of *Deep Freeze: The United States, the International Geophysical Year, and the Origins of Antarctica's Age of Science* (University Press of Colorado, 2006).

**Noel D. Broadbent** is an archaeologist and anthropologist at the National Museum of Natural History, Smithsonian Institution.

**Christopher Carter** is an independent scholar who studies geophysical sciences in the early modern period.

**Erik M. Conway** is a historian at the Jet Propulsion Laboratory, California Institute of Technology, Pasadena, California.

**Philip N. Cronenwett** is a special collections librarian *emeritus* at Dartmouth College.

**Michael Aaron Dennis** is an independent scholar who is involved in the study of cold war science and technology.

**David H. DeVorkin** is a senior curator at the National Air and Space Museum, Smithsonian Institution.

**Steven J. Dick** is the former chief historian for the National Aeronautics and Space Administration.

**James Rodger Fleming** is Professor of Science, Technology and Society at Colby College, Maine.

**Gregory A. Good** is Director, Center for History of Physics, at the American Institute of Physics.

**Adrian Howkins** is Assistant Professor of History, Colorado State University.

**Roger D. Launius** is a senior curator at the National Air and Space Museum, Smithsonian Institution.

**Cornelia Lüdecke** is Professor of Geography at the University of Munich, Germany.

**Lisbeth Lewander** is Associate Professor of political science at Gothenburg University, Sweden.

**Teasel Muir-Harmony** is a Ph.D. candidate in science and technology studies at the Massachusetts Institute of Technology.

**Allan A. Needell** is a curator at the National Air and Space Museum, Smithsonian Institution.

**Marc Rothenberg** is agency historian for the National Science Foundation.

**William R. Stevenson III** is a Ph.D. candidate in modern Japanese history at the University of Hawaii at Manoa.

**Dasan M. Thamattoor** is associate professor of chemistry at Colby College, Maine.

**Zuoyue Wang** is associate professor of history, California State Polytechnic University.

**Jiuchen Zhang** is associate professor, Institute for the History of Natural Science, Chinese Academy of Sciences.

# Permission

This material is based upon work supported by the National Science Foundation under Grant No. 0646619.

# Introduction

# Rise of Global Scientific Inquiry in the International Polar and Geophysical Years

*Roger D. Launius, James Rodger Fleming, and David H. DeVorkin*

The fiftieth anniversary of the International Geophysical Year (IGY) and the opening of a fourth International Polar Year (IPY) in 2007 provided historians and other students of science and technology a unique opportunity to assess the origins and influence of past IPYs as organized international research efforts. An interdisciplinary conference at the Smithsonian Institution in the fall of 2007, in conjunction with the annual meeting of the History of Science Society, explored ways in which these collaborative scientific activities supported the growth of professionalism (both disciplinary and interdisciplinary) in science, and how they promoted international relations of science and of governments. Participation in polar collaborative research efforts left a marked impact on the state of scientific investigation beginning in the late nineteenth century, when the first IPY took place, and more recent efforts from the 1930s to 1950s have extended its effect. They have fostered a new structure for the conduct of investigations, for the analysis of scientific data, and for the incorporation of what has been learned into the corpus of world knowledge. At the same time, these international scientific efforts evoked a healthy dose of nationalism, imperialism, and diplomatic necessity. The merger of national interest with scientific curiosity represented one of the most significant outgrowths of these collaborations. As historian Allan A. Needell wrote about the 1957–58 collaboration (although it may also be said of the earlier IPYs): "The IGY's claim for government support was substantially bolstered by its potential value as a vehicle of foreign policy, as a means of providing certain information required by the military, and as a source of scientific and technical intelligence."[1] At the same time it fostered changes in the manner in which the community conducted science, generated ideas, and transmitted understanding.

The polar years of 1882–83, 1932–33, and 1957–58 are often described by scientists as initiatives that "have had a major influence in overhauling our understanding of global processes."[2] They have also been hailed as exemplars of the universality of science; observers of science from many disciplines are quick to contend that "The truths which scientists seek to discover are not national truths; they are the same everywhere and so can be universally recognized." But some have also explored the many compromises that, leaving the ideals of science aside, tested the political realities of its practice.[3] For one, sociologist Ian Mitroff concluded that most scientists' ideas about their subject reflected to some degree their individual preconceptions in the face of evidence. They tended to cling to ideas, despite evidence to the contrary, and only grudgingly modified their theories to hold on to their ideas. This suggests that any discovery of universal scientific truths is a much more problematic issue than the practitioners are willing to admit.[4]

The first IPY resulted, as James Rodger Fleming recently noted, from "the ideas of the Austrian naval officer and polar explorer Karl Weyprecht (1838–81) and the organizational skills of Georg von Neumayer (1826–1909), director of the German Hydrographical Office, along with Heinrich Wild (1833–1902), director of the Central Physical Observatory in St. Petersburg."[5] The IPY eventually involved 11 countries placing 12 stations around the North Pole and two near the South Pole to monitor seasonal changes systematically. The second IPY, proposed during an international conference of directors of national meteorological services, was also largely Arctic and encompassed meteorology, Earth magnetism, and aurorae, applying the new technologies of radio science, and air reconnaissance and transport. The IGY was suggested at first by an informal group of elite scientists who then engaged national institutions and international scientific organizations to mount a hugely ambitious worldwide survey involving many disciplines, tens of thousands of scientists, and more than 60 countries. It was punctuated by the Soviet Union's spectacular launch of Sputnik in October 1957, which was answered by the United States in January 1958, when it launched Explorer 1. In terms of historical writing and popular media coverage, and especially in the public mind, this space race has almost totally eclipsed the extraordinary collective activities of the IGY in more traditional realms. To date, no history has adequately addressed how the eclipse of the IGY by Sputnik has influenced how humanity remembers the equally dramatic activities of global scientific investigation. Several essays in this volume explore this and other salient issues arising from the experiences gained during the IPYs and the IGY and how they relate one to another. Always, a primary goal is to gain additional historical insight into the forces that shaped the modern Earth sciences.[6]

Even though the IPYs and the IGY arose from varied motivations and interests, many of them far removed from the pursuit of basic science, they proved a true success from the standpoint of advancing scientific understanding about Earth. While the IPYs and the IGY advanced knowledge and shaped its interdisciplinary understanding, both by scientists themselves and the broader public, the details of these endeavors are less well-known. Even among historians and scientists dedicated to the study of geophysics and related disciplines, understanding of the IPYs and the IGY remains rather superficial. Accordingly, we took the occasion of the fiftieth anniversary of the IGY as an opportunity to organize, at the Smithsonian Institution, a conference at which most of the essays collected in this volume were first presented. Each essay demonstrates how the goals and methods of scientific research were shared across political and scientific boundaries, and that there was much to be gained from the exploitation of shared commitments and values. The essays also illuminate important tensions within the IPY and IGY activities. They not only focus on the science produced in each of the international efforts, but also on the nationalistic, economic, and military motivations and implications that helped propel and, in some cases, hinder those undertakings.

The overarching theme of this volume is an exploration of the role of the successive International Polar Years and the International Geophysical Year in the emergence of a global sense of geoscience, embedded as it is in transnational scientific organizations and competing national interests. A. Hunter Dupree may have been the first historian of science to recognize and articulate the inherent interest that governments of all forms have in campaigns of exploration, and his observations have been extended and refined by others who demonstrated that, whether or not they were the ones who paid the bills, governments still benefited—through such contributions as infrastructure and security—from the activity and facilitated research. The essays in this volume build on this picture by examining how the IPYs and the IGY stimulated professionalization of the various disciplines and the emergence of an interdisciplinary sense of identity within the geosciences, how they elucidated national styles, how they fostered the development of scientific instrumentation, and how they focused patronage for science.[7]

The first IPY campaign involved meteorology, oceanography, glaciology, geomagnetism, auroral observations, and latitude/longitude determinations—later expanded in the second IPY to include solar activity and studies of cosmic rays and the ionosphere and in the IGY to geodesy and magnetospherics. Some of the expeditions also returned significant collections of great ethnological importance and stimulated much new interest in that field. Synoptic data collection spread temporally and spatially as instrumental techniques and opportunities expanded, providing scientists with the means to explore the Earth as an integrated system and historians with a rich opportunity to compare and assess the sources, evolution, and application of diverse technologies in the pursuit of scientific knowledge. In part this resulted from a convergence of scientific questioning and technological prowess in developing instruments that could pursue those questions within the context of a unique political situation, and especially within the context of a *Weltanschauung* that emphasized scientific understanding as a core value in modernity. As one example, scholars now ask how the state of technology has influenced the transition of the IPY and IGY efforts from exploration to field work—from temporal excursions into the unknown to the establishment of calibrated networks of stations collecting and archiving comparable data. Doing normal science in the field requires an infrastructure to be in place, or transported, whereas exploration tends to stretch things like supply lines and communications links. Several scholars publishing in this volume ask the question: To what extent did the evolution and maturation of new technologies, such as radio and rocketry, inspire and shape the IPY and IGY efforts? Alternatively, did those campaigns stimulate new technologies for exploration? And how might one creatively situate the scientific goals in a push-pull relationship with technology?

Very different conceptions, methodologies, and technologies supported the first two IPYs from the IGY of the 1950s. Each of them took full advantage of the capabilities available. Therefore, this discussion of the IPYs and the IGY capitalize on an ideal opportunity to take a broader look at the relationship between technology and science, between physics and metaphysics, especially the question: Does technology merely enable science, or does it define science also? There is already considerable historical work on this subject, but the essays in this volume add important case studies of global cooperative scientific efforts.[8] Put another way: In what ways were the questions asked in these international campaigns time-bound and constrained by the scientific, technological, and political imagination, and in what ways were they visionary, setting goals beyond extant capabilities and perhaps overreaching? These essays also address the core issue of how the IPY and IGY campaigns contributed to knowledge-growth concerning Earth as a system and the interactions of physical, biological, and human components.

Since the first IPY in 1882–83, these collaborative scientific activities may appropriately be viewed as part of an ongoing effort by scientists to apply the latest technologies to measure and monitor remote and extreme environments; to incorporate that knowledge into a broader capability to share, archive, and compare data, and to address social problems and contribute to human welfare. The interrelatedness of science, technology, politics, economics, and culture in the IPYs and the IGY offers a uniquely valuable perspective for investigating these issues in larger contexts. The findings from this volume's essays are especially relevant because they present a manageable set of test cases—historical moments or "snapshots" that are amenable to comparison—in which to explore an array of questions concerning science and its pursuit in the industrial world. The nineteen essays in this volume, contributed by historians and historically oriented social and natural scientists, address questions relating to the IPYs and the IGY that involve the social, military, and political implications of international science, the intellectual and technological claims and aspirations of global geoscience, and (crucially important but often marginalized) the role of the ethno-sciences. The result is a search for commonalities and tensions among diverse scholars and disciplines to bring to the fore a coherent view of the historical importance of these international campaigns and their relevance to the present. These essays emphasize the interplay of individual and institutional, national and international, and scientific and technological dimensions.

This volume is divided into five major sections, each with either three or four essays therein. The first section, "Meanings, Interpretations, and Historiography," provides an overview of the volume as a whole. In the first essay, Michael Aaron Dennis offers wide-ranging comments on the meaning of scientific pursuit in a global, interlocking context. He finds the IPYs and the IGY represent a fundamental act of "disciplining of the Earth to scientific ends wherein science and scientists work to make the globe safe for both science and humanity." As Marc Rothenberg makes clear in his essay, although collaborative scientific efforts did not originate with the IPYs, they raised the bar in what could be accomplished by a succession of coordinated, related, and interlocking activities. Philip N. Cronenwett's contribution investigates the manner in which the data collected during the first IPY was disseminated, or in many cases not disseminated, and how that experience influenced later efforts to improve data-sharing and use. Finally, Roger D. Launius's essay offers an outline of the historiography of the IPYs and the IGY, seeking to place them in the context of larger historical interpretations of the industrial age.

The second section, on "National Case Studies," illuminates efforts by the UK, Sweden, Japan, and China concerning the IPYs and IGY. Christopher Carter's contribution points out that between the Congress of Vienna (1815) and the beginning of World War I (1914), and then only during periods of relative European tranquility, could scientists and planners find the necessary social and political support necessary to ensure international collaboration. The first IPY enjoyed success in no small measure because of the opening of a period of relative peace after the Congress of Berlin in 1878. A valuable lesson from this essay is the relative tenuousness of opportunities for widespread international cooperation. The essay by Lisbeth Lewander provides a much-needed gendered perspective on this discussion by focusing on the experience of Sweden in what most considered a masculine enterprise at the poles. William R. Stevenson III offers a unique set of observations about the significance of the IPYs in Japan, reporting that the IPYs "have never been free from larger historical trends." During the first IPY, Japan's national ambitions ran contrary to the core spirit of international cooperation. Having reached world-power status by the time of the second IPY, Japan sought to enhance its imperial position through its role in that collaborative effort. Another Asian power, the People's Republic of China, desired participation in the IGY, but withdrew in 1957 after Taiwan became involved. The study by Zuoyue Wang and Jiuchen Zhang presents a unique understanding of the two-China dilemma as it relates to the nature of science and transnational issues. Yet there is still much to be done, with some 50 nations not represented in this volume, most notably, perhaps, the Soviet Union.

"Networked Personalities and Programs," the third section, highlights the role of individuals in advancing international scientific efforts. Cornelia Lüdecke's essay underlines the importance of Georg Balthasar von Neumayer, who along with Karl Weyprecht made possible the first IPY in 1882–83. The essays by Gregory A. Good on Sydney Chapman and by Allan A. Needell on Lloyd Berkner profile the essential but sometimes ambiguous roles of these scientific entrepreneurs in the International Geophysical Year. Finally, James Rodger Fleming's study elevates the significance of Harry Wexler's role in the emergence of modern meteorology, concluding, "As international teams of scientists seek evidence of past and ongoing global changes, as satellite sensors and other technologies continually monitor the planet, and as global climate modelers attempt to understand it all, it is historically rewarding to follow the career of Harry Wexler as a central figure and exemplar of the processes that made meteorology the kind of global science it is today."

The fourth section, specifically on the International Geophysical Year, addresses tensions between national politics and transnational scientific efforts and agendas. The four essays in this section examine issues of sovereignty in Antarctica during the IGY and the decision to hold the continent as a preserve for science, free from imperialistic claims. As the essays by Adrian Howkins and Dian Olson Belanger reveal, the science efforts in the IGY were anything but apolitical. Teasel Muir-Harmony presents a case study of the senior partner/junior partner nature of scientific exchanges in East–West relations as it affected the tracking data of scientific satellites. Closing this section, Steven J. Dick unravels the fascinating place of the innovative but obsolescent Markowitz Moon Camera and its role in the IGY and studies of geodesy.

"Legacies of Global Science" concludes the volume with three essays that examine the long-term ramifications of the transnational efforts in the geosciences initiated by the IPYs and the IGY. Historian Erik M. Conway draws the connection between the IGY and the exploration of the planets of the solar system that began on its heels. Anthropologist Noel D. Broadbent lays out the emergence of a major area of investigation in relation to the Saami of the Arctic, which was built squarely on the work of the IPY ethnographers. Closing the volume, scientist Dasan M. Thamattoor examines the personal involvement of Nobel laureate F. Sherwood Rowland in the evolution of our understanding of stratospheric ozone depletion and greenhouse gases since the IGY.

In each of these essays, the authors have been asked to keep in mind the guiding principle that good history is the foundation for good public policy. Historical study is not for historians alone. Our desire is that this collective effort by scholars to examine the efficacy of past polar years may foster a more robust, healthy, and capable scientific infrastructure. We hope that these historical insights may prove useful to policy makers, program planners, and anyone interested in polar, international, or global science. We further hope that conferences and studies like this will lead to further investigations, not just of additional details, important as they may be, but to works of overarching synthesis and relevance.

One of the great surprises as we undertook this conference and collected work was the realization that there is no systematic overall history of the IPYs and the IGY. A serious contribution remains to be made by any scholar completing a useful synthesis of these international scientific endeavors. More surprising, perhaps, was the lack of sophisticated historical exploration—at least in English—of such critically important national efforts such as those of Russia/Soviet Union, the Scandinavian countries, Canada, the United States, and nations at the most southerly latitudes. These types of works represent the building blocks of history and are essential precursors to the completion of successful syntheses. Perhaps other scholars will take up this quest in future years. Perhaps other conferences will probe other core issues relating to the history of the IPYs and the IGY.

## Acknowledgments

Whenever scholars take on a project such as this they stand squarely on the shoulders of earlier investigators and incur a good many intellectual debts. The editors and authors would like to acknowledge the assistance of the institutions and individuals who aided in the preparation of this book. Our greatest debt is to the National Science Foundation, which provided a generous grant that made possible the conference on which this volume is based. We also wish to acknowledge the many people at the Smithsonian Institution and at Colby College who supported this endeavor. The conference was planned at the National Air and Space Museum while one of us (JRF) was a visiting scholar there, and the initial evaluation and editing of the manuscripts were accomplished by all of us during an intense week together at Colby, supported by the Goldfarb Center for Public Policy and Civic Engagement. The History of Science Society coordinated our final plenary session with the opening plenary session of its annual meeting in Crystal City, Virginia. Several individuals at the National Aeronautics and Space Administration and the National Oceanographic and Atmospheric Administration also lent support to this effort, and we thank them for their efforts.

For their many contributions in completing this project, we wish also to thank Jane Odom and her staff archivists at the NASA History Division, who helped track down information and correct inconsistencies, in addition to Steve Dick, Steve Garber, Glen Asner, and Nadine Andreassen at NASA; the staffs of the NASA Library and the Scientific and Technical Information Program, who provided assistance in locating materials; Marilyn Graskowiak and her staff at the NASM Archives; and many archivists and scholars throughout several other organizations. Patricia Graboske, head of publications at the National Air and Space Museum, provided important guidance for this project. Our deep thanks are due to all these fine people.

In addition, we wish to acknowledge the following individuals who aided in a variety of ways: Debbora Battaglia, William E. Burrows, Bruce Campbell, Tom D. Crouch, Gen. John R. Dailey, Jean DeStefano, Ronald Doel, Robert Farquhar, Jens Feeley, James Garvin, Lori B. Garver, John Grant, G. Michael Green, Doria Grimes, Barton C. Hacker, James R. Hansen, Wes Huntress, Peter Jakab, Dennis R. Jenkins, Violet Jones-Bruce, Fae L. Korsmo, Sylvia K. Kraemer, John Krige, Jennifer Levasseur, John M.

Logsdon, Ted Maxwell, W. Patrick McCray, Howard E. McCurdy, Jonathan C. McDowell, Karen McNamara, Valerie Neal, Allan A. Needell, Michael J. Neufeld, Mette Fog Olwig, Robert Poole, Stephen Pyne, Cara Seitchek, Alan Stern, Harley Thronson, and Margaret Weitekamp. Several interns provided assistance at various stages of this project and to them we offer our sincere thanks: Curtis Anderson, Mary Bergen, Meleta Buckstaff, Jessica Kirsch, Alina Naujokaitis, Claire Pope, Heather van Werkhooven, Joni Wilson, and Sam Zeitlin. We also thank Christopher Chappell and the staff of Palgrave Macmillan for their efforts in seeing this book through to publication.

These essays are offered in the spirit of scholarly debate. Not all will agree with everything in them, but we envision that this volume will consolidate and bring to the fore current understandings on this important subject.

Roger D. Launius
James Rodger Fleming
David H. DeVorkin

## Notes

1. Allan A. Needell, *Science, Cold War, and the American State: Lloyd V. Berkner and the Balance of Professional Ideals* (Amsterdam: Harwood Academic Publishers, 2000), 317.
2. See the discussion in "What Is IPY," available online at http://www.ipy.org/about/what-is-ipy.htm (accessed July 24, 2006).
3. Jean-Jacques Saloman, "The Internationale of Science," *Science Studies* 1, no. 1 (1971): 23–42; Nikolai Krementsov, *International Science between the World Wars: The Case of Genetics* (London: Routledge, 2005). A push-back on scientific objectivity is Paul R. Gross and Norman Levitt, *Higher Superstition: The Academic Left and Its Quarrels with Science* (Baltimore, Maryland: Johns Hopkins University Press, 1997).
4. Ian L. Mitroff, *The Subjective Side of Science* (New York: Elsevier, 1974), 54, 132, 164.
5. James Rodger Fleming and Cara Seitchek, "Advancing Polar Research and Communicating Its Wonders: Quests, Questions, and Capabilities of Weather and Climate Studies in International Polar Years," in I. Krupnik, M.A. Lang, and S.E. Miller, eds., *Smithsonian at the Poles: Contributions to International Polar Year Science* (Washington, D.C.: Smithsonian Institution Scholarly Press, 2009), 1–12.
6. Specifically on *Sputnik 1*, see these important works: Robert A. Divine, *The Sputnik Challenge: Eisenhower's Response to the Soviet Satellite* (New York: Oxford University Press, 1993); Rip Bulkeley, *The Sputniks Crisis and Early United States Space Policy: A Critique of the Historiography of Space* (Bloomington: Indiana University Press, 1991); Roger D. Launius, John M. Logsdon, and Robert W. Smith, eds., *Reconsidering Sputnik: Forty Years since the Soviet Satellite* (Amsterdam: Harwood Academic Publishers, 2000); Matt Bille and Erika Lishock, *The First Space Race: Launching the World's First Satellites* (College Station: Texas A&M University Press, 2004); Paul L. Dickson, *Sputnik: The Shock of the Century* (New York: Walker and Co., 2001); and Matthew Brzezinski, *Red Moon Rising: Sputnik and the Hidden Rivalries that Ignited the Space Age* (New York: Times Books, 2007). Other aspects of this are recounted in Martin Collins, ed., *After Sputnik: 50 Years of the Space Age* (New York: Collins, 2007); Asif A. Siddiqi, *Sputnik and the Soviet Space Challenge* (Gainesville: University Press of Florida, 2003); James J. Harford, *Korolev: How One Man Masterminded the Soviet Drive to Beat America to the Moon* (New York: John Wiley & Sons, 1997); Philip Nash, *The Other Missiles of October: Eisenhower, Kennedy, and the Jupiters, 1957–1963* (Chapel Hill: University of North Carolina Press, 1997); Peter J. Roman, *Eisenhower and the Missile Gap* (Ithaca, New York:

Cornell University Press, 1995); Kenneth Osgood, *Total Cold War: Eisenhower's Secret Propaganda Battle at Home And Abroad* (Lawrence: University Press of Kansas, 2006); Homer Hickam, *Rocket Boys* (New York: Delacorte, 1999); Constance McL. Green and Milton Lomask, *Vanguard: A History* (Washington, D.C.: Smithsonian Institution Press, 1971); *Brainpower for the Cold War: The Sputnik Crisis and the National Defense Education Act of 1958* (Westport, Connecticut: Greenwood Press, 1981).

7. A. Hunter Dupree, *Science in the Federal Government: A History of Policies and Activities*, 2nd ed. (Baltimore, Maryland: Johns Hopkins University Press, 1986); William Goetzmann, *Exploration and Empire: The Explorer and the Scientist in the Winning of the American West* (New York: W.W. Norton and Co., 1967); Sharon Gibbs Thibodeau, "Science in the Federal Government," *Osiris* 1 (1985): 81–96.

8. See, as examples, Merritt Roe Smith and Leo Marx, *Does Technology Drive History? The Dilemma of Technological Determinism* (Cambridge, Massachusetts: Massachusetts Institute of Technology Press, 1994); Helen Rozwadowski and David K. van Keuren, eds., *The Machine in Neptune's Garden: Historical Perspectives on Technology and the Marine Environment* (Canton, Massachusetts: Science History Publications, 2004); David H. DeVorkin, *Science with a Vengeance: How the Military Created the US Space Sciences after World War II* (New York: Springer-Verlag, 1992).

# Part One

# Meanings, Interpretations, and Historiography

# Chapter 1

# A Polar Perspective

*Michael Aaron Dennis*

April 14, 1955, began like any other day at the Barbersville, Maryland, trailer park. Located between the nation's capital and Baltimore, the trailer park was a little more than a mile from a newly established Nike antiaircraft missile battery at Fort Meade. Shortly after 12:30 p.m., the day became unusual. Trailer park residents found their quiet, rainy Saturday shattered by a Nike-Ajax missile that escaped its launcher during a drill and slammed into a hillside. A cloud of aluminum fragments from the missile's fuselage fell along the relatively new Baltimore–Washington Parkway; the clarion calls of multiple sirens from the state and military police, in addition to the fire department, became the ambient aural atmosphere. Officials followed the weapons' truncated flight path, collecting debris and storing it in the trunks of their vehicles. From her kitchen window, Mrs. Nancy Bishop watched as space-suited figures in hazardous materials garments collected the fuselage's larger pieces and toxic fuel from the hill behind her manufactured home. By nightfall, all that remained was a pair of holes, one much deeper than the other, and the memories of when the cold war had hit home.[1]

The missile's accidental target was just as dangerous and fanciful as the weapon's ostensible target—a massive Soviet bomber armada that would cross the North Pole, weave through Canadian air space, and fly below U.S. radar coverage before disgorging nuclear munitions on American cities. The cold-war American state saw the Arctic as a potential battle space; consequently, there was a massive investment by the armed services in understanding and mapping the region, whether undersea, on the ice and land, or in the air and space above the pole.[2] To the extent that the polar regions mattered in the cold-war public imagination, they existed as potential sites of conflict, in addition to being exotic and nearly unimaginable sites of adventure. The polar perspective figured most prominently in the symbol of the United Nations, a view of the globe from high atop the North Pole. Surely it is one of the great ironies that Antarctica, the most inhospitable continent, and the northern polar regions have become the most important observing stations for understanding the consequence of anthropogenic climate change.

**Figure 1.1** The Nike-Ajax was an important missile in the early cold war era, used for base defense. First operational in 1953, the Nike-Ajax served as an anti-aircraft missile system until the 1960s. Image of the Nike-Ajax on display at the National Air and Space Museum's Udvar Hazy Center, Chantilly, Virginia. (Photograph by Dane Penland, courtesy of NASM)

Polar bears and penguins, the charismatic actors in the new global theater of climatic destruction, have brought more attention to the polar regions than war or the threat of global thermonuclear war. Charismatic faunas trump individual explorers in the new political economy of polar research.

The most publicly visible polar investigation of the cold war was the collaborative research effort of 1957, the International Geophysical Year (IGY). Although polar research was an essential feature of the IGY, its history has been largely overshadowed by the launch of the first artificial Earth satellite, the Soviet *Sputnik 1*, in October 1957. My charge here is to address this tension between the history of exploration and the history of cold-war science and technology. However, unlike the IGY, the International Polar Years (IPY) of earlier years are not the subject of a vast historical literature outside the work of a small community.[3] But my charge also included presenting some understanding of the tensions involved in the title of this volume, the tension among several unspoken assumptions: that science is by definition global, that science applies the same across the globe in its totality, and that

**Figure 1.2** The logo of the International Geophysical Year, 1957–58

participation in the production of scientific and technical knowledge is, seemingly by popular acclamation, international. From this perspective, science is a sophisticated intellectual version of Esperanto or the universal languages that the heroes of the scientific revolution imagined as instruments of global communications. This title makes the claim that science must be made global, that it is through a rather complicated array of social processes that science becomes global, and that exploration, rather than being the heroic adventure of an earlier generation's historiography, is the complicated extension of state and private institutions (some with state-like interests) into environments and geographic spaces that they desperately try to shape and by which they are themselves shaped. If, as the popular media tells us, we live in the second great age of globalization, the first ending in the trenches of World War I, then we might want to understand how the set of institutions and practices we know of as science are themselves either agents of, or products of, the mysterious process of globalization. In the following remarks, I locate globalization in acts of standardization, discipline, and imperial practices. Following this we move to the transformation of the scientific practices that the cold war produced, not the least of which was a new focus on the instrument as the object of research.

## Making Knowledge Portable

What makes exploration possible? Money is certainly essential, but more often than not we jump to the explorers, such as Ernest Shackleton and Robert Perry, who form the foundation of an earlier generation's historiography. However, there is a much more mundane set of prerequisites for exploration and the production of knowledge. What allows knowledge to move around, whether in a research vessel on its way to the polar regions or in a satellite orbiting Earth, is the adherence by practitioners to standard units of measure and also to practices that are the products of disciplined action. More often than not, standards are embedded in instruments, but even the use of instruments requires a degree of standardized action, as was apparent in discussions of German Antarctic researchers using preprinted forms to record data over time. Standards are more than the material of science; nation-states expend considerable amounts to maintain centers, such as the National Institute of Standards and Technology, formerly known as the National Bureau of Standards. Metrology is both a big business and a vital state function. To use Bruno Latour's metaphor, standards are the railroad tracks on which knowledge moves; without tracks, the train of knowledge is stuck.[4] Or more crudely, knowledge is fragile, and its movement requires a vast amount of engineering—social, material, and political.

Measuring the Earth, especially for the purposes of navigation and improved timekeeping, was an integral aspect of the IGY. As Steve Dick's essay in this volume on the Markowitz Moon camera makes clear, basic issues in geodesy remained open to debate and discussion; the IGY allowed for the introduction of a new method to supplement existing geodetic solutions and produce new knowledge of the Earth's size and shape. Measurements and techniques would simultaneously use existing knowledge to refine those same figures, yielding a more precise measure of the planet. Although Markowitz's camera technique was an apparent failure, the networks in and through which the researchers worked made evident the need for, and the use of, specific standards. As Dick makes clear, the major contribution to global timekeeping from Markowitz's failed camera experiment remains central to the way the world works.

Standards do not exist in a vacuum; they require extensive institutional and political networks in which to operate. Standardization is a form of discipline, and a crucial question is where does the power come from that allows institutions to compel standard behavior and practices? The Nike accident, with which this essay began, emerged from the failure to create an ensemble of practices that would allow Nike operators, all soldiers, to practice safely in the rain. It was a "simple short" in the firing control system caused by the rain that brought the space-suited figures to Nancy Bishop's trailer park.

There are at least two essential prerequisites for such discipline, besides the educational establishments, that produce researchers. First, the presence of individuals located in an array of social and scientific networks. In nearly every essay we find ourselves confronting specific individuals, such as Georg Balthasar von Neumayer, Roald Amundsen, Harald Sverdrup, Erich von Drygalski, Lloyd Berkner, and Sydney Chapman. What unites these disparate individuals is their ability to move within and among a variety of social networks, ranging from the German state and its various bureaucracies to the various military bureaucracies that dominated the technical world of cold-war America. Drygalski's ability to secure funding for polar research in a host of political regimes, ranging from pre–World War I Germany through the Weimar era, demonstrates a level of political ability that few possess even today. Negotiating such diverse contexts meant repackaging polar research to fit new audiences and new demands. Sydney Chapman, the IGY's architect and its most peripatetic participant, also possessed this ability to make multiple nations see a common endeavor as a means of satisfying individual national goals. What is clear from all the essays published here is that Paul Forman's insights about scientific internationalism remain as relevant as ever. Forman argued that Weimar-era physicists used the rhetoric of scientific internationalism to fulfill strictly nationalistic goals; Chapman was the master of this rhetorical jujitsu, allowing nations involved in a cold war to cooperate in the IGY.[5]

As important as are the individual researchers, so, too, are the nation-states that fund and promote the research at the level of the IPY and IGY. In particular, it appears that some form of national motivation is vital to the success of promoting this research. Imperial and military ambitions are essential prerequisites to successful polar investigation. Japan's interest in the first two IPYs was at one with the outlook of the Meiji Restoration and the subsequent national effort to become a major Western power; in turn, Japanese participation in the IGY was part of the American effort to ensure that Japan, rather than China, would emerge as a major scientific power in

the Pacific. Scandinavian dominance in polar exploration was part of powerful national projects to exploit the polar regions for economic and political
power. These projects were most tangible in the form of the Bergen School
in meteorology and various polar expeditions, but also extended to the ways
in which individual members of the expeditions saw their physical conditioning as at one with their intellectual goals.[6] National political goals are not
somehow separable from the scientific research; they are constitutive of it, as
several chapters in this collection make clear.

Standardization does not occur without struggle, just as the railways that
dominated the transformation of the American economy in the nineteenth
century did not develop without significant physical violence. We saw how the
construction of satellite ground stations in India and Japan depended as much
on negotiation as on the impression of standards upon local observers, but we
might learn something from certain historians who are working on the history of the sciences in other times and places. In particular, Simon Schaffer's
work on the dialogue between the South Pacific islanders who adorned their

**Figure 1.3**   The logo of the International Polar Year, 2007–08

bodies with tattoos and the astronomers who rendered their celestial observations and drawings on paper in ink offers us a model of scholarship, linking standards and practices, instruments, inscriptions, and political cultures. Rather than see the "natives" and the British imperial enterprise as utterly dissimilar, Schaffer makes an impressive case that tattoos functioned in much the same way as the astronomer's instruments and notebooks—inscription devices that allowed for the recording and movement of particulars. As he so eloquently concluded, tattoos were part of the technology for the creation of political subjects and, hence, the reproduction of political relations. Many different kinds of inscription devices achieve this end. They were the significant concern of specialist artisans from many different cultures who both contributed to, and relied on, their long networks of astronomy, navigation, and political power.[7]

Imagine applying Schaffer's insight to the location and construction of satellite ground stations. We might come to conclude that negotiation was not the key element, but that a form of coproduction took place in which each party, the SAO and each observation site, brought something to the table and constructed something novel and unprecedented in that local context. Even more intriguing is applying Schaffer's framework to the IGY and the exploration of Antarctica; one might get a very different understanding of how the circulation of particular matters of facts make up our representations of larger entities, such as planets.

## Instruments and Understanding

Writing in *Science* on October 31, 1930, geographer Isaiah Bowman explained that

> The new ideas in polar exploration are not airplane and radio-these are but instruments of discovery. They are astonishingly reliable and useful instruments but they are of mechanical interest only, apart from the ideas they serve. Science feared for a time that they would run away with the game, because the popular mind is still on the romance of flying and the magic of communication by wireless. The really big game of the polar hunt are the scientific ideas or laws upon which the polar regions, and in some cases they alone, can throw light. Science is searching for particular things, not just anything. Real exploration has ceased to be a blind and adventurous wandering into the unknown."[8]

Bowman was concerned about a basic philosophical point dealing with the relationship between science and technology, one privileging science and its ideas over the technologies used for investigation. Although this is a point well worth arguing, our interest in this quote, as Paul Forman has demonstrated, lies in the novelty that the argument would acquire in the postwar era as science increasingly began to take on the character of a larger-scale enterprise.[9] Bowman wants to make it clear that the ideas are far more important than the tools, no matter how popular and exciting the tools. One might argue, however, that Bowman missed an important point—planes were not simply shuttles that transported researchers to the poles, but platforms that became laboratories with wings. What airplanes and radios were to the IPY of 1932, rockets and satellites were to the IGY of 1957. In his work on the use of captured V-2 rockets to explore the upper atmosphere, David DeVorkin confronted this issue when rocket researchers at government military laboratories—such as the Naval Research Laboratory (NRL) and the Johns Hopkins University Applied Physics Laboratory (APL)—attempted to enroll ground-based astronomers in the rocket-research programs. Despite their best efforts, James Van Allen and his team at APL could not protect Jesse Greenstein, a Yerkes Observatory astronomer, from the possibility that his spectrograph might suffer a catastrophic failure. When the instrument failed during its maiden flight, Greenstein retreated to the observatory; neither he nor his superiors were again willing to risk the funds that the failure entailed.[10]

Compare that attitude to one articulated by Ernst Krause, one of the rocket pioneers at NRL:

> Now, this is a good way to do some experimentation. We're going to get away from this business of having a complicated, costly set of apparatus in a physics laboratory in a basement in some university, and because it is complicated and costly, it lasts for 50 years and generation after generation grinds out theses on that same equipment because its expensive and new equipment is more expensive. We've got a setup here which by its very definition is going to get destroyed each time. How good can you have it?[11]

Krause nicely captures the excitement and novelty of a time in which instruments, despite being capital investments, became disposable. Technical and intellectual change went together with each new rocket flight. Schaffer's Astronomer Royal loathed losing valuable astronomical instruments in the South Pacific; Greenstein and ground-based astronomers continued to loathe oblivion; their astronomy remained one "designed to accumulate, account, and store data and hardware."[12]

What was new about the IGY and the science of the postwar, or cold-war, era was that the instrument had become the central focus of investigation with an important twist. Instruments had once been worked to exhaustion before being cannibalized for the next experiment; now the instrument was destroyed in the very act of performing the experiment, much as a bullet is fully consumed in its use. Instruments had become disposable commodities, part of the larger consumption-culture of the postwar era. Krause understood what Greenstein did not—the temporal dynamics of research had undergone a fundamental shift, and they were not going to revert to their prewar modes. Part of it was because of the growth of the federal government as the dominant patron of the sciences and the move from one kind of budget calendar to the federal government's annual series of appropriations, but another was built into the fabric of research itself.

## Globalizing Polar Science

What our title means is most apparent in the emblems that the IGY and the current IPY chose as their public faces. Look closely at the IGY emblem of 1957 displayed on page 15—a single, divided satellite orbits an Earth divided equally into black and white. The emblem draws attention to the IGY's most memorable event, *Sputnik*, but it also makes clear the cold-war's binary logic, although one might just as easily argue that the black-and-white division marks the day/night terminator. In case you were unaware of the satellite's path, it is etched onto the emblem and nicely captures the IGY goal of mapping the entire planet. Although there are no people on the emblem, the presence of both English and French as global languages reassures viewers that this is not an American adventure. Now compare that emblem with the symbol for the IPY of 2007–08 on page 18. The current design resembles the patch for a shuttle or a space mission rather than an emblem. And there is a person straddling the globe, although the person does not appear to have hands, feet, a neck, or gender; but it is most clearly a person. The zooming arrow that moves from the lower left to the upper right corner is a faint echo of the satellite in the original, and we might read its upward direction as one of progress. English is the only visible language, the lingua franca of our modern world. As in the 1957 emblem there are no nations present on the globe, only lines of latitude and longitude, an affirmation of the trivial point that nations are the products of politics rather than nature. It is a point at one with the comment made by spacefarers who see "spaceship Earth" from space rather than individual nations. Of course, these comments are made from platforms built by individual nations, and—as we have seen throughout this conference—the lines of latitude and longitude are the products of politics and science.

What is important about the abstract human's presence on the new logo is that the poles are now central to humanity's attempt to understand its own effect on the planet's climate. As Sheila Jasanoff argued, the image of spaceship Earth replaced the polar perspective at the outset of the environmental movement and continues to motivate an array of diverse actors as they seek to preserve the planet.[13] Here we have the meaning of making science global that deserves emphasis, the disciplining of Earth to scientific ends wherein science and scientists work to make the globe safe for both science and humanity. Surely one cannot object to such an end without being churlish, but another image, one of the density of Internet connections, brings out how much work remains to be done to make science truly global.[14] In these images, Europe, the United States, South Korea, and Japan are ablaze: beacons of technological progress and power. The southern hemisphere is dramatically dimmer, with South America and Africa nearly invisible save for isolated patches of brightness. To the extent that science is now impossible without the Internet, these domains are relatively isolated; researchers there lack access to all that we take for granted. If "Globalizing Polar Science" is the description of a process that has a history and requires substantial effort, it is also the description of a project that remains incomplete.

## Notes

1. For details on the missile accident, see the following front-page articles: William J. Boady, "Runaway Nike Bursts near D.C., Pelts Road," *Washington Post* (April 15, 1955); Anthony Leviero, "Rogue Nike Missile 'Runs Away,' Explodes in Flight," *New York Times* (April 15, 1955); also useful are Patricia Sullivan, "Chester Joswick, 79; Served in Missile Mishap," *Washington Post* (March 1, 2007): B8; and Merle T. Cole, "The One That Got Away," *Anne Arundel County History News* (January 2001): 1–2, 10–12. The latter is also available online at ed-thelen.org/rouguenike-cole.html.

2. For more on this project, see Ronald E. Doel, "Constituting the Postwar Earth Sciences: The Military's Influence on the Environmental Sciences in the USA after 1945," *Social Studies of Science* 33, no. 5 (2003): 635–666.

3. Two places to begin understanding the historiography are Roger D. Launius, John M. Logsdon, and Robert W. Smith, eds., *Reconsidering Sputnik: Forty Years since the Soviet Satellite* (Amsterdam: Harwood Academic Publishers, 2000); and Rip Bulkeley, *The Sputniks Crisis and Early United States Space Policy* (Bloomington: Indiana University Press, 1991).

4. The best introduction to metrology and its consequences remains J. O'Connell, "Metrology: The Creation of Universality by the Circulation of Particulars," *Social Studies of Science* 23 (1993): 129–173; the most accessible version of Latour is Bruno Latour, *Science in Action: How to Follow Scientists and Engineers through Society* (Berkshire, United Kingdom: Open University Press 1987).

5. Paul Forman, "Scientific Internationalism and the Weimar Physicists: The Ideology and Its Manipulation after World War I," *Isis* 64 (1972): 151–180.

6. On the Bergen School and its context of development, see Robert Marc Friedman, *Appropriating the Weather: Vilhelm Bjerknes and the Construction of a Modern Meteorology* (Ithaca, New York: Cornell University Press, 1989).

7. See Simon Schaffer, "On Seeing Me Write: Inscription Devices in the South Seas," *Representations* 97 (2007): 90–122.

8. Isaiah Bowman, "Polar Exploration," *Science* 72, no. 1870 (1930): 439–449, 441.

9. The place to begin such a discussion of the science technology relationship is Paul Forman, "The Primacy of Science in Modernity, of Technology in Postmodernity, and of Ideology in the History of Technology," *History and Technology* 23, no. 1/2 (2007): 1–153. See also the subsequent critiques of Forman's argument in the same issue.

10. On Greenstein, see David H. DeVorkin. *Science with a Vengeance: How the Military Created the US Space Sciences after World War II* (New York: Springer Verlag, 1992), 204–207.

11. DeVorkin, *Science with a Vengeance,* 214.

12. Schaffer, "On Seeing Me Write," 107.

13. Sheila Jasanoff, "Image and Imagination: The Formation of Global Environmental Consciousness," in Clark A. Miller and Paul N. Edwards, eds., *Changing the Atmosphere: Expert Knowledge and Environmental Governance* (Cambridge, Massachusetts: The Massachusetts Institute of Technology Press, 2001), 309–337.

14. Available online at http://www.chrisharrison.net/projects/InternetMap/ (accessed February 4, 2009.

# Chapter 2

# Making Science Global? Coordinated Enterprises in Nineteenth-century Science

## Marc Rothenberg

The title of the conference that inspired this collection of essays can be interpreted as implying that the International Polar Years (IPY) were the catalyst for the globalization of science. To read the title in this way can be misleading, especially if this implies the conference was an argument that the IPYs were representing a new path for science. It is the contention of this essay that there was little that was radical about the IPYs. For the most part, they were logical outgrowths of what historians have identified as two well-established traditions in Euro-American science.

Before moving forward, it will be useful to look at three ways in which the phrase "Globalizing Polar Science" can be interpreted. First, the process of globalization may be seen as the breaking down of national boundaries. "Globalizing Polar Science" may serve as a label for the transformation of science from a local to an international activity. In making science global, the community of practitioners comes to agreement on standards for observations and the free exchange of data, overcoming political boundaries and local prejudices. Any scientific discipline can become global in that sense. But the word "global" can also be shorthand for recognition by the scientific community that certain phenomena are worldwide and require a global approach if the scientific community is ever to comprehend them. The geophysical disciplines, such as terrestrial magnetism or meteorology, have become global in this sense. These two uses place "global" squarely in the context of the history of international scientific cooperation. Yet a third meaning of "global" is the expansion of local science to virgin parts of the globe. In this usage, "global" is part of the tradition of scientific exploration, of the expansion of Euro-American science across geographical frontiers on Earth, and embraces a wide range of field sciences.

Looking back from the perspective of the fourth IPY, the first IPY of 1882–83, which was the brainchild of Lieutenant Karl Weyprecht, an Austro-Hungarian Arctic explorer, was a reaction to the needs of global science in all

three usages of the term. Although it was, in the narrowest sense, a response to
the demands of meteorology, which needed to integrate data from the extreme
ends of the Earth, the first IPY, which attracted a dozen nations, with most of
the energy directed at the Arctic, drew on a well-established tradition of international scientific cooperation. Such cooperation had been a characteristic of
Euro-American science since at least the eighteenth century, if not before. The
challenges of doing science in the polar regions made international cooperation important, if not imperative. But it also drew on the traditions of scientific
exploration, particularly in the Arctic, such as practiced by Weyprecht. It was
an expansion of scientific research to new areas of the globe.

The modest objective of this essay is to propose that the history of
the first IPY—and its successors, the IPY of 1932–33, with participation
from 40 nations, which was promoted by the International Meteorological
Organization and focused primarily, although not exclusively, on the Arctic,
and the International Geophysical Year (IGY) of 1957–58, in which 67
nations conducted research at both poles and in space and greatly expanded
the disciplinary reach of the IPY[1]—can be understood only in the contexts
of the history of scientific exploration and the history of international scientific cooperation and the coming together of these two histories. By providing a sense of the historiography regarding these two distinct areas, the hope
is to offer context and framework for the essays that follow.

## International Cooperation[2]

Karl Weyprecht did not invent the concept of international scientific cooperation. International cooperation has been one of the hallmarks of Euro-
American science since at least the mid-eighteenth century. During the
subsequent two-plus centuries, researchers increasingly began taking on
problems, particularly those in astronomy or geophysics, which required
coordinated efforts among geographically scattered observers. To answer the
questions they were asking, scientists required data gathered simultaneously
by observers at different, widely scattered locations, using, in some cases,
standardized instrumentation.

The nature of that cooperation was transformed over more than two centuries, becoming more institutionalized and bureaucratic over time. The
eighteenth century was noteworthy for its international networks of scientists who kept one another abreast of their work. During the nineteenth
century, "international cooperation moved from an individual level to an
institutional one." National scientific and voluntary international societies
took leadership roles. Finally, in the twentieth century, cooperation among
nations became evident.[3]

Observations of the 1753 transit of Mercury are an early example of
international cooperation based on a mixture of personal correspondence
and impersonal publication, and driven by one concerned scientist. Only
through geographically scattered and coordinated observations could the
problem of the distance between the Earth and Sun be solved.[4] Although

the resulting observations failed to solve the scientific question asked, they did serve as a practice run, so to speak, for the transits of Venus of 1761 and 1769, which used a different set of observations to solve the same problem. Coordinated through an international correspondence network among astronomers and scientific societies, these observations have been credited with marking a major milestone in the history of science. As Harry Woolf concluded in the classic historical study of these events, the scientific problem of the Earth-Sun distance, to be solved through observations of the transit of Venus, "brought to a common focus men of almost every national background with an abiding concern for the advancement of knowledge. In doing so, it helped to shape the growing international community of science and to demonstrate with striking clarity what cooperation and good will might achieve in the peaceful pursuit of science."[5] Although recent scholarship has challenged the true extent of international cooperation, at least at the level of government policy—pointing to the negative impact of the Seven Years' War (1756–63) on some of the observation plans—there seems little question that the American and European scientific communities increased their level of collaboration across international boundaries in the wake of efforts to observe the transits of Venus.[6]

**Figure 2.1** Exploration had ranged to the poles before the International Polar Year of 1882–83, but never before was it so well coordinated. Here the 1842 voyage of the French vessels Astrolabe and Zelee toward Antarctica is depicted from *Voyage au pole sud et dans l'Oceanie* (1842) as they encounter an ice field. (Courtesy of National Oceanic and Atmospheric Administration, NOAA Library Collection, Washington, D.C.)

This collaboration accelerated in the nineteenth century, as more reliable and faster steamships, the laying of the Atlantic cable, and the development of extensive railroad and telegraph systems in North America and Europe made interaction among scientists and the sharing of data much easier and, eventually, routine. In the late nineteenth century, the discovery of a transitory astronomical phenomenon could be communicated in a matter of hours from rural New York State to all the major European observatories.[7] Just as crossing the Atlantic or the Alps—major endeavors as late as the 1820s and 1830s—were taken for granted, the coming together at international congresses to exchange information and attempt to establish standards was becoming commonplace.

It is important to recognize that international cooperation was an integral part of Victorian-era, Euro-American society, not simply a phenomenon limited to the scientific community. International congresses were a manifest expression of the urge to gather and solve common problems that transcended national boundaries. To take one example, 32 international congresses met in conjunction with the 1878 Exposition in Paris. On the various congress agendas were cooperation, coordination, standardization, exchange of information, best methods for the gathering of statistics, and/or efforts at common solutions for common problems. Among the scientific fields to hold congresses in Paris that year were anthropology, botany, geology, and meteorology, each gathering with the hope of establishing agreement on simultaneity of observations or uniform nomenclature. Science and commerce were both invoked in congresses revisiting the long-debated issue of an international system of weights, measures, and coinage, and the congress for the unification and numbering of threads. Other congresses focused on legal issues, such as international copyright and patent rights, or social issues, such as prevention of cruelty to animals, the treatment of alcoholism, guidelines for military ambulance service, or aid to the blind and deaf.[8]

A number of nineteenth-century collaborative or cooperative international scientific efforts are important in understanding the prehistory of the IPYs. These efforts included the Magnetic Crusade, a collaborative endeavor to solve fundamental questions in terrestrial magnetism; a variety of plans for international cooperation in the gathering of meteorological data; and the many expeditions sent out throughout the world to observe the transits of Venus. The level of cooperation ranged from simply improving communication among scientists to establishing common standards for recording observations. When we look at these efforts, it will also be important to remember that the urge for cooperation did not always translate into true collaboration. This century marked by collaboration was also marked by the growth of nationalism, the unification of Italy and Germany, and intense international rivalry among European nations in the wake of the Franco-Prussian War of 1870. The line between international cooperation and rivalry was narrow.[9]

The first great international effort at coordinating physical science research in the nineteenth century was the Magnetic Crusade, which focused on the gathering of terrestrial magnetic observations.[10] Early nineteenth-century scientists recognized that the variations of Earth's magnetic field were extremely complex, requiring the gathering of data over the entire Earth, and they began creating an informal system of contacts to exchange information and provide some uniformity to the observations. Although both French and German physicists established formal networks, it was the system established by the British Association for the Advancement of Science in 1838 that was truly global. Funded by the British government and the private sector (coverage was expanded to India with the cooperation of the East India Company), the British system also coordinated with 23 other observatories scattered throughout the Russian Empire, Asia, North America, North Africa, and Europe. All these other observatories were funded by their respective governments, except for those in the United States, which were funded by academic institutions. The Magnetic Crusade even included polar terrestrial-magnetic observations in its program. The British, French, and Americans all sent ships to Antarctica, while observations were made in the Arctic region by Germans, Russians, and Norwegians.[11]

The Magnetic Crusade was, however, more of a limited international cooperative venture rather than a true collaboration. Although the British system synchronized its observations with the Germans, so that data could be compared, there was no formal collaboration, while the Paris Observatory acted independently of its other European counterparts. And like many subsequent large-scale cooperative programs, much of the data was never published or, if

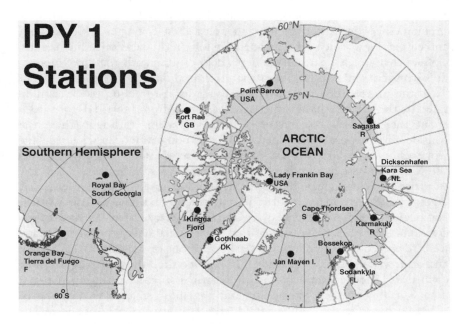

**Figure 2.2** The scientific stations established during the first International Polar Year. (Courtesy of National Oceanic and Atmospheric Administration, NOAA Library Collection, Washington, D.C.)

published, never properly analyzed. Nonetheless, by the time the Crusade formally ended in 1848, there was a firmly established network of magnetic observatories in Europe, throughout the British Empire, and in the United States, which continued to make observations and exchange data. Most important, as John Cawood argued, the Magnetic Crusade demonstrated "that large-scale operations could be organized and carried through."[12] Even C.J. Taylor, who has argued that the IPY served as the great demonstrator of the possibilities of large-scale international cooperation, admitted that the Magnetic Crusade "provided many precedents for subsequent global scientific endeavours."[13]

Weather does not respect political boundaries, and many meteorologists realized the need for cooperation. During the mid–nineteenth century, national weather networks or services had been organized throughout western and central Europe, in addition to imperial Russia and the United States. German meteorologists were the first to try to coordinate observations across national boundaries, with such organizations as the Süddeutsche Meteorologische Verein (1841) and the Königlich Preussische Meteorologische Institut (1847). These organizations had relatively limited geographical coverage, however, and were international only because of the political fragmentation of the German scientific community. Other efforts followed, ranging from the exchange of storm warnings among nations to international gatherings.[14] There was, however, one major issue confronting scientists wishing to collaborate internationally: the lack of a standard system of measurement. As Paul Edwards has argued, scientific "standards are socially constructed tools." They are the result of "negotiations that are simultaneously technical, social, and political in character." Meteorology and climatology, which are dependent on long-term data collecting, are disciplines that are also particularly dependent on standards that have more to do with social and political issues than technical ones.[15]

In the nineteenth century, nationalism was a stumbling block, as indicated by the debate among American scientists—from the 1850s forward—about the possible adoption of the metric system. Although there was considerable support for the metric system among American scientists (and Congress even legalized it for use in the United States in 1866), neither the American Association for the Advancement of Science nor the National Academy of Sciences would endorse it. In 1865 Joseph Henry, then secretary of the Smithsonian Institution and soon to become president of the National Academy of Sciences, rejected the metric system as an artificial creation of the French. He argued that since England and the United States were "destined to control the political operations of the world"—in science and otherwise—a true international system required the leadership of those countries.[16] Henry would repeatedly return to this issue during the next decade, calling on the United States and England, perhaps joined by Russia, to establish standards. And he would repeatedly attack any efforts to extend the French metric system in the United States.[17]

The issue of standard measurement arose most frequently in meteorology, where there was a continuing conflict between the metric and English systems of measurement. For example, the 1873 congress in Vienna, which ultimately attracted representatives from 20 nations, was unable to adopt the metric system as the standard because of opposition by the English delegate. His resistance reflected the fear that this might be the first step in efforts to force the wider acceptance of the meter in Great Britain.[18]

Another problem facing meteorologists was that they were often government-supported by large investments in their national meteorological programs. Collaboration might mean acknowledging that another country's system was superior. In 1845 an international meeting of scientists interested in terrestrial magnetism and meteorology was held in conjunction with the meeting of the British Association for the Advancement of Science. Efforts to establish some sort of coordination of meteorological observations, akin to the Magnetic Crusade, ran into this serious obstacle. Government meteorologists of the various European nations had too much invested in their own systems to lay them aside for some common system. As Edward Sabine remembered two decades later, the government meteorologists had "manifested so marked a disposition... to adhere to their respective arrangements in regard to instruments, times of observation, and modes of publication," as to make it clear the time for a uniform system "had not then arrived."[19]

Perhaps the most successful international approach to meteorological observations during the mid-nineteenth century was the Smithsonian Institution's meteorological network. Joseph Henry, who became the first secretary of the Smithsonian Institution in 1846, sought a unique niche for the new organization. Henry made coordination of large-scale research projects a major element of his program, something he felt no other institution in the United States was equipped to do. An experienced meteorological observer and participant in the Magnetic Crusade, he was aware of the power of coordination with regard to throwing light on geophysical questions. The first such project Henry embraced was the development of what Elias Loomis, one of his consultants, characterized as "a grand meteorological crusade" for collecting meteorological observations.[20]

The system devised by Henry had two distinct, but interrelated, components, one requiring international cooperation, one not. To provide data for long-term climatological conditions and weather tendencies, Henry created a system of observers who—using standard apparatus, techniques, and forms to the greatest extent possible—maintained monthly logs of weather conditions that were sent to the Smithsonian for reduction. From the onset, it was recognized that "to give this system its greatest efficiency, the co-operation of the British government and of the Hudson's Bay Company [in Canada] is absolutely indispensable."[21] Both the British government and the private Hudson's Bay Company quickly agreed to cooperate. To extend coverage to the Caribbean and the Gulf of Mexico, observers were recruited in Bermuda, Mexico, all the Central American countries, and throughout the West Indies, frequently drawing, in the latter two cases, on Americans residing there.[22]

The second component of the Smithsonian system was strictly domestic. The telegraph was used to forward data on weather in real time to the Smithsonian.

In the 1850s, this network of telegraphic observers provided data for the publication of the first scientifically based weather forecasts in newspapers and the first publicly posted weather maps. Unlike the long-term data gathering, the forecasting only lasted a few years because of funding issues.

The great strength of Henry's international system was that there were no government meteorologists involved who felt protective of their own national systems. Instead, Henry relied on an international network of independent observers, most of whom were not scientists. The Smithsonian, which provided the coordination, prided itself on being an independent scientific institution, not a component of the U.S. government. This was a cooperative venture of scientists, not governments.

Some historians have argued that only when the collaboration efforts bypassed the official government meteorological organizations and became "a pattern of voluntary cooperation between meteorologists on international problems" were these problems solved.[23] Edith Brown Weiss dates this cooperation to 1879 and the second international meteorological congress, held in Rome; she argues that this pattern continued until at least 1950.[24]

The last examples of pre-IPY international scientific cooperation we shall examine are the transits of Venus 1874 and 1882. Like the eighteenth-century events, the transits attracted an international observation corps. In some countries government commissions were established to oversee the efforts, and among members of the astronomical community there was considerable communication regarding observing protocols and coordination of observing sites.[25]

Because Antarctica was among the desirable observing locations for these transits, there was, for the first time, serious discussion of establishing an astronomical observing site in Antarctica. Scientists were divided as to whether it was possible to establish and maintain an observing party and whether the potential rewards were commensurate with the risks—both those to the lives of the astronomers and those to the success of the mission because of the possibility of bad weather. Englishman George Airy, British Astronomer Royal, who had begun planning as early as 1857, was uncertain and called for a reconnaissance ahead of time.[26] J.E. Davis, a British naval officer and Arctic explorer, was optimistic, although realistic as to the difficulties such an effort would present. Davis developed a plan in 1869 for observations from Antarctica of the 1882 transit, but noted in his presentation to the Royal Geographical Society that the observing parties would have to over-winter. Davis argued that that they should be landed in late 1881 with sufficient supplies to last two years, even though the plan was to have them picked up in about a year. It was necessary to leave a margin of error. He did warn of the problematic weather conditions, describing the weather as "either very bad or very delightful."[27] To Davis, who was not an astronomer, it was a gamble worth taking. In contrast to Davis and Airy, Simon Newcomb, the leader of American astronomy and a member of the American Transit of Venus Commission, rejected out of hand the idea of astronomical observations from "the Antarctica continent and the neighboring islands."

Not only did he fear that "a party can neither be landed nor subsisted there," but perhaps more important from the perspective of the working astronomer, "if they could, the weather would probably prevent any observations from being taken."[28]

In the end, no observations were made from the continent of Antarctica, although the 1874 transit was observed by parties from the United Kingdom, Germany, France, and the United States from stations on islands within the Antarctic convergence, including Kerguelen (Newcomb apparently thought Kerguelen sufficiently north not to be considered "a neighboring island") and Saint-Paul. The 1874 observations were only moderately successful because of adverse weather (and a planned American station on the Crozet Islands was aborted because of weather conditions). Although the combination of the uncertainty of the weather and the difficulties, dangers, and expense of sending parties to Antarctica seemed to have discouraged most further efforts in that direction for the 1882 transit, Germany sent an expedition to South Georgia for the dual purpose of conducting transit-of-Venus observations and other observations as part of the IPY.[29]

The seeds had been planted, however, for more extensive, cooperative scientific research in the Antarctic. Subsequently scientists have acknowledged that the transit-of-Venus expeditions established a precedent for later cooperative research in Antarctica, including both the IPYs[30] and the IGY.[31]

## Scientific Exploration

Scientific exploration had its halcyon days during what William Goetzmann called the "Second Great Age of Scientific Discovery," that period stretching from the publication of great insights into the laws of motion by Isaac Newton in the late seventeenth century until the turn of the twentieth century.[32] In Goetzmann's characterization, scientific exploration was a mixture of geographical discovery and scientific research. Contributions were made to a broad range of disciplines, including botany, zoology, geology, meteorology, oceanography, and terrestrial magnetism. Early on in the period, with the question of survivability of the expedition uppermost in many minds, and with great geographical blanks to be filled, exploration took precedence over scientific research. The explorer had to overcome a hostile environment, hostile inhabitants, or both. A trained scientist, let alone a scientific party, was a dangerous luxury for most exploring expeditions. It was a member of the party, often not a trained scientist, who made the observations and collected the data as a sideline or incidental activity. As survivability became less of an issue and more blanks were filled in on maps, the balance shifted. Progressively less time was spent on geographical discovery and more on cataloging nature's resources and data gathering. The scientist gradually replaced the military or civilian explorer on center stage. Using American exploration as an example, one sees a progression from teaching an explorer some basic science (Lewis and Clark Expedition, 1804–06) to allowing a naturalist to accompany an essentially military exploring expedition (Long

Expedition, 1819–20) to sending out mixed parties of military surveyors and scientists (the Navy's U.S. Exploring Expedition, 1838–42, and the Army's Railroad Surveys, 1853–55) and finally sending out a civilian, scientific endeavor (U.S. Geological and Geographical Survey of the Territories, 1867–68). Nonetheless, even at the end of this period, the scientific community depended on the complex infrastructure of the military supply system to provide logistical support. Perhaps most important, scientific exploration was not an isolated activity, at least not in the United States. Throughout the nineteenth century, there was continuous interplay between scientific exploration and the development of American science, scientific institutions, and public policy.[33]

In the history of exploration, the polar regions are the last frontier, where the hostile environment continued to represent a challenge long overcome elsewhere. Goetzmann concludes his study of American involvement in the Second Great Age of Discovery with an examination of polar exploration. The American contributions to the first IPY fit neatly into his construction of history. By the end of the Second Great Age of Discovery, polar research has the potential at least to become as unproblematic as geological research in the Rocky Mountains. The success of Lieutenant Henry P. Ray's IPY expedition to the Alaskan Arctic circle could be contrasted with the disaster and death that accompanied Lieutenant Adolphus Greeley's IPY expedition to Ellesmere Island. Thanks to proper planning, "[Ray] and his men...made survival in the Arctic wastes...look routine to Americans for the first time."[34] This was a major step forward.

It would be incorrect, however, to assume that polar research truly became routine with better planning and the subsequent technological innovations of the twentieth century. As the emphasis shifted from the Arctic to the Antarctic, another frontier was opened and another age of discovery began. Goetzmann is vague about when the "Third Great Age of Discovery" began, although it is clearly post–World War I and it appears he dates it around the time of the second IPY.[35] Perhaps its initiation can be linked to Richard E. Byrd's second expedition to Antarctica, 1933–35, the major component of the second IPY to focus on the South. The expedition established the first inland Antarctic research station and harvested many scientific specimens.[36] In contrast to Goetzmann, Stephen Pyne clearly dates the Third Age of Discovery to the second half of the twentieth century, arguing that the launching of the IGY marked the beginning of this Age of Discovery.[37] Pyne also proposes that Antarctica was a different environment to contend with than those faced by earlier generations of explorers. The lack of a native population to supply guides or survival skills, and the limited ecosystem, among other characteristics, made it "an almost extraterrestrial presence."[38] It was exploration truly on a new level.

In any case, the challenges facing the participants in Byrd's expedition or those involved in the IGY were extensive. This was still scientific exploration in the earlier sense of the term. As Dian Olson Belanger makes evident in her study of American participation in the IGY, even as late as the 1950s the military and civilian staff that provided infrastructure and logistical support was crucial to the success of the scientific endeavor.[39] For the United States the reliance of scientific research on a support structure that was primarily provided by the military harked back to a distant past. Antarctica in the second half of the twentieth century recalled an approach to the scientific endeavor in the American West a century or more earlier.

## Conclusion

In emphasizing historical continuum in the background of the IPYs, I am not arguing for an unproblematic progression from coordination through scientific correspondence, to take the example of the transit of Mercury, to organization through international treaty for Antarctic activities, which was accomplished in the wake of the IGY. Nor has scientific exploration, even in just the American context, remained unchanged over two centuries. But recognizing precedents is important, both for the practicing scientists and for the historians who are studying them. It should be comforting, both for the practitioner and those who fund the research, to know that similar challenges confronted earlier generations and that there were precedents upon which to draw.

## Notes

1. The historical literature for the IPYs is extremely thin, although this will no doubt change during the next few years as the fruits of the historical research being conducted on the occasion of the fiftieth anniversary of the IGY become available. For the first IPY, the three essential studies are F.W.G. Baker, "The First International Polar Year, 1882–1883," *Polar Record* 21 (1982): 275–285; Cornelia Lüdecke, "The First International Polar Year (1882–1883): A Big Science Experiment with Small Science Equipment," *History of Meteorology* 1 (2004): 55–64; James Rodger Fleming and Cara Seitchek, "Advancing Polar Research and Communicating Its Wonders: Quests, Questions, and Capabilities of Weather and Climate Studies in International Polar Years," in I. Krupnik, M.A. Lang, and S.E. Miller, eds., *Smithsonian at the Poles: Contributions to International Polar Year Science* (Washington, D.C.: Smithsonian Institution Scholarly Press, 2009), 1–12. For the IGY, ignoring the literature on Sputnik and the space race, there is agreement that the essential publication is the account by the journalist Walter Sullivan, *Assault on the Unknown: The International Geophysical Year* (New York: McGraw-Hill, 1961).

2. I have previously explored the issue of the context of polar research and international cooperation, but with a different thrust, in "Cooperation at the Poles? Placing the First International Polar Year in the Context of Nineteenth-Century Scientific Exploration and Collaboration," in Krupnik, Lang, and Miller, eds., *Smithsonian at the Poles*, 13–21.

3. Jessica Ratcliff, *The Transit of Venus Enterprise in Victorian Britain* (London: Pickering & Chatto Publishers, 2008), 18.

4. Ratcliff, *Transit of Venus*, 13.

5. Harry Woolf, *The Transits of Venus: A Study of Eighteenth-Century Science* (Princeton: Princeton University Press, 1959), 197.

6. Ratcliff, *Transit of Venus*, 18.

7. Marc Rothenberg et al., eds., *The Papers of Joseph Henry* (Sagamore Beach, Massachusetts: Science History Publications, 2007), 11:447n–448n; Bessie Zaban Jones and Lyle Gifford Boyd, *The Harvard College Observatory: The First Four Directorships, 1839–1919* (Cambridge, Massachusetts: Harvard University Press, 1971), 194–198.

8. United States Commission to the Paris Universal Exposition, *Reports of the United States Commissioners to the Paris Universal Exposition, 1878* (Washington, D.C.: U.S. Government Printing Office, 1880), 1:455–464.

9. John Cawood, "The Magnetic Crusade: Science and Politics in Early Victorian Britain," *Isis* 70 (1979): 518.

10. Cawood, "Magnetic Crusade," 493–518; John Cawood, "Terrestrial Magnetism and the Development of International Collaboration in the Early Nineteenth Century, *Annals of Science* 34 (1977): 551–587.

11. Ratcliff, *Transit of Venus*, 23.

12. Cawood, "Magnetic Crusade," 516.

13. C.J. Taylor, "First International Polar Year, 1882–83," *Arctic* 34 (1981): 370.

14. James Rodger Fleming, *Meteorology in America, 1800–1870* (Baltimore, Maryland: Johns Hopkins University Press, 1990), 165–166; James Rodger Fleming, "Meteorological Observing Systems before 1870 in England, France, Germany, Russia, and the USA: A Review and Comparison," *World Meteorological Organization Bulletin* 46 (1997): 249–258.

15. Paul N. Edwards, "'A Vast Machine': Standards as Social Technology," *Science* 304 (May 7, 2004): 827–828, quotations on p. 827.

16. Henry to Hubert Anson Newton, June 15, 1865, in Rothenberg, *Papers*, 10:524.

17. Henry to Charles Davies, July 29, 1872, in Rothenberg, *Papers*, 11:415.

18. Katharine Anderson, *Predicting the Weather: Victorians and the Science of Meteorology* (Chicago: University of Chicago Press, 2005), 246.

19. Edward Sabine, "Note on a Correspondence between Her Majesty's Government and the President and Council of the Royal Society Regarding Meteorological Observations to Be Made by Land and Sea," *Proceedings of the Royal Society of London* 15 (1866): 29–38.

20. *Annual Report of the Board of Regents of the Smithsonian Institution for 1847* (Washington, D.C.: U.S. Government Printing Office, 1848), 207. Fleming, *Meteorology in America*, 75–93.

21. *Annual Report for 1847*, 207.

22. Fleming, *Meteorology in America*, 123; *Annual Report of the Board of Regents of the Smithsonian Institution for 1868* (Washington, D.C.: U.S. Government Printing Office, 1872), 68–69.

23. Edith Brown Weiss, "International Responses to Weather Modification," *International Organizations* 29 (1975): 809.

24. Weiss, "International Responses," 809–810.

25. Hilmar W. Duerbeck, "The German Transit of Venus Expeditions of 1874 and 1882: Organizations, Methods, Stations, Results," *Journal of Astronomical History and Heritage* 7 (2004): 8–17; Steven J. Dick, "The American Transit of Venus Expeditions of 1874 and 1882," in *Proceedings of IAU Colloquium No. 196* (Cambridge, England, 2004), 110–111.

26. George B. Airy, "On the Means which will Be Available for Correcting the Measure of the Sun's Distance, in the Next Twenty-Five Years," *Monthly Notices of the Royal Astronomical Society* 17 (1857): 216.

27. J.E. Davis, "On Antarctic Discovery and Its Connection with the Transit of Venus," *Journal of the Royal Geographical Society* 39 (1869): 91–95, quotation on p. 93.

28. [Simon Newcomb,] "The Coming Transit of Venus," *Harper's New Monthly Magazine* 50 (1974): 30.

29. Duerbeck, "German Transit of Venus Expeditions."

30. Julian Dowdeswell, Scott Polar Research Institute, International Polar Year Outreach Project.

31. Kenneth J. Bertrand, *Americans in Antarctica, 1775–1948* (New York: American Geographical Society, 1971), 255.

32. William H. Goetzmann, *New Lands, New Men: America and the Second Great Age of Discovery* (New York: Viking, 1986), 1.

33. The standard study of scientific exploration in the United States is William H. Goetzmann, *Exploration and Empire: The Explorer and the Scientist in the Winning of the American West* (New York: Alfred A. Knopf, 1968).

34. Goetzmann, *New Lands, New Men,* 428.

35. Goetzmann, *New Lands, New Men,* 453.

36. Bertrand, *Americans in Antarctica,* 313–361.

37. Stephen J. Pyne, *The Ice: A Journey to Antarctica* (Iowa City: University of Iowa Press, 1986), 108.

38. Pyne, *The Ice,* 88.

39. Dian Olson Belanger, *Deep Freeze: The United States, The International Geophysical Year, and the Origins of Antarctica's Age of Science* (Boulder: University of Colorado Press, 2006).

# Chapter 3

# Publishing Arctic Science in the Nineteenth Century: The Case of the First International Polar Year

*Philip N. Cronenwett*

The seeds of the first International Polar Year (IPY) were sown by Karl Weyprecht, an Austro-Hungarian naval officer and Arctic explorer. In his essay "Fundamental Principles of Scientific Arctic Investigation," delivered to the Academy of Sciences in Vienna in January 1875, Weyprecht argued that cooperation by several nations could and would provide useful data that could be compared across a vast stretch of the Arctic. The presentation was repeated at a conference of naturalists and physicians in Graz later in the year. Quickly translated into English and other European languages and published in Vienna as a pamphlet, Weyprecht's lecture posited the need for a multinational, cooperative year of scientific polar research using similar equipment, similar instruments, and similar data-gathering methods.[1] Within a few years, the first IPY conference met in Hamburg in 1879, the second in Berne in the following year, and the third, and perhaps the most important, met in St. Petersburg in 1881.

At the St. Petersburg conference, the permanent International Polar Commission set the dates for the first IPY as August 1, 1882, to August 31, 1883. Equally important, at least in the context of this essay, the St. Petersburg conference set a 39-point program for research and data gathering. The final three points treated the organization and publication of the data amassed by the various expeditions. These included summaries of observations to be published in a uniform manner as soon as possible, publication in extenso of observations "when their discussion is complete," and the use of metric and centigrade scales in all measurements and reports.[2] The report of the fourth commission meeting, held in Vienna in 1884, provided extensive, but broadly written, direction for the publication of the scientific results of the expeditions.[3] The final meeting was held in Munich in 1891.

The publications subsequent to the completion of the expeditions of the cooperating nations and the data gathered by the IPY research stations form the basis of this essay. After a brief review of each of the publications, the total IPY record will be considered in the context of what may be called the *Challenger* paradigm, and some conclusions will be drawn.[4] It should be noted that the so-called auxiliary expeditions and the several expeditions to the southern hemisphere will not be considered.

## Expeditions and Reports

The Austro-Hungarian contribution to the IPY was primarily a privately funded expedition to Jan Mayen Island. Johann Nepomuk and Graf von Wilczek provided much of the resources necessary to send Captain Emil von Wohlgemuth and his crew to obtain a full array of meteorological and magnetic data, to gather biological and mineral specimens, and to begin glaciological studies on the island. The report of the expedition was published by the Imperial Academy of Sciences in 1886.[5]

The three volumes of the Austro-Hungarian report begin with a lengthy introduction by the commander, Wohlgemuth, and continue with specific sections carefully describing all aspects of data collection. Illustrations, including plates and maps, complete the volumes. Of interest to the history of publishing is the fact that many of the illustrations within the text are individually described as a *Holzschnitt*, woodcut, rather than a *Stich*, engraving. Given the inability of woodcuts to withstand the pressure and battering of a late nineteenth-century mechanical press, it is hard to imagine how these illustrations were completed without a significant amount of hand work. The answer is, I believe, that the original illustrations were woodcuts that were recut as engravings that could be printed on a text page on a press that could run at high speeds.[6] This did nothing more than make the images muddy and useless as scientific illustrations.

Great Britain, in a joint expedition with its former colony, Canada, mounted an expedition to Fort Rae, a Hudson's Bay Company post on the Great Slave Lake. This was a half-hearted effort, to say the least, by the Royal Society and the Canadian Meteorological Office, with observers from the British military. Only the barest minimum of data, including meteorological and magnetic data, was gathered and then published in a single volume in 1886.[7] The volume—containing sketch maps, tables of data, and nearly three dozen folding sheets of diagrams—was edited by Henry Dawson, commander of the expedition.

Denmark's contribution to the IPY was an expedition to Godthåb, Greenland, and to Nanortalik, in eastern Greenland, to gather meteorological and magnetic data. The data was gathered into two volumes, edited by Adam Paulsen, the commanding officer of the expedition, and published in1886 and 1894.[8] The volumes, replete with charts and diagrams, include a section of data (in the second volume) gathered from Andreas Hovegaard's expedition to the Kara Sea aboard the *Dijmphna*. Hovegaard also published several stand-alone articles on his 1884 expedition. His colleague, C. F. Lütken, published an extensive monograph analyzing the zoological and botanical specimens several years later.[9]

While the Dutch government expedition failed to reach its designated harbor in the Kara Sea, it managed to be quite successful in the work it did achieve. The loss of the research vessel *Varna*, beset in the ice near the Danish vessel *Dijmphna,* should have spelled disaster for the Dutch expedition. It did not, however, stop the Dutch from gathering a significant quantity of meteorological data, ice-drift data, and oceanographic studies. There were obviously no terrestrial magnetic observations to report, but the remainder of the data was important. Maurits Snellen, the commanding officer of the expedition, published a brief narrative of the voyage in 1886 in Dutch, but it was not until 1910 that the data gathered was finally digested and published.[10]

Finland mounted a government-sponsored observation station at Sodankylä, in Finnish Lapland, to gather meteorological and magnetic observations in 1882 and 1883. One of the more interesting aspects of the scientific work conducted by the Finns was the study of auroras and the propagation of artificial auroras at Kultala, a site farther north. Because of the success of this work with auroras, the program was extended for a second year, and the men returned to Helsinki in 1884, a year later than planned. The data gathered during these two winters was published in three volumes over a 12-year period.[11]

After being briefly stymied by ice, the German Polar Commission's expedition to Cumberland Sound and Baffin Island established a station in Kingua Fjord. Because of its location, the expedition was regularly visited by the Inuit, allowing for ethnographic studies and observations. As a consequence, the German IPY publications are among the few that contain a significant quantity of ethnological data. In addition to meteorological, magnetic, and ethnological studies, the reports contain botanical and geological studies. The main German publications were released in four volumes: two in 1886 and two more in 1890–91.[12] Both sets of publications are heavily illustrated with color lithographs—as frontispieces to the first volume of each set—and color and black-and-white engravings. Interestingly, the 1890–1891 volumes are printed in a Fraktur typeface making it extremely difficult to read today. Perhaps because of the amount of data gathered, this expedition was one of the few in which individual members published data separately.[13]

The Norwegian IPY contribution was a government-sponsored observation site in Bossekop, in northern Norway. Organized by the Meteorological Institute in Christiania (Oslo), the small team of Norwegians gathered large amounts of data on astronomical observations, magnetism, meteorology, and auroras. This data was published in two volumes, in multiple parts, with all the requisite charts and data sets, in 1887 and 1888. Of particular interest is a historical introduction that provides a glimpse into the decision-making process regarding Norway's participation in the IPY.[14]

The first of two Russian IPY expeditions was sent to the Lena River delta under the aegis of the Imperial Russian Geographical Society. The expedition was sponsored by the army, navy, and the hydrographic department. In addition to the required magnetic and meteorological observations, the team studied auroras, gathered hydrographic data, and made natural-history observations. At the end of the first year, the Russian Polar Commission asked that they remain for another winter. In this second winter, the team made a series of sledge journeys significantly extending their observation range. The reports of the research were published in parts, ultimately occupying two volumes, over a period of nine years.[15] The volumes were enhanced with maps, plates, diagrams, and tables.

Russia's second expedition, again under the aegis of the Imperial Russian Geographical Society and funded by the government, was sent to Novaya Zemlya to complete the magnetic and meteorological data gathering. Whenever and wherever possible, the team also gathered botanical and geological specimens for later study. The result of this work was another important publication, edited by the commander, K. P. Andreyev, and published over a period of six years in parts gathered into two volumes. The volumes include charts, plates, maps, and tables.[16]

Sweden's contribution to the IPY was a series of observations from Svalbard. Unable to reach their primary site at Mosselbukta because of ice, the party settled in at Kap Thordsen for the year. The work was sponsored by the Royal Academy of Sciences and the government, and had private financial support from a Stockholm businessman. In addition to amassing magnetic and meteorological observations, the team surveyed the area, gathered hydrographic data, and conducted a variety of studies on auroras and other optical phenomena. The data was published in parts over a six-year period and gathered into two volumes. Each of the volumes contains maps and charts as well as data.[17] The plates that accompany these volumes are quite spectacular, particularly the images depicting aurora borealis.

The first of two U.S. IPY expeditions sailed from San Francisco to Point Barrow, Alaska, in 1881. Under the command of Lt. Patrick H. Ray of the U.S. Signal Corps, the small body of men gathered data for more than two years. Note that this expedition began before the final decision was made regarding the form of data collection and recording. Nevertheless the Point Barrow expedition managed to gather an unbroken series of tidal, magnetic, and meteorological data. They also gathered floral and faunal specimens and prepared a significant analysis of the ethnography of the Inuit in the Point Barrow area. The Inuit vocabulary gathered and published in the report remains of interest to this day. The publication of their results— some 700 pages of data and analysis—early in 1885, made this the first of the formal IPY reports to become available.[18] To complicate matters bibliographically, the report was not only issued as a response to a congressional resolution, but also as one of the U.S. Signal Office's Arctic Series, and as a House of Representatives document with a congressional serial number! Of particular interest in the development of scientific publications on the Arctic is that the Point Barrow expedition record is illustrated with the usual color lithographs and engravings, but also with phototypes, an early technical term for metal blocks prepared by photogravure to reproduce a photograph for printing.

The final U.S. expedition was the ill-fated expedition to Ellesmere Island. In what is often referred to as the Greely Expedition, or the Lady Franklin Bay Expedition, the research ship sailed to Lady Franklin Bay in 1881, again before the International Polar Commission determined the details of data gathering and dissemination. While there was a significant amount of data collected—including meteorological, magnetic, hydrological, botanical, zoological, ethnological, and geological data—the expedition was not resupplied by relief ships for two summers. As a consequence of this failure to resupply the expedition and perhaps because of a less than stellar command, 18 of the 25 men died. Most starved or died of scurvy, one committed suicide, and one—and this is the only case in the annals of polar exploration that I am aware of—was executed for stealing food from his fellow starving crew members. There is also evidence of cannibalism. Seven men were rescued in June 1884; one of these men died on the voyage home.

Possibly as a result of the harrowing experiences of the crew and the loss of life, this expedition not only has an official publication of its data with a historical introduction, but also a two-volume narrative prepared by A. W. Greely, the commanding officer, which saw print two years before the official record was released.[19] The official record of the expedition was illustrated in typical fashion of the day: engravings made after photographs. In this case, the photographs were taken by a member of the expedition, George Rice, who had been designated official photographer. In addition, writer Charles Lanman produced a life and times of Lt. James Lockwood, both to solidify Lockwood's claim to the farthest north record and to ensure that Lockwood's reputation was not sullied by the horrors of the expedition. Finally, near the end of his life, David Brainard, one of expedition's heroic figures, published his version of events, including a description of the last desperate weeks before rescue.[20] There have also been a number of modern books on the expedition, none of which has made use of all the records in federal repositories or publicly accessible libraries.[21]

## The *Challenger* Paradigm

There are roughly three periods of exploration in the Arctic during the nineteenth century. The first, in the decades following the Napoleonic era, was essentially an era of geographic exploration in service of empire, employing, in the British case, the surplus vessels of the Royal Navy and a multitude of senior officers on half-pay. This was followed by nearly two decades of searching for Sir John Franklin, and the stirrings of a race to the North Pole. Finally, due in large measure to the circumnavigation of HMS *Challenger*, 1872–1875, with its strong focus on science and the gathering of data and specimens, a new age of polar exploration began.

While some of the early data from the *Challenger* was published while the expedition was under sail and immediately thereafter,[22] the first formal volume of the *Challenger* report was not published until 1880. The final volume was not published until 1895. In those 15 years, 50 volumes containing more than

30,000 pages, 200 maps, and 3,000 lithographs were published. What set the *Challenger* publication record apart from previous expeditions was the five-year grant from the British government to underwrite the salaries and expenses of the editors and scientists. Previously the onus of digesting and editing data fell to the scientific members of an expedition, often at their own expense. While the publishing itself might be underwritten by a government or a sponsor, the salaries of the individuals doing the work was not. While the cost of editing and publishing was a small part of the overall *Challenger* expedition cost, which totaled £170,000, it is significant that the work of digesting and disseminating scientific results was considered a part of the cost of the expedition.[23]

The publication of the *Challenger* expedition data profoundly influenced the publication of all subsequent exploration records. Not only did the quantity of data published increase dramatically, but the quality of the analysis and the analyzed data also rose. The first inkling of this can be found in the report of the fourth IPY Commission meeting in Vienna in 1884. This report contained instructions for the preparation and publication of the IPY data.[24] The *Challenger* paradigm is clearly in evidence in the IPY expedition publications, the first of which reached the public in 1885 with P. H. Ray's Point Barrow report. The *Challenger*'s influence went far beyond what Her Majesty's Parliament ever could have envisioned.

## Conclusion

The publications of the IPY expeditions marked a signal change in the manner in which Arctic science was presented and made available. While the *Challenger* paradigm has much to do with this change, a major factor must also be the international cooperation engendered by both Weyprecht's call for a joint program and the work of the International Polar Commission, particularly in the third meeting in St. Petersburg in 1881 and the fourth meeting in Vienna in 1884. Some specific points to note:

- One of the major issues was the opportunity to share and compare data. This was not successful in the first IPY. The publication of the data from each of the national expeditions did occur, but few nations had a complete set of published results from all the expeditions in one institute or in one library. There is no concrete evidence of exchanges of publications among the nations. Then, too, illustrations, maps, and charts often disappeared from these volumes. It is only recently, with the on-line publication of all the first IPY data by the National Oceanographic and Atmospheric Administration's Arctic Research Office, that the published data is readily available in one place.[25]
- Even with the instructions given by the commission, publication of the data was inconsistent. This is due, in some measure, to the commission's instructions in the third and fourth meetings. The requirements were broadly written and lacked the specificity that would have made both data gathering and publication uniform.

- The reports all contained maps, charts, tables, and illustrations. These were reproduced in a variety of formats, including woodcuts recut as engravings, engravings themselves, lithographs in both color and gray-scale, and phototypes (an early form of photographic reproduction). The varied formats for illustrations meant that some illustrations were useless for scientific purposes.
- Funding for publication was rarely built into the budget of the nongovernmental expeditions. While there was funding for the U.S. expeditions, because they were both organized by the U.S. Signal Corps, other expeditions that were only in part federally or privately funded had to seek publication subvention. This does not appear to have had an impact on the quality of the publications, the length of the works, or the number of illustrations. It does, however, remind us that publication and dissemination of results was often a neglected aspect of scientific research in the nineteenth century.
- All the results were published in a western European language that scholars were expected to know. English, French, and German were accepted as languages in which scientific results could be published and read. In some cases, such as the Dutch expedition, the first report was in Dutch, while the final report with much of the data was published in French.
- The attention paid to instrumentation in all of the reports is worth noting. Prior to the IPY, little was reported about the instruments used to gather data. They were, to be sure, listed in the past and, in some cases, illustrations were provided to show the use of the instruments. In nearly every case of the IPY expeditions, care was taken to identify the instruments and their use and to describe the methods of calibration used to keep the instruments accurate.
- The pattern of waiting for the commander of the expedition to publish his narrative or memoirs before publishing any other materials, scientific or otherwise, is broken by the IPY. Before this joint scientific venture, it had been a custom and sometimes a contractual agreement that the commanding officer of the expedition would publish his narrative before any other expedition-related publications in order for him to reap the glory and sometimes the financial reward of publishing first. Only in the case of the U.S. Lady Franklin Bay expedition, an anomaly in every sense of the word, was the commanding officer's extensive narrative published before the scientific reports.

Finally in a period of much interest in the Arctic and its history, few critical studies have yet been written either about the individual IPY expeditions and their successes and failures or the scientific results of the first IPY.

## Notes

1. See Karl Weyprecht, *An Address Delivered by Lieutenant Charles Weyprecht of the I. R. Austrian Navy, Before the 48th Meeting of German Naturalists and Physicians at Graz,*

of the 18th September1875, Fundamental Principles of Scientific Arctic Investigation (Vienna: W. Stein, 1875); Charles Weyprecht, Discours prononcé devant la 48e assemblée des naturalistes et médecins allemands à Graz (Vienna: W. Stein, 1875).

2. The instructions are conveniently printed in Niels H. Heathcote and Angus Armitage, "The First International Polar Year," Annals of the International Geophysical Year 1 (1959): 9–14. It is unfortunate that there is no general history of Arctic science similar to G.E. Fogg's History of Antarctic Science (Cambridge, United Kingdom: Cambridge University Press, 1992), although Trevor Levere's Science and the Canadian Arctic, A Century of Exploration, 1818–1918 (Cambridge, United Kingdom: Cambridge University Press, 1993) comes close to filling this void. The magisterial work of Helen Roswadowski, particularly "Small World: Forging a Scientific Maritime Culture for Oceanography," Isis 87, no. 3 (1996): 406–429, must be read in order to understand the childhood and adolescence of a number of maritime scientific disciplines in the nineteenth century. There are also a number of studies that deal with specific national Arctic scientific work including Willem F.J. Mörzer Bruyns, "The Dutch in the Arctic in the late 19th Century," Polar Record 23, no. 142 (1986): 15–26.

3. Vorläufige Mittheilung über die Wichtigeren Ergebnisse der Internationalen Polar-Conferenz, Wien, 1884, Minimal-Umfang von jeder Station zu publicirenden Materials (Vienna: K.K. Hof-u. Staatsdruckerei, 1884). The Dartmouth copy of this report is annotated, emended, and signed by A.W. Greely, who had only returned recently from commanding the ill-fated U.S. International Polar Year expedition to Ellesmere Island.

4. This essay does not review the history of the first International Polar Year as this has been ably done by a number of scholars. These include William Barr, The Expeditions of the First International Polar Year, 1882–83, AINA Technical Paper, 29 (Calgary, Canada: Arctic Institute of North America, 1985); William Barr, "Geographical Aspects of the First International Polar Year, 1882–1883," Annals of the Association of American Geographers 73, no. 4 (December 1983): 463–484; Heathcote and Armitage, "First International Polar Year," 6–99; and F.W.G. Baker, "The First International Polar Year, 1882–83," Polar Record 21, no. 132 (1982): 275–285. The First International Polar Year publication record can be conveniently found in the first three volumes of Arctic Bibliography (Montreal: McGill University Press, 1953–55).

5. Die Internationale Polarforschung 1882–1883, Die österreichische Polarstation Jan Mayen, Beobachtungs-Ergebnisse, 3 vols. (Vienna: K.K. Hof- und Staatsdrükerei, 1886). It is not possible to ascertain the print runs or the scientific impact of any of the reports of the First International Polar Year. We do know that there were exchanges of volumes among several of the scientific bodies involved, but the true impact of the research and the publications must wait for a history of science in the polar regions in the nineteenth century.

6. I wish to thank my colleague, Patti L. Houghton, sometime rare book specialist, Dartmouth College Library, for her assistance and advice on this issue and with the question of phototypes.

7. Observations of the International Polar Expeditions, 1882–83, Fort Rae, H.P. Dawson, ed., (London: Eyre and Spottiswoode for Trübner & Co., 1886). A less well-known publication relating directly to the Fort Rae program is Great Britain, Meteorological Office, Contributions to Our Knowledge of the Meteorology of the Arctic Regions, Meteorological Office Official Publication No. 34, 2 vols. in 1 (London: HMSO, 1885–88).

8. Observations faites à Godthaab, 2 vols. (Copenhagen: Institut météorologique de Danemark, 1886–94).

9. A.P. Hovegaard, "Die Eiszustände im Karischen Meere," Petermanns Geographische Mitteilungen 30 (1884): 253–259; A.P. Hovegaard, Dijmphna expeditionen 1882–83, Rapporter til Dijmphna's rheder, Herr grosserer, fabrikeier Augustin Gamél (Copenhagen: Forlagsbureauet in Kjøbenhavn, 1884); and C.F. Lütken, Dijmphna-togtets zoologisk-botaniske udbytte (Copenhagen: H. Hagerup, 1887).

10. Maurits Snellen, *De Nederlandische Pool-Expeditie 1882–83* (Utrechet: L.E. Bosch en Zoon, 1886); Maurits Snellen and H. Ekama, *Rapport sur l'expédition néerlandaise qui a hiverené dans la mer de Kara en 1882/83* (Utrecht: J. Van Boekhoven, 1910).

11. Selim Lemström and Ernest Biese, *Observations faites aux stations de Sodankylä et de Kultala*, 3 vols. (Helsingfors: Imprimerie des héritiers de Simelisu, 1886–98).

12. *Die Internationale Polarforschung 1882–1883, Die Beobachtungs-Ergebnisse der deutschen Stationen*, 2 vols. (Berlin: Verlag von A. Asher, 1886); *Die Internationale Polarforschung 1882–1883, Die Beobachtungs-Ergebnisse der deutschen Stationen*, 2 vols. (Berlin: Verlag von A. Asher, 1890–91).

13. See G.J. Pfeffer, "Mollusken, Krebse und Echinodermen von Cumberland-Sund nach der Ausbeute der deutschen Nordexpedition 1882 und 1883," *Jahnburch der Hamburgischen Wissenschaftlichen Anstalten* 3 (1886): 23–50; H. Abbes, "Die deutschen Nordpolar-Expedition nach dem Cumberland-Sunde," *Globus* 46 (1884): 294–298, 312–315, 328–331, 343–345, and 365–368; and H. Abbes, "Die Eskimos des Cumberland-Sundes," *Globus* 46 (1884): 198–201 and 213–218.

14. *Die Internationale Polarforschung, 1882–1883, Beobachtungs-Ergbnisse der Norwegischen Polarstation Bossekop in Alten, Im Auftrage des Königl. norwegischen Cultus-Ministeriums, herausgegeben von Aksel S. Steen*, 2 vols. (Christiania [Oslo]: Gröndahl & Sons for the Cultus-Ministerium, 1887–88). The historical introduction is in part 1 of the first volume.

15. *Ekspeditsiya k ust'yu reki eny, 1882–84, Trudy Russkoy polyarnoy stantsii na ust'ye leny. Beobachtungen der russischen Polarstation an de Lenamündung*, 2 vols. (St. Petersburg: Tipografiya Imperatorskoy Akademii Nauk, 1886–95).

16. *Trudy Russkoy polyarnoy stantsii na Novoy Zemle, Beobachtungen der russischen Polarstation auf Nowaja Semija*, 2 vols. (St. Petersburg: Tipografiya Imperatorskoy Akademii Nauk, 1886–91).

17. *Observations faites au cap Thordsen, Spitzberg, par l'expédition suédoise, publiées par l'Academie Royale des Sciences de Suède*, 2 vols. (Stockholm: P.A. Norsted & Söner for l'Academie Royale des Sciences de Suède, 1886–91).

18. *Report of the International Polar Expedition to Point Barrow, Alaska, in Response to the Resolution of the [U.S.] House of Representatives of December 11, 1884* (Washington, D.C.: U.S. Government Printing Office, 1885); U.S. Signal Office, Arctic Series of Publications 1 (1885); and U.S. Congress, House of Representatives, ex. doc. 44, 48th Congress, 2nd session, vol. 23, no. 2298.

19. *International Polar Expedition, Report on the Proceedings of the United States Expedition to Lady Franklin Bay, Grinnell Land*, U.S. Congress, House of Representatives, Miscellaneous Document No. 393, 2 vols. (Washington, D.C.: U.S. Government Printing Office, 1888); and Adolphus W. Greely, *Three Years of Arctic Service, An Account of the Lady Franklin Bay Expedition of 1881–84 and the Attainment of the Farthest North* (New York: Charles Scribner's Sons, 1886).

20. Charles Lanman, *Farthest North; or, the Life and Explorations of Lt. James Booth Lockwood, of the Greely Arctic Expedition* (New York: D. Appleton & Co., 1885); and David L. Brainard, *Six Came Back: The Arctic Adventure of David L. Brainard* (Indianapolis: Bobbs-Merrill Co., 1940). Brainard's diary of the final weeks of the expedition, when many died and Charles Henry was executed, purports to be printed in extenso in the book. It is not; the version in the book is heavily expurgated and a comparison of that text with the original at Dartmouth is revealing. In addition to many of Brainard's papers, Dartmouth also holds a body of papers of A.W. Greely, the diaries of P.W. Johnson, George Rice, and Frederick Hoadly, and Henry Howgate's notes on the genesis of the expedition. Most of these papers have not been thoroughly utilized by modern researchers. In addition, the remaining volumes of the Fort Conger library are in the collections of the Explorers Club in New York City. I am grateful to Clare Flemming, sometime curator of Research Collections at the Explorers Club for this information.

21. Among the more recent publications are Pierre Gauroy, *Les affamés de la banquise* (Paris: A Bonne, 1964); Leonard Guttridge, *The Ghosts of Cape Sabine: The Harrowing True Story of the Greely Expedition* (New York: G.P. Putnam's Sons, 2000); A.L. Todd, *Abandoned: the Story of the Greely Arctic Expedition, 1881–1884*, with a foreword by Terrence Cole and introduction by Vilhjalmur Stefansson, 2nd ed. (Fairbanks: University of Alaska Press, 2001), first published in 1961; and Geoffrey E. Clark, *Abandoned in the Arctic: Adolphus W. Greely and the Lady Franklin Bay Expedition, 1881–1884* (Portsmouth, New Hampshire: Portsmouth Athenaeum, 2007). The last-named volume accompanies a documentary film of the same name.

22. See, for example, *Reports on Ocean Soundings and Temperatures, New Zealand to Torres Strait, Torres Strait to Manila and Hong Kong* (London: Her Majesty's Stationary Office, 1874) and *Report on Ocean Soundings and Temperatures in the Atlantic Ocean* (London: Her Majesty's Stationary Office, 1876).

23. Margaret Deacon, *Scientists and the Sea, 1650–1900, A Study of Marine Science*, 2nd ed. (Brookfield: Ashgate Publishing Co, 1997) remains the best study of the work of the Challenger expedition. Aside from the importance of the analysis found in chapter 15 of the work, pp. 333–365, her footnotes and bibliography provide a wealth of information regarding contemporary papers and publications, official and otherwise.

24. *Vorläufige Mittheilung über die Wichtigeren Ergebnisse der Internationalen Polar-Conferenz, Wien, 1884, Minimal-Umfang von jeder Station zu publicirenden Materials.*

25. Available on-line at http://www.arctic.noaa.gov/aro/ipy-1/Data-P1.htm (accessed October 24, 2007).

## Chapter 4

# Toward the Poles: A Historiography of Scientific Exploration during the International Polar Years and the International Geophysical Year

*Roger D. Launius*

Historians around the world have paid insufficient attention to understanding both the Arctic and the Antarctic. While there may be considerable historical attention paid to these polar regions from explorers, geographers, surveyors, anthropologists, archaeologists, ethnologists, ethnographers, geologists, botanists, climatologists, ornithologists, zoologists, journalists, and novelists—and some of this work is excellent—each has been molded by the perspectives and methodologies of his or her discipline.[1] As Michael Bravo and Sverker Sörlin recently concluded: "Arctic [and Antarctic] exploration and research was for a long time part of public memory and national myth-making rather than of professional historiography; not until very recently has this been considered a worthwhile undertaking by professional scholars."[2] Too much of this writing about the poles has been anecdotal—some of it with a propensity for sensationalism—and cannot thus be considered an authoritative source of history. In the end, there has been a scarcity of historically oriented, intellectually rigorous historical analysis of this theme by those best suited to write it.

More than this, the histories of the first and second International Polar Years (IPYs)—1882–83 and 1932–33—and the International Geophysical Year (IGY) of 1957–58 has failed to elicit much in the way of serious scrutiny from scholars probing the past. The IPYs and the IGY have largely attracted historical attention when the historian has been interested in some other subject. For example, the IGY is always a part of the historical record when discussing the origins of the space age and the Sputnik crisis of 1957–58. Likewise, there are a few general histories of Arctic and Antarctic exploration that contain sections on the IPYs and the IGY. There are also popular accounts of these events, but only a few scholarly discussions exist. There are

TAKING AN OBSERVATION AT THE POLE.

**Figure 4.1**   Polar explorer Roald Amundsen arrived in the general vicinity of South Pole on the afternoon of December 14, 1911. This image is from *The South Pole* (1913) by Roald Amundsen. (Courtesy of National Oceanic and Atmospheric Administration, NOAA Library Collection, Washington, D.C.)

documentary and biographical accounts concerning people and events associated with these scientific efforts, but as yet there is neither a sophisticated literature of the IPYs and the IGY, nor of the major questions that these activities engendered: the origins of these efforts, their political dimensions, the scientific questions and conclusions, and their consequences. Major themes not fully addressed in what does exist in the scholarly literature include the place of the poles in the human imagination; the place of these explorations in scientific-discipline formation, cultural nationalism, politics, and transnationality; the emergence of the modern geosciences; the uses of new technologies to explore the poles; changing assessments of the nature of human cultures in high latitudes; and polar contributions to environmental awareness. This essay will explore some of this literature and assess its place in creating a historiography of the IPYs and the IGY and its place in the larger story of human concern about the poles and the natural aspects of Earth as a global system.

## General Works

There is a range of general histories on the Arctic and Antarctic regions and of the efforts of those living elsewhere in the world to understand them more effectively. Perhaps the most useful general overview of Arctic history is Richard Vaughan's *The Arctic: A History*, published in 1994. After a survey of indigenous Arctic peoples and cultures, Vaughan then concentrates on exploration and empire in the north country and includes sections on the first and second IPYs.[3] Syntheses are never easily assembled, and Vaughan's is about as sophisticated as one could hope for in a general overview.

Although originally published in 1934 and updated only in 1970, another useful popular account of the Arctic region and its exploration remains Jeannette Mirsky's *To the Arctic: The Story of Northern Exploration from Earliest Times to the Present*.[4] It contains a lengthy discussion of the first IPY (calling it the International Circumpolar Year) and concentrates on the Greely expedition to Lady Franklin Bay in northwestern Greenland to establish one of a series of meteorological, astronomical, and polar magnetic observation stations for the IPY. Sponsored by the U.S. Army Signal Corps, this expedition was stranded for more than two years and, when finally rescued, 18 of the 25-person crew had died from starvation, drowning, hypothermia, and in one case a firing squad ordered by its commander, Lt. Adolphus Greely. One more man died after the rescue.[5] This is a dramatic episode to which Mirsky and virtually all other writers about the first IPY have been drawn like a moth to a flame.

There are, of course, other somewhat interesting and useful general histories of Arctic exploration. These include, by publication date: L.P. Kirwan, *A History of Polar Exploration* (New York: W.W. Norton and Co., 1960); L.H. Neatby, *Conquest of the Last Frontier* (Athens: Ohio University Press, 1966); Ragnar Thorén, *Picture Atlas of the Arctic* (New York: Elsevier, 1969); Louis Rey, ed., *Unveiling the Arctic* (Anchorage: University of Alaska Press, 1984); Fred Bruemmer, William E. Taylor, and Ernest S. Burch, *The Arctic World* (Toronto, Canada: Key Porter Books, 1987); Mick Conefrey and Tim Jordan, *Icemen: A History of the Arctic and its Explorers* (New York: T.V. Books, 1998); Charles Officer and Jake Page, *A Fabulous Kingdom: The Exploration of the Arctic* (New York: Oxford University Press, 2001); Fergus Fleming, *Ninety Degrees North: The Quest for the North Pole* (New York: Grove Press, 2002); and Richard Sale, *Polar Reaches: The History of Arctic and Antarctic Exploration* (Seattle, Washington: Mountaineers Books, 2005). Most of these are little more than descriptive narratives; many are popular in tone and content. Some, such as Steven B. Young, *To the Arctic: An Introduction to the Far Northern World* (New York: John Wiley, 1989), is a more scholarly guide focusing on natural history. All these works are readable and offer a broad overview as a starting point for the history of the IPYs and the IGY as they relate to the larger history of the Arctic. Derek Hayes has produced an especially useful *Historical Atlas of the Arctic* (Vancouver, Canada: Douglas & McIntyre, 2003), which contains more than 300 maps relating to all phases of Arctic exploration and activities. Finally David Sugden, *Arctic and Antarctic: A Modern Geographical Synthesis* (New York: Rowman and Littlefield Publishers, Inc., 1982), makes available an introduction to the geography of the polar regions.

When considering Antarctica, there are a range of general historical works that provide an overview of the subject. Western civilization has spent much less time at Terra Australis Incognita—the unknown southern land, as Antarctica was called. While Captain James Cook sailed into the region briefly in the seventeenth century, in 1819 the first seafarers of record spied the Antarctic Peninsula when a merchant captain named William Smith sailed around Cape Horn on his way to Valparaiso, Chile, only to detour southward to avoid unfavorable winds. On February 20, he fixed the position of an uncharted landfall at 62°17'S and 60°12'W, further south than anyone had ever before recorded. Thereafter seal hunters from throughout Western civilization initiated the virtual extermination of South Shetland Island seals. Seal hunting dominated Antarctic enterprise throughout the remainder of the nineteenth century and drove settlements on outlying islands north of Antarctica.[6]

Other voyages of discovery followed, but only in 1898 did the first explorers begin travels overland on the continent in search of the South Pole. The so-called heroic age of exploration of Antarctica followed, as Roald Amundsen, Ernest Shackleton, Erich von Drygalski, Robert Falcon Scott, Otto Nordenskjld, Jean-Baptiste Charcot, and others sought the South Pole. Amundsen eventually arrived there in 1912 after a race to the pole against Scott. While these expeditions had a cache of scientific pursuit surrounding them, they were largely undertaken by adventurers in search of fame in the drawing rooms of Europe and North America. The enhancement to science, therefore, was some knowledge of the geography of the Antarctic continent. These efforts in Antarctica, while there had been some incursions, were not central to the first and second IPYs, but during the IGY in 1957–58 they became a major goal for scientific exploration.

General histories of Antarctica are not numerous, and none are authoritative. The only work of substance is Stephen Martin, *A History of Antarctica*, a judicious account of scientific discovery and recent international cooperation. Concerned especially with the evolution of the Antarctic Treaty, Martin avoids much discussion of the so-called "heroic age" of Antarctic exploration when famous explorers such as Shackleton, Amundsen, Scott, and Douglas Mawson led their expeditions to the South Pole.[7] In Ian Cameron's *Antarctica: The Last Continent*, the emphasis is on dramatic stories of exploration from the earliest Polynesians in about 600 CE until the time of the book's publication in 1974.[8] Even so, none of these works approaches the sophistication of Richard Vaughan or Jeannette Mirsky on the Arctic as full-blown histories of Antarctica.

For the earliest encounters of the West with Antarctica, perhaps the best work is Michael H. Rosove's *Let Heroes Speak: Antarctic Explorers, 1772–1922*.[9] Focusing on 150 years of polar exploration, Rosove begins with the efforts of Captain James Cook, the great emissary of the British Empire, and closes with Ernest Shackleton's final expedition in 1922. In between, he offers a wide-ranging narrative emphasizing these core questions: What lured these explorers? What was their motivation? How did they measure success? How harsh was the natural environment, and how did they cope with it?

Finally, what had they wrought? Almost as useful is *A Chronology of Antarctic Exploration: A Synopsis of Events and Activities until the International Polar Years, 2007 to 2009*, compiled by Robert Headland.[10] This is the second edition of an authoritative sourcebook on Antarctic discovery and exploration.

There are several quite good national overviews of Antarctic activities. *Americans in Antarctica, 1775–1948* by Kenneth J. Bertrand is a case in point.[11] It contains an authoritative narrative of more than 500 pages on the United States on the continent, all well-researched and well-written, but offers little in the way of transnational relations or analysis of the United States in relation to other exploration efforts. Dean Beeby, *In a Crystal Land: Canadian Explorers in Antarctica*; Douglas Mawson, *Home of the Blizzard: The Story of the Australasian Antarctic Expedition, 1911–1914*; Lars Christensen. *Such Is the Antarctic*, about the Norwegian whaling industry and Antarctic exploration; and David Thomas Murphy, *German Exploration of the Polar World, 1870–1940*, are also authoritative for the nations they chronicle.[12] T.H. Baughman, *Before the Heroes Came: Antarctica in the 1890s*, details the history of Antarctic exploration during a single decade, one characterized by rising interest from the West but before the continent was assaulted by the adventurers who would come near the turn of the twentieth century. He concentrates on the work of Matthew F. Maury and Georg von Neumayer, both of whom undertook meteorology and physical-science observations inside the Antarctic Circle. Additionally the scientifically spectacular mission of Adrien de Gerlache in 1897 stimulated new interest in the region and sparked the onslaught that would follow during the "heroic age" of the first part of the twentieth century.[13] Finally, *Ice: Stories of Survival from Polar Exploration* edited by Clint Willis contains excerpts from the famous accounts of Antarctic exploration by Apsley Cherry-Garrard, Ernest Shackleton, Douglas Mawson, and others.[14]

## The Poles in the Human Imagination

Joseph Campbell once wrote, "Myths are public dreams, dreams are private myths."[15] For humanity the poles have been mythological in both concept and fact. This obsession has been discussed in several outstanding books that explore the place of the Arctic and Antarctica in the human imagination. An excellent place to start is Eric G. Wilson, *The Spiritual History of Ice: Romanticism, Science, and the Imagination*.[16] Wilson asserts that near the end of the eighteenth century, scientists, writers, mystics, and others combined to reinterpret ice, glaciers, and cold places as something beyond a traditional wasteland, indeed, a place of romance and spiritualism. While much of this reinterpretation represented wishful or even wrongheaded thinking, the romantic imagination inspired by it energized Western civilization's quest to explore it. Focused on the literary heritage of ice as mythical place, Wilson offers a convergence of crystallography, esoteric glaciology, group psychology, and polar exploration into a portrait of an age and its fascination with frozen phenomena.

Following closely on the metaphysical attributes of Wilson's investigation of *The Spiritual History of Ice* is Barry Holstun Lopez's *Arctic Dreams*. Lopez observes that "it is possible to live wisely on the land, and to live well. And in behaving respectfully toward all that the land contains, it is possible to imagine a stifling ignorance falling away from us." He finds that the Arctic is "rich with metaphor, with adumbration." There are three themes at the center of his narrative: "the influence of the arctic landscape on the human imagination. How a desire to put a landscape to use shapes our evaluation of it. And, confronted by an unknown landscape, what happens to our sense of wealth. What does it mean to grow rich?" In the end Lopez thinks that the land is like poetry, "it is inexplicably coherent, it is transcendent in its meaning, and it has the power to elevate a consideration of human life."[17]

If Lopez approaches the Arctic as poetry, Robert McGhee interprets the experience of Western civilization there in *The Last Imaginary Place: A Human History of the Arctic World*, even as it uses powerful language to make his points.[18] As the curator of Arctic archaeology at the Canadian Museum of Civilization, McGhee has a wealth of experience in the region, and he uses it to relate history with personal memoir to offer a realistic view of the region, its challenges, and its place in world history. Beau Riffenburg's *The Myth of the Explorer: The Press, Sensationalism, and Geographical Discovery* and John McCannon's *Red Arctic: Polar Exploration and the Myth of the North in the Soviet Union 1932–1939* are two excellent books that examine the process whereby sensation-hungry Americans and Europeans responded to the accounts of explorers in the Arctic. These books reveal the subterfuge and the bravery of those undertaking the various expeditions, especially the polar quest. A similar work analyzing the myth of exploration in Antarctica is Stephanie Barczewski's *Antarctic Destinies: Scott, Shackleton, and the Changing Face of Heroism.*[19]

Four key books in the historiography of the polar imagination require the attention of all who seek to understand this subject. In *The Coldest Crucible: Arctic Exploration and American Culture*, Michael F. Robinson focuses on the public perception of American Arctic explorations in order to illuminate developments in the political and cultural history of the United States between 1850 and 1909. He approaches the Arctic as "a faraway stage on which explorers played out dramas that were unfolding close to home." He unpacks the political and cultural demarcations of American culture and uses the polar explorers as a means of illuminating society. Most important, Robinson finds that the explorers offered an unequivocal statement of American exceptionalism that all could embrace. In the end, he asserts that Americans used the Arctic as a surrogate for other controversies and dramas played out far from the snow-covered region.[20]

Two books on the British fascination with the poles are Francis Spufford, *I May Be Some Time: Ice and the English Imagination*, and Robert G. David, *The Arctic in the British Imagination, 1818–1914.*[21] Spufford's work seeks to show the relationship between polar exploration and English literature. He asks why British polar explorers willingly placed their lives in jeopardy in the harsh polar environment; was it gold or glory or something else? The answer, Spufford believes, rests not with the explorers themselves but with the English imagination as expressed in the writings of such authors as the Brontës, Edmund Burke, Samuel Taylor Coleridge, Charles Dickens, and others. As one reviewer commented:

> Spufford's study is about the power of ideas formed in one geographical place to distort perceptions of other environments and how the experiences of the early polar explorers redefined heroism in ways that allowed abandoned wives, sisters, and female admirers to identify with their brave and suffering kinsmen and compatriots. His book shows how the concept of the sublime articulated by Burke and Coleridge, in which nature simultaneously inspires and terrifies, was corrupted by Markham and Scott into a nostalgic identification of the poles with bygone English glory. Conquest of the polar regions by Englishmen was inevitable, but so was failure.[22]

In contrast, Robert David seeks to answer why the British government pressed so doggedly for the Northwest Passage and a route through the Arctic, despite disaster. For David the Arctic was a cultural constant for the British people and it energized excitement through a variety of representations. These not only solidified support for public expenditures but also ritualized the sacrifice of those lost in the effort.

## Before the IPYs and the IGY: The "Heroic Age" of Exploration

Although the IPYs and the IGY were not much associated with the "heroic age" of exploration, the earlier efforts shaped much that followed in the scientific undertakings of those international endeavors.[23] The literature on the heroic age is massive, and the following recounts only a few of the best works on the subject. Much of the heroic age was predicated on reaching one or the other of the poles, which achievement involved fame, fortune, bragging rights, and in some cases imperial claims of one expedition leader over another. All this was played out in the first quarter of the twentieth century, as the Arctic and Antarctic represented the last parts of the global map to be filled in with detail.

The stories of Arctic discovery and scientific exploration for the heroic age typically fall into three broad categories. The first is the sometimes coordinated, at other times competitive, effort to find the Northwest Passage, a far-north water route from Europe to Asia.[24] Ironically, because of global warming this is now beginning to open up, with Arctic sea ice melting. Sought for centuries by various European nations as a possible trading route, only in 1903–06 was it finally navigated by the Norwegian Roald Amundsen. Even so, he found the Arctic icepack prevented regular marine operations. That might have been to the good because contested claims of sovereignty could have sparked international crises had there been navigable waterways.[25] In the twenty-first century, furthermore, as the ice is thinning

**Figure 4.2** A trail through the ice serving as the gangway to the research vessel *Gauss*. From *Deutsche Südpolar-Expedition 1901–1903 Meteorologie I* (1911) by Erich von Drygalski. (Courtesy of National Oceanic and Atmospheric Administration, NOAA Library Collection, Washington, D.C.)

and the Arctic Ocean is more navigable than ever before, this issue has come to the fore.[26] The place to start in understanding this quest is Pierre Berton, *The Arctic Grail: The Quest for the North West Passage and the North Pole, 1818–1909*.[27] Berton, a Canadian who has evinced a lifelong fascination with the poles, offers a narrative account of the efforts of British, American, and Scandinavian explorers seeking the Northwest Passage through Canada's northern archipelago and the North Pole. He emphasizes not only the romance motivating this effort, but also the poor planning and administrative bungling that torpedoed so many of the expeditions sent there. The British Admiralty, especially, sent a succession of failed missions in search of the Northwest Passage. At a remarkable level the arrogance and romance of Western civilization combined for more than a century of fruitless search for a presumed economic trade route that did not exist. Also comprehensive, and more up-to-date, are Glyn Williams's *Voyages of Delusion: The Quest for the Northwest Passage* and Ann Savours's *The Search for the North West Passage*, which both cover much the same ground.[28] An indispensable reference work, *Historical Dictionary of the Discovery and Exploration of the Northwest Passage*, compiled by Alan Edwin Day, provides detailed accounts of expeditions and issues associated with the subject.[29]

The Russians also sought the Northwest Passage in their own way and over a long period of time. L.M. Starokadomskiy's *Charting the Russian Northern Sea Route: The Arctic Ocean Hydrographic Expedition 1910–1915,* after translation into English in 1976, quickly became the standard account of Imperial Russia's attempt to survey a Northeast Passage route. Likewise, Terence E. Armstrong, *The Northern Sea Route: Soviet Exploitation of the North East Passage,* carried this story forward until World War II, demonstrating that interest in the quest did not wane with the Bolshevik Revolution and the creation of the Soviet Union. Concluding this Russian/Soviet story, Lawson W. Brigham has edited *The Soviet Maritime Arctic,* which offers an up-to-date treatment of the subject from the standpoint of environmental, economic, legal, and geopolitical issues.[30]

Of course, no expedition in search of the Northwest Passage has garnered more scrutiny than the futile efforts of Sir John Franklin, the British Royal Navy officer who was on a sustained mission to chart the Northwest Passage in the Canadian Arctic when his entire expedition perished in 1847. Icebound, Franklin's party suffered the slow decimation of starvation, hypothermia, exposure, and cannibalism. Numerous rescue and recovery expeditions sought to learn the fate of the expedition, but no definitive tale has yet been pieced together to explain all of what happened. Many books have been written on this party and the search for its remnants. Patricia D. Sutherland edited a solid collection of essays, *The Franklin Era in Canadian Arctic History, 1845–1859,* that present the basic story.[31] Richard J. Cyriax, *Sir John Franklin's Last Arctic Expedition: A Chapter in the History of the Royal Navy,* offered an early analysis of the fate of Franklin and his men and the many searches for them that followed their disappearance in addition to an evaluation of the Arctic environment of the time.[32] David Woodman's *Unraveling the Franklin Mystery: Inuit Testimony* offered new perspectives on the fate of the expedition using accounts from native inhabitants to reconstruct what might have happened.[33]

The second major category of Arctic exploration involves the race to the North Pole.[34] *True North: Peary, Cook, and the Race to the Pole,* by Bruce Henderson, tells the story of intense hatred and jealousy between Americans Robert E. Peary, the supposed discoverer of the North Pole in 1909, and his former colleague, Frederick Cook, who claimed to have reached it a year earlier. Attacks on the generally accepted Peary account began in 1911 with a book by the rival claimant, Frederick Cook in his *My Attainment of the Pole,* a dispute that has played out ever since.[35] In *True North* the author marshals evidence to support the claim of Cook that he reached the pole in April 1908, a full year before Peary.[36] There have been many other books dealing with the debate about who was the first to the reach the North Pole, all of them making a case for one or the other of the rival explorers. Sometimes the prose is laced with vitriol. Definitively concluding which explorer, if either, was first to reach the North Pole continues to be debated in the popular media. Although Peary received great acclamation in his lifetime for his polar exploit, most current professionals view with skepticism the claims of both Peary and Cook to have reached the North Pole. More than a century after the controversy first erupted it seems that neither claim holds up well.[37]

Finally a third category of Arctic exploration of the heroic age involves international efforts to learn about the region. Samuel Elliot Morison, *The*

*European Discovery of America*, volume 1, *The Northern Voyages, A.D. 500–1600*, is a seminal work that includes discussions of all early efforts to learn about the Arctic before the Enlightenment, while Alan Cooke and Clive Holland, *The Exploration of Northern Canada, 500 to 1920*, offer a detailed chronology that fleshes out and extends Morison's work.[38] Nancy Fogelson, *Arctic Exploration & International Relations, 1900–1932*, emphasizes the international character of many of the explorations of the Arctic, particularly relating to American-Canadian relations.[39] Finally Jean Malaurie, *Ultima Thule: Explorers and Natives in the Polar North*, offers a heavily illustrated collection of explorers' tales excerpted from their accounts of the efforts emphasizing relations with the native peoples of the Arctic.[40]

The Antarctic experience during the heroic era is every bit as fascinating as it is for the northern reaches of the planet. Like the experience at the North Pole, Antarctic explorers sought first to reach the South Pole and then to cross the continent, and numerous books have been written to tell the story.[41] Norwegian explorer Roald Amundsen,[42] already well-known for his efforts in the Arctic, raced Englishman Robert Falcon Scott[43] to the geographic South Pole, arriving there on December 11, 1911, a month before Scott. Along with four others, Scott reached the pole a month later but died of starvation and extreme cold during the return trip.

Roland Huntford's *Scott & Amundsen*, issued in various editions under several titles beginning in 1979, is a benchmark analysis that accompanied a successful television documentary on the race to the South Pole. It presents a convincing case for Amundsen to have reached the pole before Scott—an issue then in dispute among dilettantes—and finds that Scott was both an unsavory character and a poor organizer. The failure of logistics that led to Scott's defeat in the race, and to the loss of his life in the process, was the opposite of the strong organizational imperative demonstrated by Amundsen. Huntford celebrated Amundsen as a hero for a technological and organizational age while assessing Scott as "a suitable hero for a nation in decline," finding that the British Empire was declining in part because of its fixation on the worldview of a previous era, unaware that its time at the top of the community of nations was nearing its end.[44]

The British explorer Ernest Shackleton's Imperial Trans-Antarctic Expedition of 1914–17 has gained notoriety for heroic efforts to survive and return home after being icebound and all but lost for more than a year. Over time, Shackleton's willingness to forego his quest for a continental crossing in favor of ensuring that all the men on his expedition made it home safely gained for him later acclaim far beyond what he enjoyed during his life. He is now routinely invoked as the model of a proper English explorer.[45] Finally, in the 1920s, as virtually the last act of the heroic age, U.S. Admiral Richard Evelyn Byrd employed the new capabilities offered by the airplane to complete flights over the poles.[46] At sum, the nature of public adulation—coming as it did to these various adventurers and explorers, for reasons not fully understood—has been well explored in several works.[47]

## Science at the Poles

The polar regions have fascinated scientists and others from the first time they encountered them. How is it possible to understand an environment that tolerates virtually no life, one that is derived from a single substance, and one that is for the most part a single color? The level of both horror and enthrallment experienced simultaneously at the poles' awesome desolation was something everyone commented on when first encountering it. But horror and enthrallment, despite their reasonableness as human reactions, were insufficient for scientists. The nineteenth century witnessed the rise of a scientific order that tackled the systemization of knowledge about the planet.[48] Several ambitious projects resulted, including the Magnetic Crusade, a cooperative endeavor to solve fundamental questions in terrestrial magnetism; international and systematic efforts to collect meteorological data; and multiple networks to record astronomical observations. It was logical that this systematizing of knowledge would include the polar regions and inhabited parts of the globe, but it was not until the latter part of the century that the ability to undertake large-scale observations, especially in terms of safety and survival, extended to the poles. Once accomplished, the results transformed the Western world and set in motion the major elements of the twentieth century's reliance on science and technology for its welfare.[49]

More than this, polar science became a useful tool for delineating empire. At a sublime level, it represented the "pursuit of power through the pursuit of knowledge."[50] Scholars have argued that scientific colonialism dominated the pursuit from at least the eighteenth century on; as Roy McLeod concluded, "[T]he central issue becomes no longer science in imperial history, but science as imperial history."[51] Of course there were eminently practical reasons for pursing knowledge about the polar regions. The most knowledge of geography, meteorology, geomagnetism, oceanography, and a host of other disciplines enabled empires to survive in these regions. Learning about the flora and fauna, in addition to the lifestyles of the native peoples, also enabled survival. Besides these practical efforts, those engaged in this gathering of information pursued studies that excited them, some of which were pathbreaking. The more they learned, the more the nations that supported them were able to thrive at the poles, enhancing the imperial stature of the nation involved.[52]

The poles have repeatedly served as a vehicle by which to focus national aspirations; competition among international rivals played out in these settings. In the process the science to be conducted could be co-opted by the larger imperial design, just as it was in other settings and times and disciplines. "Debate about the origins of arctic science may be of interest to scholars," according to Cynthia Lamson; "however, there is general consensus that political and economic circumstances were—and remain today—principal factors in promoting scientific enterprise."[53] For the Arctic, this might be called a commitment to "Nordicity," a shorthand explaining the region's place in shaping national self-image and the role of science in promoting it.[54]

Specifically on the question of the history of Arctic science, the starting point is a general overview of the early era, *Unveiling the Arctic: Conference on the History of the Discovery of the Arctic Regions as Seen through the Descriptions of Travellers and the Work of Cartographers from Early Antiquity to the 18th Century.* Edited by Louis Rey, Marvin W. Falk, and Claudette Reed Upton, this landmark collection of essays emerged in 1986 from a conference held at the Vatican Library in Rome to assess the state of scientific knowledge and its formation in the region through the eighteenth century.[55] As one reviewer noted, "*Unveiling the Arctic* presents a fascinatingly detailed picture of struggle and growing technical achievement in one of the world's harshest environments."[56]

For a contemporary discussion of scientific knowledge about the Arctic, a solid, jubilant narrative is offered in Barry Lopez's *Arctic Dreams: Imagination and Desire in a Northern Landscape.* It offers a portrait of the landscape, animals, and people of the contemporary Arctic, eloquently capturing the beauty and peril of the region, its harsh climate, and its place in shaping the more temperate parts of the world teeming with modern, urbanized life.[57]

A strong book on the role of one scientific institution in the Arctic may be found in Debra J. Lindsay, *Science in the Subarctic: Trappers, Traders, and the Smithsonian Institution.*[58] It tells the story of the mid-nineteenth-century collecting adventures of the Smithsonian's Spencer Baird and Robert Kennicott in the subarctic and the Arctic. Baird, the Smithsonian Institution's second secretary, set out to improve knowledge of the North American flora and fauna. His efforts to increase the number of specimens gathered and to ensure their quality led to a revolution in the nature of scientific data collection—how specimens were collected and processed, by whom, and why. Seeking to end ignorance of northern regions of North America, Baird sent Robert Kennicott, a skilled young naturalist, to the Mackenzie River in northwestern Canada in search of natural history specimens. Kennicott eventually brought home more than 12,000 natural history specimens, several volumes of field notes, ethnographic accounts from the region, and more than 500 indigenous artifacts that were among the earliest anthropological specimens at the Smithsonian. Accomplished between 1859 and 1868, Kennicott's pioneering efforts became the foundation of the Smithsonian Institution's subarctic and Arctic scientific studies.

Another work on Arctic science, this one focused on a single national effort, is *Science and the Canadian Arctic: A Century of Exploration, 1818–1918*, by Trevor H. Levere.[59] Beginning with the British Royal Navy's expeditions in Arctic Canada in 1818, the author assesses both scientific exploration and nationalistic aspirations in this important study of how knowledge was created about the polar region. Contained therein are discussions of such episodes as the Franklin expedition, the Hudson's Bay Company's practical approach to creating knowledge for private gain, and a host of other scientific expeditions. The author views the British effort as central to this story, followed by the Americans and Canadians, but he does give the Norwegians, Swedes, and Germans high marks for their efforts late in the nineteenth century. The central themes of the search for a Northwest Passage, science and colonialism, imperialism and nationalism, and scientific internationalism press the analysis forward into the twentieth century. As Levere makes clear, during the period discussed in this work science moved from a national undertaking to "one of nondependence and integration into an international scientific culture."[60]

A second national perspective is offered in the edited work of Michael Bravo and Sverker Sörlin, *Narrating the Arctic: A Cultural History of Nordic Scientific Practices*.[61] It offers a sober collection of seminal essays concerning the history of Arctic exploration. Making a comparative study of episodes in the history of science and Arctic exploration, editors Bravo and Sörlin make clear that far from there being a united Nordic or Scandinavian history of Arctic science, narratives of colonial science about the Arctic are each different from the other. These narratives move from Inuit exploration to Danish Arctic research to Swedish Arctic exploration. At center is the national character of these scientific efforts. Accordingly, taking center stage are the stories of encounters between Danes and Greenlanders, Swedish naturalists and the Saami, and ethnographers and Inuit. Bravo and Sörlin conclude with a plea for historians of exploration to do more comparative work and be more attentive to the deep narrative structures in exploration historiography.

Finally William F. Althoff, *Drift Station: Arctic Outposts of Superpower Science*, carries to the present this story of Arctic science propelled by imperial ambitions.[62] Floating on the icecap, these stations operated by a multitude of nations have been both the intentional and the unintended consequences of national scientific efforts. For example, Althoff begins his narrative with a discussion of the science conducted during the three-year drift (1893–96) of the research vessel *Fram*, caught in the floe during an expedition led by Fridtjof Nansen. The Soviet Severnyy Polyus-1 (North Pole-1) or SP-1 expedition of 1937–38 brought drift stations into the modern era and, after an interruption for World War II, they remained in continuous existence thereafter. The first American drift station was established in 1950 and others followed thereafter. Until the last one finally shut down in 2004, the Soviet Union/Russia had created 32 such drift stations over the years. The Americans tended toward shorter-term encampments but also were always present in the Arctic. Beginning with the USS *Nautilus* in the 1950s, this presence was extended by nuclear submarines, all of which conducted certain types of experiments built on the needs of the U.S. military. In all cases, these scientific facilities were military in operation; nevertheless, their place as key outposts collecting data on meteorology, oceanography, and other scientific disciplines held consequences beyond the geopolitical and national-security interests that had sparked them in the first place. At the same time, as Althoff makes clear, the strategic considerations of the North Pole for the cold war transformed the region into a theater for military operations.

In terms of Antarctic science, G. E. Fogg's *A History of Antarctic Science* is the recognized starting point for any study of the subject.[63] His central thesis is that Antarctic science is just as important as the glamorous adventures of exploration on the continent. From Edmond Halley's travels in 1700 to the rise of technologically sophisticated and international efforts to understand the region that came after World War II, scientific enterprises represented a fundamental justification—and an appropriate one—for the broad investigation of

**Figure 4.3**  Portrait of Ernest H. Shackleton in *The Heart of the Antarctic*, Volume I (1909), by E.H. Shackleton. (Courtesy of National Oceanic and Atmospheric Administration, NOAA Library Collection, Washington, D.C.)

Antarctica. Fogg divides the scientific history of Antarctica into three stages: (1) the "heroic era," (2) the period between about 1930 and the IGY, and (3) the post–IGY era. The result is a historically sophisticated account of the rise of big science, which Fogg insists was necessary to marshal the financial, political, and logistical support necessary to be successful in the polar regions.

Nationalism dominated the scientific efforts in Antarctica from the beginning, and even in the cooperative ventures of the IGY and later, nationalism played a key part in the story.[64] One attribute of Fogg's account is that it situates Antarctic science and its evolution in the twentieth century in the larger context of international relations. He makes clear that there were also national styles to the science conducted in Antarctica. The United States, for example, concentrated on large, splashy short-term expeditions and projects, while the British committed to a few small and less well-known bases, where scientists undertook observations over many years. The American effort succeeded in organizing huge teams of researchers, but without those observations taken over time, made possible by such encampments as those operated by the British, the change in the ozone hole above Antarctica would not have been as rapidly understood.[65] Overall this approach has been a positive development for science, as the nations involved have made available the latest in technological advances to further science in this hostile environment. One important element in Fogg's study is the central role played in this period by the Scientific Committee for Antarctic Research, planning for and conducting the IGY, and thereafter. *A History of Antarctic Science* admirably describes the state of science in the region and the hypotheses explored, but especially the observations and results therefrom.

An ambitious book by Stephen J. Pyne approached the massive ice sheet of Antarctica from all possible angles and was hailed as a pathbreaking study on its publication in 1986. *The Ice: A Journey to Antarctica* remarks on the sublime nothingness of the continent.[66] Pyne touches on everything about Antarctica, emphasizing its history, landscape, literature, geology, and biology. He argues that for all its geology and geography, geomagnetism and weather, and its biology and boredom, Antarctica remains at sum a diminished location in which water has been transformed into mineral. There are no cultural studies beyond those done on the explorers themselves. As many have commented, the ice of Pyne's characterization is a study in nihilism. According to one reviewer, "The appeal of this rich and awestruck book lies in its author's strenuous attempts to come to terms with the sheer negativity and materialism of Antarctica."[67] Another reviewer noted, "Pyne uses two metaphors throughout the book. Antarctica is an information sink, requiring the input of huge amounts of information before it will give anything in return; it is a reductionist, abstract environment, both physically and intellectually alienating. Second, Antarctica is a distorted mirror, reflecting back what each individual and culture brings to it." Pyne offered "a mystical mood in this book that hints that human endeavors in Antarctica will never really touch *The Ice*."[68] Environmental historian Donald Wooster commented about Pyne's Antarctica: "To penetrate it scientifically required airplanes, remote sensors, and advanced crystallography. To apprehend it aesthetically took the modernist evolution in the arts, which has emphasized abstraction, subjectivity, and minimalism."[69]

The book by Richard S. Lewis, *A Continent for Science: The Antarctic Adventure,* offers an accessible account of scientific efforts in the region.[70] It argues that the only sound reason that may be offered for expending the vast sums of governmental money in Antarctica is for the creation and dissemination of scientific knowledge, an expenditure that Lewis believes is fully justified by the results reported. He explores the development of Antarctic glaciology, geology, glacial chronology, weather, geophysics, and biology to understand how Antarctica affects the rest of the globe. This is popular history that opens discussions of each aspect with a captivating story and stunning scientific results before moving on to a dialogue, a kind of detective story, of how it all came about. While showing partiality toward the efforts of Americans in Antarctica, Lewis provided a useful overview of the subject just as the results from the IGY effort were starting to become apparent.

In *Antarctica,* editor Trevor Hatherton encapsulates the totality of Antarctic science as understood in the middle part of the 1960s.[71] Discussing for a specialized audience a broad range of Antarctic science, from the surrounding seas to the ice, the land beneath the ice, and the atmosphere, it focuses on the scientific efforts of English-speaking observers from New Zealand, Australia, the United States, and the United Kingdom. As one reviewer commented, "While reading the various sections one is continually made aware that the Antarctic and its scientific problems are now indissolubly linked to the general body of scientific investigation of our planet. This is a far cry from the situation in 1952 when it was not even possible to draw a detailed outline of the continent, and the interior physiography could only be guessed at."[72]

More recent reviews of Antarctic science have followed. In 1987 D.W.H. Walton assembled *Antarctic Science* to capture the state of the art in scientific knowledge about the region.[73] It bespeaks the vigor of the sustained efforts under way in Antarctica and offers box scores on which nation's record has been the most prestigious. For example, scientists in the United States contributed to 18 percent of all scholarly publications on Antarctic science during the 1979–82 timeframe, while second place was taken by the United Kingdom with 17 percent. Overall it provides an excellent review of what was learned during the preceding 25 years in Antarctic life sciences, Earth sciences, and atmospheric sciences. The most recent example of a historical review is a specialized volume, *Smithsonian at the Poles: Contributions to International Polar Year Science,* edited by Igor Krupnik, Michael Lang, and Scott Miller.[74] The result of a conference focused on depicting, with a Smithsonian twist, the state of the art in polar science, the 31 papers published in the proceedings discuss disparate topics ranging from balloon experiments to cultural studies, from biology to cosmology, marine science, and under-ice research.

## International Scientific Cooperation and the
## Two International Polar Years

One area that has failed to garner the historical attention it deserves encompasses the IPYs of 1882–83 and 1932–33. The first IPY, which ultimately had 11 Western nations, sponsored 14 separate expeditions to the poles. The Arctic region dominated with 12 expeditions but two also went to Antarctica. Those undertaking this endeavor blazed a trail every bit as significant as the dramatic explorations around the world; it succeeded in organizing a large-scale, coordinated, international effort to collect a set of observations on meteorology, geography, geology, biology, and a host of other scientific disciplines; it transformed the competitive polar undertakings of the past into the cooperative ventures that would follow in the twentieth century; and it demonstrated the power and authority of the modernity that these endeavors portended. In the latter case, it represented the rise of a worldview that celebrated science and technology and gave wheels to the idea of progress.[75]

The IPY was the brainchild of Austrian explorer Karl Weyprecht, whose "drive, ambition, and connections" brought it to fruition. During an Austrian expedition to Franz Josef Land in 1872–74, Weyprecht concluded that an international, cooperative approach to solving the scientific questions brought to the fore at the poles had the most probability of success. This led to his belief: "Decisive scientific results can be attained only through a series of synchronous expeditions, whose task it would be to distribute themselves over the Arctic regions, and to obtain one year's series of observations made according to the same method."[76] Not the first to come to this conclusion, he certainly carried the idea to a new level with the IPY. Although he died in 1881, before the first IPY was officially launched, Weyprecht deserves credit as the godfather of systematic, broad-based polar science. Even so, it is unlikely that Weyprecht would have gotten far without the moral support, and in some cases political and organizational excellence, of such individuals as Georg von Neumayer and Count Wilczek. Most significantly this vision of scientific cooperation during the IPY demonstrated that scientists and scientific organizations could work in partnership across a broad spectrum despite national rivalries.[77] As historian Marc Rothenberg has concluded of Weyprecht's initiative: "What occurred with the first IPY was not a revolution in international science, but the transformation of Polar science; it began to more closely resemble the norm in international science."[78]

The starting point for any study of the first IPY is William Barr's 1985 seminal work, *The Expeditions of the First International Polar Year, 1882–83.*[79] It shows clearly that Weyprecht's foresight and drive made real the efforts to undertake synchronous observations using the same methods and instruments at several locations. Initially conceived to concentrate on meteorology, geomagnetism, and auroras, the expeditions ended up doing much more in disciplines ranging from biology to astronomy. Even when expeditions went awry, as happened with the U.S. Greely expedition and the Dutch expedition to the northern Siberian coast, they still collected important scientific data. In an irony of the first magnitude, as Barr points out, all this scientific harvest was contained in a series of expedition reports published by the various nations, and in different languages—French, German, Russian, Swedish, Dutch, and English.[80] For all the farsightedness of Weyprecht in conceiving and carrying out an international scientific effort, once the expeditions were completed there was essentially no thought given to ensuring that the results were properly shared by polar scientists around the globe. Barr's historical study is a valuable digestion of these disparate efforts, including a chapter on each major expedition and the fate of their scientific results. It represents the first, and so far the only, time anyone has sought to encapsulate the history of the first IPY in a single book-length study. There are, however, several article-length works that offer a short introduction to the first IPY.[81]

There are also specialized works exploring aspects of the first IPY. William Barr's 1983 article on geographical knowledge resulting from the first IPY is one such example. In terms of the geographical spinoff from the project, several expeditions made notable contributions in the area of exploration and cartography, particularly an American expedition to Ellesmere Island and a French one to Cabo de Hornos (Cape Horn). Although ethnography was not a part of the official program, several scientists made extensive observations of the indigenous peoples in their respective areas, and Barr highlights the studies of the natives on the North Slope by the American expedition to Point Barrow, of the Inuit of Cumberland Sound by a scientist with the German expedition to Baffin Island, and of the Yahgan Indians of Tierra del Fuego by the French expedition to Cabo de Hornos.[82]

So, too, the specialized account by Kevin R. Wood and James E. Overland, "Climate Lessons from the First International Polar Year," has important implications not only for history but for scientific understanding. The authors noted that the combined climate records for the first IPY offer a significant counterpoint to the data collected in the twentieth century. Wood and Overland "found that surface air temperature (SAT) and sea level pressure (SLP) observed during 1882–83 were within the limits of recent climatology, but with a slight skew toward colder temperatures, and showed a wide range of variability from place to place over the course of the year, which is a feature typical of the Arctic climate today." They also found that "Observations recorded after the Krakatau eruption in 1883 show cooling effects, notably at Sodanklyä, Finland, where residents regarded the change as quite exceptional."[83]

R.G. Barry's historical analysis using IPY scientific data, "Arctic Ocean Ice and Climate: Perspectives on a Century of Polar Research," is equally useful. Barry surveys the data collected over time to consider the complexities introduced by snow cover, summer melt ponds, and polynyi (an area of open water in sea ice). He concludes that while "ice anomalies are pronounced on a regional scale, in response to atmospheric circulation patterns, there are also strong correlations between high-latitude temperature trends and total ice area."[84] Important studies of geomagnetism and associated cooperative disciplines in the poles also have appeared.[85] There are also several short studies of individual expeditions, usually built around national teams. For example, a German expedition to Clearwater Fiord, Baffin Island, set an unbroken record of 359 daily observations of hourly meteorological and magnetic readings.[86]

Finally Cornelia Lüdecke's intriguing essay, "The First International Polar Year (1882–83): A Big Science Experiment with Small Science Equipment," offers a model study in a merger of social-construction theory with the history of science and technology. Lüdecke argues that "each participating state of the IPY was in charge of its own expedition being part of a big experiment to investigate the meteorology and earth's magnetic field in high latitudes." There were dramatic differences between the ways in which the various participants organized their efforts, but the assembly of them into a meaningful whole represented the biggest type of science yet undertaken in global history. If internationality represented the "big science" of the first IPY, small science ensured useful results. Individuals on the ground at key locations at the poles made possible the creation of data sets still being mined for scientific knowledge. That data became the benchmark on the scientific measurements undertaken in later expeditions.[87]

The importance of the first IPY for scientific understanding cannot be underestimated. Because of this success, scientists pursued a reprise in 1932–33, one that was less productive in large part because of the international economic depression that limited national commitment to the endeavor.[88] This IPY was proposed by the International Meteorological Organization. Like Weyprecht in the earlier era, the driving force before this IPY was D. la Cour, director of the Danish Meteorological Institute. At some point during the 1932–33 period, a total of 49 nations participated in the second IPY, although no nation's contributions were quite as extensive as those made in the first IPY. In the end 22 countries sent special expeditions, largely to the poles. Coupled with existing scientific posts in more temperate zones, a worldwide portrait of scientific characteristics of the globe emerged. Accordingly, the effort yielded important new knowledge about meteorology, magnetism, atmospheric science, and ionospheric phenomena. One of the keys to the second IPY was the attempt to take vertical measurements in the atmosphere and also to use radio communications to explore geomagnetism and the aurora.[89] The participants established 40 observation stations in the Arctic, any of which became more or less permanent, and for the first time Antarctica became a major research arena.[90] As R.S. Patton, director of the U.S. Coast and Geodetic Survey, commented: "While the fundamental activities, magnetic and meteorological observations, are the same for the First and Second Polar Years, there is a great contrast in the associated activities, practically all of which were undreamed of at the time of the First Polar Year. Our knowledge of the interrelations of the different phenomena that are to be observed is still inadequate both from the scientific and from the practical viewpoint." With the cooperation of all partners in the effort, Patton concluded, this IPY will begin filling "important gaps in observations."[91]

The U.S. contribution to the second IPY included a second Byrd Antarctic expedition that led to the establishment of a winter-long meteorological station on the Ross Ice Shelf at the southern end of Roosevelt Island about 125 miles south of Little America Station. This was the first research station inland from Antarctica's coast and a major outpost for future scientific efforts. In the north, the U.S. conducted auroral studies, ionospheric physics, ice studies, and the preparation of Northern Hemisphere weather charts. Emerging from this effort, American policy makers came to understand, as A.L. Washburn stated: "Thus, in both the Arctic and Antarctic, scientific research has political and economic implications, and the resulting financial support has given impetus to research, especially in the Antarctic. Success in this opportunity to advance both science and national self-interest through international cooperation in Antarctica is a challenge, both for governments and individual scientists, that has far-reaching implications for both polar hemispheres."[92]

Because of the difficult economic times in which the second IPY took place and World War II, 1939–45, the results of the second IPY were also not as available as the planners intended. J.A. Fleming of the Carnegie Institution of Washington and a force in the development of the IPY program, recognized this as a necessity from the beginning. He wrote in 1932:

> Upon completion of the Polar Year and publication of the resulting data by the individual governments and organizations occupying stations, it is essential that provision be made to coordinate, reduce, and compile the data under the direction of the International Polar Year Commission. After the publication of results at individual stations it is proposed to distribute the data for study and discussion to specialists who will report to the Commission, thus permitting general coordination of findings, and insuring their consideration in the broadest aspects.[93]

Unlike in the first IPY, some reports were not published at all, but like that earlier effort those that appeared were issued in a haphazard manner and only later did science teams seek to compile data into a useful compendium.[94] Perhaps because of situation and circumstance, the second IPY was not the substantive scientific effort that had been intended.

### International Scientific Cooperation and the International Geophysical Year

The historical literature on the International Geophysical Year (IGY) is massive, in part because of its fundamental success in kick-starting both an ongoing, aggressive science collection program at the poles and a space-based science program, both of which have been glamorous to most observers since they began. The genesis of the IGY took place at a dinner party in the home of James A. Van Allen in Silver Spring, Maryland, in the summer of 1950. This event has taken on legendary status and serves a range of purposes from a nearly mystical birth for the IGY to the reaffirming of the authority of science in modern life. Pressed by American science entrepreneur Lloyd

V. Berkner, the International Council of Scientific Unions (ICSU) agreed in 1952 to pursue a comprehensive series of global geophysical activities to span the period July 1957–December 1958.[95] The IGY, as it was called, was timed to coincide with the high point of the 11-year cycle of sunspot activity. There are several accounts of the genesis of the IGY. The most thorough in English is contained in Constance McLaughlin Green and Milton Lomask, *Vanguard: A History.* More historically sophisticated in analysis are several publications by Fae L. Korsmo, Rip Bulkeley, and Allan A. Needell.[96]

The IGY, taking place from July 1, 1957, to December 31, 1958, eventually involved 67 nations undertaking research in 11 major scientific areas: Earth sciences, aurora and airglow, cosmic rays, geomagnetism, gravity, ionospheric physics, precision mapping, meteorology, oceanography, seismology, and solar activity. Managed by the International Council of Scientific Unions, more than 4,000 research stations, either already in operation or established as part of the IGY, participated in the cooperative endeavor. While efforts were concentrated in the polar regions, terrestrial stations along the equator and at several geographic lines north to south also yielded valuable scientific data. For example, scientists defined the mid-ocean ridges (developing the theories of plate tectonics and the nature of Earth's crust that has defined modern geology), discovered the Van Allen radiation belts using data from *Explorer 1* and *Explorer 3*, charted ocean depths and ocean currents, and measured a range of terrestrial phenomena from the magnetic field to upper atmospheric winds to the geophysical nature of the planet.[97]

In the run-up to the beginning of the IGY, Berkner described the excitement it offered in the pursuit of science:

> Perhaps the most important aspect of the IGY is the simultaneity of all the related scientific programs, which will make possible a synthesis of the results of complementary observations in related sciences. Continuous programs of geophysical observation are expensive and burdensome, but concentration on simultaneous programs during 1 year will be feasible and productive, particularly when emphasis is placed on observation during regular and special World Days. The powerful tools now available for observation in themselves insure outstanding contributions to our understanding and comprehension of the environment in which man now finds himself.

Berkner's closing words were stirring to be sure but half-true at best: "Tired of war and dissension, men of all nations have turned to 'Mother Earth' for a common effort on which all find it easy to agree."[98]

There are many specialized works on the history of IGY, but no sophisticated book-length overviews. In 1959 one of the godfathers of the IGY, Sydney Chapman, wrote a short overview of the effort, but it was more of a booster's pamphlet than a history. *New York Times* journalist Walter Sullivan's popular history from 1965 does not offer anything more than a general narrative of the major elements of the IGY.[99] So to, there are other general accounts in article form that offer snippets about the history of IGY and call for a redoubling of collaborative big science efforts using government funds.

Often outlining the number of research stations, the contributions of the partnering nations, and the proposed results of the endeavor, these works are helpful but decidedly not comprehensive accounts.[100] A work of substance charting the length and breadth of these activities for the United States, and a host of scientific results, is Frank M. Marson and Janet R. Terner, compilers, *United States IGY Bibliography, 1953–1960: An Annotated Bibliography of United States Contributions to the IGY and IGC (1957–1959)*, a compendium well worth using as an entrée into the literature of the IGY.[101] Innumerable reports emerging from the IGY have been published; much of the data collected has also been stored in the conglomeration of World Data Centers established by ICSU as a part of the effort to ensure the preservation and use of scientific data from the IGY.[102] A useful starting point for understanding, and locating, these materials is in K.O. Murra, *International Scientific Organizations: A Guide to Their Library, Documentation, and Information Services*.[103] Finally the *Annals of the International Geophysical Year*, published by ICSU in 46 volumes between 1957 and 1970 is an invaluable resource for all aspects of the IGY.[104] There were also films and popularizations for schools

Two overwhelmingly significant events occurred as a result of the IGY. The first was the occupying of Antarctica by several nations through an international treaty regime that maintained it as a reserve for science. The establishment of this peaceful, collaborative occupation of Antarctica is the subject of several important books and articles. An early attempt to understand the evolution of this effort came in 1973 with the publication of a document by the U.S. House of Representatives: "The Political Legacy of the International Geophysical Year." This work laid out in stilted bureaucratese the evolution of the treaties and conventions holding Antarctica as a refuge for science.[105] Four books dealing specifically with international politics in the region are Peter Beck, *The International Politics of Antarctica*; Klaus Dodds, *Geopolitics in Antarctica: Views from the Southern Oceanic Rim*; P. A. Berkman, *Science into Policy: Global Lessons from Antarctica*; and M. J. Peterson, *Managing the Frozen South: The Creation and Evolution of the Antarctic Treaty System*.[106] A fifth work, Donald R. Rothwell's *The Polar Regions and the Development of International Law*, is authoritative.[107] Other works have explored the politics of international scientific collaboration and aspects of public policy in Antarctica.[108] Two books on the American effort in Antarctica, known as Operation Deep Freeze, which began as part of the IGY but has continued to the present, are George J. Dufek, *Operation Deepfreeze*, and Dian Olsen Belanger, *Deep Freeze: The United States, the International Geophysical Year, and the Origins of Antarctica's Age of Science*.[109] There are also personal accounts of activities on the ice during and since the IGY.[110]

The second major event emerging from IGY was the space age, begun on October 4, 1957, when the Soviet Union launched *Sputnik 1*, the first artificial Earth satellite. The United States followed soon thereafter with *Explorer 1*, on January 31, 1958. There is an enormous quantity of literature

on the history of the birth of the space age.[111] The classic work on this subject remains Walter A. McDougall, *The Heavens and the Earth: A Political
History of the Space Age*, winner of the 1985 Pulitzer Prize in history.[112]
It lays out well the IGY origins of the space age but condemns those origins as "the institutionalization of technological change for state purposes,
that is, the state-funded and managed R&D explosion of our time."[113] As
McDougall wrote:

> In these years the fundamental relationship between the government and
> new technology changed as never before in history. No longer did state and
> society react to new tools and methods, adjusting, regulating, or encourag-
> ing their spontaneous development. Rather, states took upon themselves the
> primary responsibility for generating new technology. This has meant that
> to the extent revolutionary technologies have profound second order con-
> sequences in the domestic life of societies, by forcing new technologies, all
> governments have become revolutionary, whatever their reasons or ideologi-
> cal pretensions.[114]

Emphasizing the effect of the space race on society, this critique focused on
the role of the state as a powerful promoter of technological progress—to
the detriment of civilization.

## Conclusion

Volume upon volume has been written on the history of the Arctic and
Antarctica, but relatively few raise critical historiographical questions deserv-
ing of sustained and sophisticated attention. This essay has surveyed some
of the more prominent examples of this literature, but much remains to be
considered. At one level, historians of science must pursue several core questions when considering this subject. Some of these include:

- To what extent did the evolution and maturation of new technologies
  inspire these scientific efforts at the poles?
- Did those campaigns stimulate new technologies for exploration?
- How might one situate the IPYs and the IGY in a larger context of trans-
  national science?
- Over time, expeditions were mounted specifically to answer scientific
  questions and were designed and led by scientists. How did this happen?
  When did it? Why?
- Does technology merely enable science, or does it define science also?
- Were the questions asked in these international campaigns limited to what
  technologies, methodologies, and processes could answer; or were the
  goals beyond extant technologies, methodologies, and processes?
- How did the politics of science shape the efforts in the IPYs and the IGY,
  and how was politics shaped by them?
- Is there such as thing as a global science? What does it entail? How did it
  arise, and how has it been sustained?

The polar years' campaigns initially involved meteorology, oceanography, glaciology, geomagnetism, auroral observations, latitude/longitude determinations, and ethnography (IPY1), and later added solar activity and studies of cosmic rays and the ionosphere in addition to other disciplines. The expeditions themselves also returned significant collections of great ethnological importance and stimulated much new interest in that field. Observations and data collection spread in both depth and breadth as instrumental techniques and opportunities expanded, providing historians with a rich opportunity to compare and assess the sources, evolution, and application of diverse technologies in the pursuit of scientific knowledge. In the early modern era, long before the first IPY, the ships of James Cook did more than transport his crews—they were "scientific instruments in their own right," mediating "the complex interplay between representation and reality."[115] More recently, technology has made possible the enormous capabilities marshaled to understand the poles.[116] How did this happen? When did it? And why? How did this form disciplines and shape scientific discovery? All of these and many more are possibilities for polar science. They remain significant avenues for explorations into the future.

## Notes

1. J.A. Fleming and V. Laursen, "International Polar Year 1932–1933," *Science* 110 (September 23, 1949): 308–209.
2. Michael Bravo and Sverker Sörlin, eds., *Narrating the Arctic: A Cultural History of Nordic Scientific Practices* (Sagamore Beach, Massachusetts: Science History Publications, 2002), 7. See also Shelagh D. Grant, "Arctic Historiography: Current Status and Blueprints for the Future," *Journal of Canadian Studies* 33 (Spring 1998): 145–153.
3. Richard Vaughan, *The Arctic: A History* (Phoenix Mill, United Kingdom: Sutton Publishing Ltd., 1994).
4. Jeannette Mirsky, *To the Arctic: The Story of Northern Exploration from Earliest Times to the Present* (New York: The Viking Press, 1934), 185–195. The 1970 revised edition was published by the University of Chicago Press.
5. Primary records of this event are available in Adolphus Greeley, *Three Years of Arctic Service: An Account of the Lady Franklin Bay Expedition of 1881–1884 and the Attainment of Farthest North*, 2 vols. (New York: Charles Scribner's Sons, 1886); *Proceedings of the Proteus Court of Inquiry on the Greeley Relief Expedition of 1883* (Washington, D.C.: Government Printing Office, 1884). The most definitive history of this expedition is Alden Todd, *Abandoned: The Story of the Greely Arctic Expedition 1881–1884*, 2nd ed. (Fairbanks: University of Alaska Press, 2001). On the expedition's rescue, see Stephen K. Stein, "The Greely Relief Expedition and the New Navy," *International Journal of Naval History* 5 (December 2006), available on-line at http://www.ijnhonline.org/volume6_number1_apr07/article_stein_greely_dec06.html.
6. See Anthony Brandt, ed., *The South Pole: A Historical Reader* (Washington, D.C.: National Geographic, 2004), 1–136; Alan Gurney, *Below the Convergence: Voyages toward Antarctica 1699–1839* (New York: W.W. Norton & Co., 1997); Alan Gurney, *The Race to the White Continent: Voyages to the Antarctic* (New York: W.W. Norton & Co., 2002); Kieran Mulvaney, *At the Ends of the Earth: A History of the Polar Regions* (Washington, D.C.: Island Press, 2001).
7. Stephen Martin, *A History of Antarctica* (Sydney, Australia: State Library of New South Wales Press, 1996). See also Marilyn J. Landis, *Antarctica: Exploring the Extreme, 400 Years of Adventure* (Chicago, Illinois: Chicago Review Press, 2001).

8. Ian Cameron, *Antarctica: The Last Continent* (London: Cassell & Co., 1974). See also Walker Chapman, *The Loneliest Continent: The Story of Antarctic Discovery* (New York: New York Graphic Society, 1964); Frank Debenham, *Antarctica: The Story of a Continent* (New York: The Macmillan Co., 1961); Gordon Elliott Fogg and Michael Smith, *The Exploration of the Antarctic: The Last Unspoilt Continent* (London: Cassell & Co., 1990); Laurence M. Gould, "Antarctic Prospect," *Geographical Review* 47 (January 1957): 1–28.

9. Michael H. Rosove, *Let Heroes Speak: Antarctic Explorers, 1772–1922* (Annapolis, Maryland: Naval Institute Press, 2000).

10. Robert Headland, comp., *A Chronology of Antarctic Exploration: A Synopsis of Events and Activities until the International Polar Years, 2007 to 2009* (London: Bernard Quaritch Ltd., 2009).

11. Kenneth J. Bertrand, *Americans in Antarctica, 1775–1948* (New York: American Geographical Society, 1971). An earlier study useful for the period that it covers is Philip I. Mitterling, *America in the Antarctic to 1840* (Urbana: University of Illinois Press, 1959).

12. Dean Beeby, *In a Crystal Land: Canadian Explorers in Antarctica* (Toronto: University of Toronto Press, 1994); Douglas Mawson, *The Home of the Blizzard: Being the Story of the Australasian Antarctic Expedition, 1911–1914* (London: Ballantyne Press, n.d.); Lars Christensen, *Such Is the Antarctic,* translated by E.M.G. Jayne (London: Hodder and Stoughton, 1935); David Thomas Murphy, *German Exploration of the Polar World, 1870–1940* (Lincoln: University of Nebraska Press, 2002).

13. T.H. Baughman, *Before the Heroes Came: Antarctica in the 1890s* (Lincoln: University of Nebraska Press, 1993).

14. Clint Willis, ed., *Ice: Stories of Survival from Polar Exploration* (Cambridge, Massachusetts: Da Capo Press, 1999).

15. Joseph Campbell with Bill Moyers, *The Power of Myth* (New York: Anchor Books, 1988), 48.

16. Eric G. Wilson, *The Spiritual History of Ice: Romanticism, Science, and the Imagination* (New York: Palgrave/Macmillan, 2003).

17. Barry Holstun Lopez, *Arctic Dreams: Imagination and Desire in a Northern Landscape* (New York: Charles Scribner & Sons, 1986), xxviii, xxix, 13, 274.

18. Robert McGhee, *The Last Imaginary Place: A Human History of the Arctic World* (New York: Oxford University Press, 2005).

19. Beau Riffenburg, *The Myth of the Explorer: The Press, Sensationalism, and Geographical Discovery* (London: Belhaven Press, 1993); John McCannon, *Red Arctic: Polar Exploration and the Myth of the North in the Soviet Union 1932–1939* (New York: Oxford University Press, 1998); Stephanie Barczewski, *Antarctic Destinies: Scott, Shackleton, and the Changing Face of Heroism* (New York: Continuum Books, 2007), Bruce Hevly, "The Heroic Science of Glacier Motion," *Osiris* 11 (1996): 66–86.

20. Michael F. Robinson, *The Coldest Crucible: Arctic Exploration and American Culture* (Chicago, Illinois: University of Chicago Press, 2006), 3.

21. Francis Spufford, *I May Be Some Time: Ice and the English Imagination* (New York: St. Martin's Press, 1997); Robert G. David, *The Arctic in the British Imagination, 1818–1914* (Manchester: Manchester University Press, 2000).

22. Bernard Mergen, review of *I May Be Some Time,* in *Isis* 89 (December 1998): 750–751.

23. The so-called "heroic age" is usually considered from about 1899 to the 1920s, although there is no hardness to these dates. It also allows for national differences. For example, American efforts in 1920, with aviation and its use in exploration, was much different than England's in 1920, after its more brutal experience in World War I.

24. An excellent article laying out the contours of the debate about the Northwest Passage may be found in D.B. Quinn, "The Northwest Passage in Theory and Practice," in J.L. Allen, ed., *North American Exploration, I, A New World Disclosed* (Lincoln: University of Nebraska Press, 1997), 292–343.

25. Office of Naval Research, Naval Ice Center, Oceanographer of the Navy, and the Arctic Research Commission, *Naval Operations in an Ice-Free Arctic Symposium, 17–18 April 2001, Final Report* (Washington, D.C.: Office of Naval Research, Arctic Research Commission, and Naval Ice Center, 2001), 5.

26. The policy history and implication of this issue are outlined in Franklyn Griffiths, ed., *Politics of the Northwest Passage* (Kingston, Canada: McGill-Queen's University Press, 1987).

27. Pierre Berton, *The Arctic Grail: The Quest for the North West Passage and the North Pole, 1818–1909* (New York: Penguin Books, 1988), with many reprints. On the same subject, see Ann Savours, *The North West Passage in the Nineteenth Century: Perils and Pastimes of a Winter in the Ice* (London: Hakluyt Society, 2003). The work on this subject written by a Canadian scholar is Leslie H. Neatby, *In Quest of the North West Passage* (Toronto: Longmans, Green, 1958).

28. Glyn Williams, *Voyages of Delusion: The Quest for the Northwest Passage* (New Haven, Connecticut: Yale University Press, 2003); Ann Savours, *The Search for the North West Passage* (New York: St. Martin's Press, 1999). In addition, Glyndwr Williams, *The British Search for the Northwest Passage in the Eighteenth Century* (New York: Longmans, 1962), presents a fine account of little-known eighteenth-century efforts.

29. Alan Edwin Day, comp., *Historical Dictionary of the Discovery and Exploration of the Northwest Passage* (Lanham, Maryland: Scarecrow Press, 2006).

30. L. M. Starokadomskiy, *Charting the Russian Northern Sea Route: The Arctic Ocean Hydrographic Expedition 1910–1915,* translated and edited by William Barr (originally published in Russian, 1946; 3rd Russian ed., 1959, Montreal: Arctic Institute of North America/McGill-Queen's University Press, 1976); Terence E. Armstrong, *The Northern Sea Route: Soviet Exploitation of the North East Passage* (Cambridge, United Kingdom: Scott Polar Research Institute, 1952); Lawson W. Brigham and Ellen M. Gately, eds., *The Soviet Maritime Arctic: Proceedings of a Workshop Held May 10–13, 1987 by the Marine Policy Center of the Woods Hole Oceanographic Institution* (Woods Hole, Massachusetts: Woods Hole Oceanographic Institution, 1988).

31. Patricia D. Sutherland, ed., *The Franklin Era in Canadian Arctic History, 1845–1859* (Ottawa, Canada: National Museums of Canada, 1985).

32. Richard J. Cyriax, *Sir John Franklin's Last Arctic Expedition: A Chapter in the History of the Royal Navy* (1939, reprint West Midlands, United Kingdom: The Arctic Press, 1997). Many other books have explored this subject subsequently: Owen Beattie and John Geiger, *Frozen in Time: Unlocking the Secrets of the Franklin Expedition* (Saskatoon, Canada: Western Producer Prairie Books, 1989); Ken McGoogan, *Lady Franklin's Revenge: A True Story of Ambition, Obsession, and the Remaking of Arctic History* (New York: HarperCollins, 2005); Ken McGoogan, *Fatal Passage: The Untold Story of John Rae, the Arctic Adventurer Who Discovered the Fate of Franklin* (New York: HarperCollins, 2001); David Murray, *The Arctic Fox: Francis Leopold McClintock, Discoverer of the Fate of Franklin* (Cork: The Collins Press, 2004); Vilhjálmur Stefánsson, *Unsolved Mysteries of the Arctic* (1938, reprint New York: Collier Books, 1962); Martyn Beardsley, *Deadly Winter: The Life of Sir John Franklin* (Annapolis, Maryland: Naval Institute Press, 2002); W. Gillies Ross, "The Type and Number of Expeditions in the Franklin Search 1847–1859," *Arctic* 55, no. 1 (2002): 57–69.

33. David C. Woodman, *Unraveling the Franklin Mystery: Inuit Testimony* (Toronto: McGill-Queen's University Press, 1992).

34. Perhaps the most entertaining of all stories about the quest for the North Pole is Chauncy Loomis, *Weird and Tragic Shores: The Story of Charles Francis Hall, Explorer* (New York: Alfred A. Knopf, 1971). The earliest attempt to tell this story in its totality is J. Gordon Hayes, *The Conquest of the South Pole: Antarctic Exploration, 1906–1931* (London: Thornton Butterworth, 1932).

35. Frederick A. Cook, *My Attainment of the Pole: Being the Record of the Expedition that First Reached the Boreal Center, 1907–1909* (New York: Mitchell Kennerley, 1912),

with many subsequent editions. The Peary case is made in Rear Admiral Robert E. Peary, *The North Pole: Its Discovery in 1909 under the Auspices of the Peary Arctic Club* (New York: Frederick A. Stokes Co., 1910); Matthew Henson, *A Black Explorer at the North Pole*, foreword by Robert E. Peary and introduction by Booker T. Washington (New York, 1912, reprint Lincoln: University of Nebraska Press, 1989).

36. Bruce Henderson, *True North: Peary, Cook, and the Race to the Pole* (New York: W. W. Norton and Co., 2005).

37. Some of these other works include: John Edward Weems, *Peary: The Explorer and the Man* (Boston, Massachusetts: Houghton Mifflin, 1967); Theon Wright, *The Big Nail: The Story of the Cook–Peary Feud* (New York: John Day Company, 1970); Dennis Rawlins, *Peary at the North Pole, Fact or Fiction?* (Washington, D.C.: Robert B. Luce, 1973); William R. Hunt, *To Stand at the Pole: The Dr. Cook–Admiral Peary North Pole Controversy* (New York: Stein & Day, 1981); Larry Schweikart, "Polar Revisionism and the Peary Claim: The Diary of Robert E. Peary," *The Historian* 48 (May 1986): 341–358; Clive Holland, ed., *Fartherest North: The Quest for the North Pole, 1818–1909* (New York: MacLelland and Stewart, 1988); Robert M. Bryce, *Cook & Peary: The Polar Controversy, Resolved* (Mechanicsburg, Pennsylvania: Stackpole Books, 1997); Wally Herbert, *The Noose of Laurels: The Discovery of the North Pole* (Garden City, New York: Doubleday and Co., 1989); Fergus Fleming, *Ninety Degrees North: The Quest for the North Pole* (London: Granta Books, 2001); Randall J. Osczevski, "Frederick Cook and the Forgotten Pole," *Arctic* 56, no. 2 (2003): 207–217.

38. Samuel Elliot Morison, *The European Discovery of America*, vol. 1, *The Northern Voyages, A.D. 500–1600* (New York: Oxford University Press, 1971); Alan Cooke and Clive Holland, *The Exploration of Northern Canada, 500 to 1920, a Chronology* (Toronto: The Arctic History Press, 1978).

39. Nancy Fogelson, *Arctic Exploration & International Relations, 1900–1932* (Fairbanks: University of Alaska Press, 1992).

40. Jean Malaurie, *Ultima Thule: Explorers and Natives in the Polar North* (New York: W.W. Norton & Co., 2003).

41. On the race, see Apsley Cherry-Garrard, *The Worst Journey in the World: Antarctic 1910–13* (1965, New York: Penguin, 1970); Charles Neider, *Antarctica: Firsthand Accounts of Exploration and Endurance* (New York: Cooper Square Publishers, 2000); Roland Huntford, *The Last Place on Earth: Scott and Amundsen's Race to the South Pole* (London: Pan Books, 1985); David Thomson, *Scott, Shackleton, and Amundsen: Ambition and Tragedy in the Antarctic* (New York: Basic Books, 2002); Rainier K. Langner, *Scott and Amundsen: The Race to the Pole* (Berlin: Haus Publishers Ltd., 2007).

42. On Amundsen, see Roald Amundsen, *South Pole: An Account of the Norwegian Antarctic Expedition in the Fram* (London: Murray, 1912); J. Gordon Vaeth, *To the Ends of the Earth: The Explorations of Roald Amundsen* (New York: Harper and Row, 1962); Roald Amundsen, *The Amundsen Photographs* (New York: Atlantic Monthly Press, 1987); Roald Amundsen, *Belgica Diary: The First Scientific Expedition to the Antarctic*, Hugo Decleir, ed., (Norwich, United Kingdom: Erskine Press/Bluntisham, 1999); Tor Bormann, *Roald Amundsen* (London: Sutton Publishing Ltd., 2006).

43. On Scott, see Elspeth Huxley, *Scott of the Antarctic* (New York: Atheneum, 1987); Diana Preston, *A First Rate Tragedy: Captain Scott's Antarctic Expeditions* (London: Constable, 1999); Susan Solomon, *The Coldest March: Scott's Fatal Antarctic Expedition* (New Haven, Connecticut: Yale University Press, 2001); Ranulph Fiennes, *Captain Scott* (London: Hodder & Stoughton, 2003); Max Jones, *The Last Great Quest: Captain Scott's Antarctic Sacrifice* (New York: Oxford University Press, 2003); David Crane, *Scott of the Antarctic: A Life of Courage, and Tragedy in the Extreme South* (New York: HarperCollins, 2005).

44. Roland Huntford, *Scott & Amundsen* (London: Hodder & Stoughton, 1979).

45. A few recent books on the subject include Francis Spufford, *I May Be Some Time: Ice and the English Imagination* (London: Faber & Faber, 1997); Roland Huntford, *Shackleton* (London: Hodder & Stoughton, 1985); Beau Riffenburg, *Nimrod: Ernest*

*Shackleton and the Extraordinary Story of the 1907–09 British Antarctic Expedition* (London: Bloomsbury Publishing, 2005); Kelly Tyler-Lewis, *Lost Men: The Harrowing Saga of Shackleton's Ross Sea Expedition* (New York: Viking, 2006).

46. Eugene Rodgers, *Beyond the Barrier: The Story of Byrd's First Expedition to Antarctica* (Annapolis, Maryland: Naval Institute Press, 1997); Lisle Rose, *Explorer: The Life of Richard E. Byrd* (Columbia: University of Missouri Press, 2008).

47. The hero question in Antarctic exploration is the specific task of the "heroic era" of exploration, for the South Pole has generated enormous investigation. See Barczewski, *Antarctic Destinies.*

48. R.W. Home and Sally G. Kohlstedt, eds., *International Science and National Scientific Identity* (Dordrecht, The Netherlands: Springer, 1991), 1–5, 32, 50.

49. James Rodger Fleming and Cara Seitchek, "Advancing Polar Research and Communicating Its Wonders: Quests, Questions, and Capabilities of Weather and Climate Studies in International Polar Years," in I. Krupnik, M.A. Lang, and S.E. Miller, eds., *Smithsonian at the Poles: Contributions to International Polar Year Science* (Washington, D.C.: Smithsonian Institution Scholarly Press, 2009), 1–12; Katharine Anderson, "Mapping Meteorology," in James Rodger Fleming, Vladimir Jankovic, and D.R. Coen, eds., *Intimate Universality: Local and Global Themes in the History of Weather and Climate* (Sagamore Beach, Massachusetts: Science History Publications), 69–92; Katharine Anderson, *Predicting the Weather: Victorians and the Science of Meteorology* (Chicago, Illinois: Chicago University Press, 2005); John Cawood, "Terrestrial Magnetism and the Development of International Collaboration in the Early Nineteenth Century," *Annals of Science* 34 (1977): 551–587; John Cawood, "The Magnetic Crusade: Science and Politics in Early Victorian Britain," *Isis* 70 (1979): 493–518; Steven J. Dick, "The American Transit of Venus Expeditions of 1874 and 1882," *Proceedings of IAU Colloquium No. 196* (Cambridge: Cambridge University Press, 2004), 100–110; Hilmar W. Duerbeck, "The German Transit of Venus Expeditions of 1874 and 1882: Organizations, Methods, Stations, Results," *Journal of Astronomical History and Heritage* 7 (2004): 8–17; William H. Goetzmann, *New Lands, New Men: America and the Second Great Age of Discovery* (New York: Viking, 1986); Helge Kragh, *An Introduction to the Historiography of Science* (New York: Cambridge University Press 1990); Jan Golinski, *Making Natural Knowledge: Constructivism and the History of Science*, 2nd ed. with a new preface (Princeton, New Jersey: Princeton University Press, 2005).

50. Quoted in Robert Tignor, Jeremy Adelman, Stephen Aron, Stephen Kokin, Suzanne Marchand, Gvan Prakash, and Michael Tsin, *Worlds Together, Worlds Apart: A History of the Modern World (1300 to the Present)* (New York: W.W. Norton and Co., 2002), chapter 14.

51. Roy MacLeod, "On Visiting the 'Moving Metropolis': Reflections on the Architecture of Imperial Science," in Nathan Reingold and Marc Rothenberg, eds., *Scientific Colonialism: A Cross-Cultural Comparison* (Washington, D.C.: Smithsonian Institution Press, 1986), 217–249.

52. The most extensive treatment of scientific imperialism may be found in Lewis Pyenson, *Cultural Imperialism and Exact Sciences: German Expansion Overseas, 1900–1930* (New York: Peter Lang, 1985). See also these shorter introductions on this subject: Lewis Pyenson, "Science and Imperialism," in R.C. Olby, G.N. Cantor, J.R.R. Christie, and M.J.S. Hodge, eds., *Companion to the History of Modern Science* (New York: Routledge, 1996), 920–933; Paolo Palladino and Michael Worboy, "Science and Imperialism," *Isis* 84 (1993): 91–102; C. Jami Petitjean and A.M. Moulin, eds., *Science and Empires: Historical Studies about Scientific Development and European Expansion* (Dordrecht, The Netherlands: Kluwer Academic Press, 1992).

53. Cynthia Lamson, "In Pursuit of Knowledge: Arctic Shipping and Marine Science," in David L. VanderZwaag and Cynthia Lamson, eds., *The Challenge of Arctic Shipping: Science, Environmental Assessment, and Human Values* (Toronto: McGill-Queen's University Press, 1990), 5.

54. See Louis-Edmond Hamelin, *Canadian Nordicity: It's Your North Too* (Montreal: Harvest House, 1979).

55. Louis Rey, Marvin W. Falk, and Claudette Reed Upton, eds., *Unveiling the Arctic: Conference on the History of the Discovery of the Arctic Regions as Seen through the Descriptions of Travellers and the Work of Cartographers from Early Antiquity to the 18th Century* (Fairbanks: University of Alaska Press, 1984).

56. Review of *Unveiling the Arctic*, in *The Geographical Journal* 152 (1986): 267.

57. Lopez, *Arctic Dreams: Imagination and Desire in a Northern Landscape.*

58. Debra J. Lindsay, *Science in the Subarctic: Trappers, Traders, and the Smithsonian Institution* (Washington, D.C.: Smithsonian Institution Press, 1992).

59. Trevor H. Levere, *Science and the Canadian Arctic: A Century of Exploration, 1818–1918* (New York: Cambridge University Press, 1993). See also Trevor H. Levere, "Vilhjalmur Stefansson, the Continental Shelf, and a New Arctic Continent," *The British Journal for the History of Science* 21 (June 1988): 233–247.

60. Levere, *Science and the Canadian Arctic*, 1.

61. Bravo and Sörlin, eds., *Narrating the Arctic.*

62. William F. Althoff, *Drift Station: Arctic Outposts of Superpower Science* (Dulles, Virginia: Potomac Books, 2006).

63. G.E. Fogg, *A History of Antarctic Science*, Studies in Polar Research (New York: Cambridge University Press, 1992).

64. The politics of science in Antarctica are discussed in three important articles: Aant Elzinga, "Antarctica: The Construction of a Continent by and for Science," in E.T. Crawford, T. Shinn, and S. Sörlin, eds., *Denationalizing Science: The Contexts of International Scientific Practice* (Dordrecht, The Netherlands: Kluwer Academic Publishers, 1994), 73–106; Aant Elzinga and Ingemar Bohlin, "The Politics of Science in Polar Regions," *Ambio* 18, no. 1 (1989): 71–76, reprinted in Aant Elzinga, ed., *Changing Trends in Antarctic Research* (Dordrecht, The Netherlands: Kluwer Academic Publishers, 1993), 7–27; Robert K. Headland, "Antarctic Odyssey: Historical Stages in Development of Knowledge of Antarctica," in Aant Elzinga, T. Nordin, D. Turner, and U. Wråkberg, eds., *Antarctic Challenges: Historical and Current Perspectives on Otto Nordenskjöld's Antarctic Expedition 1901–1903* (Göteborg, Sweden: Royal Society of Arts and Sciences in Gothenburgh, 2004), 15–24.

65. Eric L. Mills, "Southerly Enterprises," *Science* 260 (May 21, 1993): 1175–1176; L.B. Quartermain, *South to the Pole: The Early History of the Ross Sea Sector, Antarctica* (New York: Oxford University Press, 1967).

66. Stephen J. Pyne, *The Ice: A Journey to Antarctica* (New York: Ballantine Books, 1988). See also Paul Simpson-Housley, *Antarctica: Exploration, Perception, and Metaphor* (New York: Routledge, 1992).

67. Kirkus Review, available on-line at http://www.amazon.co.uk/Ice-Antarctica-Stephen-J-Pyne/dp/1842126741.

68. Richard Gillespie, review of *The Ice*, in *Isis* 78 (September 1987): 456–457.

69. Donald Wooster, review of *The Ice*, in *Environmental Review* 11 (Winter 1987): 307–309.

70. Richard S. Lewis, *A Continent for Science: The Antarctic Adventure* (New York: Viking Press, 1965).

71. Trevor Hatherton, ed., *Antarctica* (London: Methuen, 1965). A report on British science in Antarctica may be found in Sir Raymond Priestley, Raymond J. Adie, and G. de Q. Robin, eds., *Antarctic Research: A Review of British Scientific Achievement in Antarctica* (London: Butterworth & Co., 1964).

72. J.A. Heap, review of *Antarctica*, in *The Geographical Journal* 132 (March 1966): 119–120.

73. D.W.H. Walton, ed., *Antarctic Science* (New York: Cambridge University Press, 1987).

74. Igor Krupnik, Michael Lang, and Scott Miller, eds., *Smithsonian at the Poles: Contributions to International Polar Year Science* (Washington, D.C.: Smithsonian Institution Scholarly Press, 2009).

75. On modernity, see Anthony Giddens, *The Consequences of Modernity* (Stanford, California: Stanford University Press, 1990); Frederic Jameson, *Postmodernism, or, The Cultural Logic of Late Capitalism* (Durham, North Carolina: Duke University Press, 1991); John Krige and Dominique Pestre, eds., *Science in the Twentieth Century* (Amsterdam: Harwood Academic Publishers, 1997); Saskia Sassen, *Globalization and Its Discontents: Essays on the New Mobility of People and Money* (New York: New Press, 1998); Paul Forman, "The Primacy of Science in Modernity, of Technology in Postmodernity, and of Ideology in the History of Technology," *History and Technology* 23 (March/June 2007): 1–152.

76. Karl Weyprecht, "Scientific Work of the Second Austro–Hungarian Polar Expedition, 1872–4," *Journal of the Royal Geographical Society of London* 45 (1875): 19–33, quote from 33. This point is also made in R.E. Bell, "Antarctic Earth System Science in the International Polar Year 2007–2008," in A. Cooper, C. Raymond, and the ISAES editorial team, eds., *Antarctica: A Keystone in a Changing World* (Santa Barbara, California: USGS, 2007), tenth SCAR international symposium on Antarctic Earth sciences (ISAES) August 16–September 1, 2007. USGS open file report 2007–1047.

77. W.F. Budd, "The Scientific Imperative for Antarctic Research," in J. Jabour-Green and M. Haward, eds., *The Antarctic: Past, Present, and Future* (Antarctic CRC Research Report, no. 28, 2001), 41–59; C.J. Taylor, "First International Polar Year, 1982–83," *Arctic* 34 (1981): 370–376; Clark A. Miller, " 'An Effective Instrument of Peace': Scientific Cooperation as an Instrument of U.S. Foreign Policy, 1938–1950," *Osiris* 21 (2006): 133–160.

78. Marc Rothenberg, "Cooperation at the Poles? Placing the First International Polar Year in the Context of Nineteenth-Century Scientific Exploration and Collaboration," in Krupnik, Lang, and Miller, eds., *Smithsonian at the Poles*, 14.

79. William Barr, *The Expeditions of the First International Polar Year, 1882–83* (Calgary, Canada: The Arctic Institute of North America, Technical Paper No. 29, 1985).

80. The reports in order of publication include: *War Department, Proceedings of the Proteus Court of Inquiry on the Greely Relief Expedition of 1883* (Washington, D.C.: Government Printing Office, 1884); P.H. Ray, *Report of the International Polar Expedition to Point Barrow, Alaska* (Washington, D.C.: Government Printing Office, 1885); H. Dawson, *Observations of the International Polar Expedition, Fort Rae* (London: Eyre and Spottiswood for Trübner and Co., 1886): Adolphus W. Greely, *Report on the Proceedings of the United States Expedition to Lady Franklin Bay, Grinnell Land,* 2 vols. (Washington, D.C.: Government Printing Office, 1886); Adolphus W. Greely, *Three Years of Arctic Service* (London, England: Richard Bentley and Son, 1886); R. Lenz, ed., *Beobachtungen der Russischen Polarstation auf Nowaya Semlja. Expedition der Kaiserl. Russischen Geographischen Gesellschaft,* 2 vols. in 1 (1886); G. Neumayer and Börgen, *Die Beobachtungs-Ergibnisse der Deutschen Stationen,* 2 vols. (Berlin: Verlag von A. Asher & Co., 1886); E.E. von Wohlgemuth, *Österreichische Polarexpedition nach Jan Mayen. Beobachtungs-Ergebnisse,* 2 vols. (Wien, Germany: Der Kaiserliche-Königliche Hof -und Staatsdruckerei, 1886); S. Lemström and E. Biese, eds., *Observations faites aux stations de Sodankylä et de Kultala,* 3 vols. in 1 (Helsingfors: L'Imprimerie des Hértiers de Simelius, 1886–1898); W.S. Schley, *The Greely Relief Expedition of 1884* (Washington, D.C.: Government Printing Office, 1887); A. S. Steen, *Die inter-nationale Polarforschung, 1882–1883. Beobachtungs-Ergebnisse der Norwegischen Polarstation Bossekop in Alten,* 2 vols. (Christiania, Germany: Grödahl & Sons, 1887); L'Institut Météorologique de Danemark, *Exploration Internationale des Régions Arctiques, 1882–1883: Expédition danoise. Observations faits à Godthaab,* 2 vols. (København: Chez G.E.C. Gad, Librarie de L'Université, 1889–1893); N.G. Ekholm, *Observations faites au Cap Thorsden, Spitzberg, par l'expédition suédoise,* 2 vols. (Stockholm: Kongl, Boktryckeriet, P.A. Norstedt & Söner, 1890); M. Snellen and H. Ekama, *Rapport sur l'Expédition Néerlandaise qui a hiverné dans la Mer de Kara en 1882/83* (Utrecht: J. Van Boekhoven, 1910).

81. Sydney Chapman, "Introduction to the History of the First International Polar Year," in International Council of Scientific Unions, *Annals of the International Geophysical Year* (New York: Pergamon Press, 1957–1970), vol. 1, 3–5; N. Heathcote and A. Armitage, "The First International Polar Year," *Annals of the International Geophysical Year*, vol. 1, 6–105; D.C. Martin, "The International Geophysical Year," *The Geographic Journal* 124, no. 1 (1958): 18–29; F.W.G. Baker, "The First International Polar Year, 1882–1883," *Polar Record* 21, no. 132 (1982): 275–285; G.A. Corby, "The First International Polar Year (1882/83)," *World Meteorological Organization Bulletin* 31, no. 3 (1982): 197–214; C.J. Taylor, "First International Polar Year, 1882–83," *Arctic* 34, no. 4 (1981): 370–376; E.F. Roots, "Anniversaries of Arctic Investigation: Some Background and Consequences," *Transactions of the Royal Society of Canada* 20 (1982): 373–390; G.D. Garland, "Another Centenary: Poles of the Unknown in Our Earth," *Transactions of the Royal Society of Canada* 20 (1982): 359–368; T.H. Levere, "From Nationalism to Internationalism in Science: The International Polar Year 1882–1883," *Science and the Canadian Arctic: A Century of Exploration, 1818–1918* (New York: Cambridge University Press, 1998), 307–337.
82. William Barr, "Geographical Aspects of the First International Polar Year, 1882–1883, *Annals of the Association of American Geographers* 73, no. 4 (1983): 463–484.
83. Kevin R. Wood and James E. Overland, "Climate Lessons from the First International Polar Year," *Bulletin of the American Meteorological Society* 87, no. 12 (2006): 1685–1697. See also the much earlier assessment: William Herbert Hobbs, "The Climate as Viewed by the Explorer and Meteorologist," *Science* 108 (August 27, 1948): 193–201.
84. R.G. Barry, "Arctic Ocean Ice and Climate: Perspectives on a Century of Polar Research," *Annals of the Association of American Geographers* 73, no. 4 (1983): 485–501. See also W.R. Piggott, "The Importance of the Antarctic in Atmospheric Sciences," *Philosophical Transactions of the Royal Society of London, Series B, Biological Sciences* 279 (May 26, 1977): 275–285.
85. H. Nevanlinna, "Geomagnetic Observations at Sodankyla during the First International Polar Year (1882–1883)," *Geophysica* 35 (1992): 15–22; L.R. Newitt and E. Dawson, "Magnetic Observations at International Polar Year Stations in Canada," *Arctic* 37, no. 3 (1984): 255–266; Wilfred Schröder, "The First International Polar Year (1882–1883) and International Geophysical Cooperation," *Earth Sciences History* 10, no. 2 (1991): 223–226; C. Summerhayes, B. Dickson, M. Meredith Dexter, and K. Alverson, "Observing the Polar Oceans during the International Polar Year and Beyond," *WMO Bulletin* 5, no. 4 (2007): 270–283.
86. William Barr and Chuck Tolley, "The German Expedition at Clearwater Fiord, 1882–83," *The Beaver* 3 (1982): 36–45. See also other national contributions: C. Eamer, "Canada and the International Polar Years," *Meridian* (Fall/Winter 2004): 12–14; R. L. Christie, "Fort Conger: Crossroads of the High Arctic," *Musk-ox* 34 (1984): 28–34.
87. Cornelia Lüdecke, "The First International Polar Year (1882–83): A Big Science Experiment with Small Science Equipment," *History of Meteorology* 1 (2004): 55–64.
88. J. A. Fleming, "Twelfth Annual Meeting of the American Geophysical Union," *Science* 73 (June 26, 1931): 705–707.
89. Fleming and Seitchek, "Advancing Polar Research and Communicating Its Wonders," 1–12.
90. V. Laursen, "The Second International Polar Year, 1932–33," *Annals of the International Geophysical Year*, vol. 1, 211–234; V. Laursen, "Bibliography for the Second International Polar Year, 1932–33," *Temporary Commission on the Liquidation of the Polar Year 1932–33* (Copenhagen: International Meteorological Organization, 1951); J.A. Fleming and V. Laursen, "International Polar Year 1932–1933," *Science* 110 (September 23, 1949): 308–309; Isaiah Bowman, "Polar Exploration," *Science* 72 (October 31, 1930): 439–449; J.A. Fleming, "The Proposed Second International Polar

Year, 1932–1933," *Geographical Review* 22 (January 1932): 131–134; J.A Fleming, "The Jubilee International Polar Year," *Scientific Monthly* 33 (October 1931): 375–380.

91.  R. S. Patton, "Second International Polar Year," *Science* 76 (August 26, 1932): 187–188.

92.  John C. Reed, "The United States Turns North," *Geographical Review* 48 (July 1958): 321–335; John Edwards Caswell, "Materials for the History of Arctic America," *Pacific Historical Review* 20 (August 1951): 219–226, A.L. Washburn, "Focus on Polar Research," *Science* 209 (August 8, 1980): 643–653.

93.  Fleming, "Proposed Second International Polar Year," 134.

94.  Examples of this are the Soviet efforts: USSR Committee of the Second International Polar Year at the Central Administration of the Hydro-Meteorological Service, *Transactions of the Glacial Expeditions*, vol. I; *Pamirs, the Northern Pamirs and Fedshenko Glacier*, vol. II; *Tianshan Head Waters' of the Great Naryn River*, vol. III; *Zeravshan, Head Waters of Zeravshan and Fandaria Rivers*, vol. IV; *Ural, the Subpolar Region*, vol. V; *Caucasus, the Glacier Regions*, vol. VI, 1935–1936. See also the discussions of the national efforts, R. Samoilovich, "Exploration of the Polar Part of U.S.S.R. in 1934 and the Sedovexpedition," *Geografiska Annaler* 17 (1935): 663–668; "The British Polar Year Expedition, 1932–33," *Nature* 129 (1932): 385–386; N.N. Zubov, "The Circumnavigation of Franz Josef Land," *Geographical Review* 23 (July 1933): 394–528; "Canadian Scientists Attack Unknowns of Arctic Frontier," *The Science News-Letter* 22 (October 22, 1932): 262; E.V. Appleton, R. Naismith, L.J. Ingram, "British Radio Observations during the Second International Polar Year 1932–33," *Philosophical Transactions of the Royal Society of London, Series A, Mathematical and Physical Sciences* 236 (April 10, 1937): 191–259; *British Polar Year Expedition, Fort Rae, N.W. Canada, 1932–33* (London: British National Committee for the Polar Year, The Royal Society, Burlington House, 1937); Brian Roberts, "The Cambridge Expedition to Scoresby Sound, East Greenland, in 1933," *Geographical Journal* 85 (March 1935): 234–251.

95.  On ICSU, see L. Ernster et al., "ICSU: The First Sixty Years," *Science International*, special issue (1991): 1–69.

96.  Constance McLaughlin Green and Milton Lomask, *Vanguard: A History* (Washington, D.C.: Smithsonian Institution Press, 1971), 1–24; Fae L. Korsmo, "The Genesis of the International Geophysical Year," *Physics Today* (June 2007): 38–43; Fae L. Korsmo, "Shaping Up Planet Earth: The International Geophysical Year (1957–1958) and Communicating Science through Print and Film Media," *Science Communication* 26, no. 2 (2004): 162–187; Rip Bulkeley, *The Sputniks Crisis and Early United States Space Policy: A Critique of the Historiography of Space* (Bloomington: Indiana University Press, 1991), 89–103; Allan A. Needell, *Science, Cold War, and the American State: Lloyd V. Berkner and the Balance of Professional Ideals* (Amsterdam: Harwood Academic Publishers, 2000), 297–323.

97.  Ann Ewing, "International Look at Earth," *The Science News-Letter* 67 (January 15, 1955): 42–43; National Academy of Sciences, *Antarctic Research: Elements of a Coordinated Program* (Washington, D.C.: National Academies of Sciences, 1949).

98.  Lloyd V. Berkner, "International Scientific Action: The International Geophysical Year 1957–58, *Science* 119 (April 30, 1954): 569–575, quotes from 575.

99.  Sydney Chapman, *IGY: Year of Discovery, The Story of the International Geophysical Year* (Ann Arbor: University of Michigan Press, 1959); Walter Sullivan, *Assault on the Unknown: The International Geophysical Year* (New York: McGraw-Hill, 1961).

100. Sydney Chapman, "The International Geophysical Year and Some American Aspects of It," *Proceedings of the National Academy of Sciences* 40, no. 10 (1954): 924–926; Joseph Kaplan, "The Science Program of the International Geophysical Year," *Proceedings of the National Academy of Sciences* 40, no. 10 (1954): 926–931; Werner Buedeler, *The International Geophysical Year: UNESCO and its Programmes*

(Paris: United Nations Educational, Scientific and Cultural Organisation (UNESCO), 1957); Gould, "Antarctic Prospect," 25–28; D.C. Martin, "The International Geophysical Year," *Geographical Journal* 124 (March 1958): 18–29; Sydney Chapman, "From Polar Years to Geophysical Year," *Studia Geophysica et Geodaetica* 4 (December 1960): 313–324; Marcel Nicolet, "Historical Aspects of the IGY," *Eos* 64, no. 19 (1983): 369–370; Marcel Nicolet, "The International Geophysical Year 1957/58," *WMO Bulletin* 31, no. 3 (1982): 222–231; Marcel Nicolet, "The International Geophysical Year (1957–1958): Great Achievements and Minor Obstacles," *GeoJournal* 8 (December 1984): 303–320; Ronald Fraser, *Once around the Sun: The Story of the International Geophysical Year* (New York: Macmillan, 1957); J.T. Wilson, *IGY: The Year of the New Moons* (New York: Alfred A. Knopf, 1961).

101. Frank M. Marson and Janet R. Terner, comps., *United States IGY Bibliography, 1953–1960: An Annotated Bibliography of United States Contributions to the IGY and IGC (1957–1959)* (Washington, D.C.: National Academy of Sciences-National Research Council, 1963).

102. The World Data Centers have a massive on-line presence. They are accessible through the National Oceanic and Atmospheric Administration: http://www.ngdc.noaa. gov/wdc/ (accessed January 30, 2009).

103. K.O. Murra, *International Scientific Organizations: A Guide to Their Library, Documentation, and Information Services* (Washington, D.C.: Library of Congress, 1962).

104. ICSU, *Annals of the International Geophysical Year*. Rip Bulkeley wrote of the IGY's *Annals* in 1991: "It is appropriate to comment at this point that the *Annals* are frequently unreliable or inconsistent on minor points of organizational detail." (Bulkeley, *The Sputniks Crisis and Early United States Space Policy*, 237). The Geophysical Institute at the University of Alaska, Fairbanks, also maintains an exhaustive bibliography on-line concerning the IGY's scientific efforts, http://www. gi.alaska.edu/services/library/IGY.pdf (accessed January 30, 2009).

105. H. Bullis, "The Political Legacy of the International Geophysical Year," Committee on Foreign Affairs, U.S. House of Representatives (Washington, D.C.: Government Printing Office, 1973).

106. Peter Beck, *The International Politics of Antarctica* (London: Croom Helm, 1986); Klaus Dodds, *Geopolitics in Antarctica: Views from the Southern Oceanic Rim* (New York: John Wiley, 1997); A. Berkman, *Science into Policy: Global Lessons from Antarctica* (San Diego, California: Academic Press, 2002); M.J. Peterson, *Managing the Frozen South: The Creation and Evolution of the Antarctic Treaty System* (Berkeley: University of California Press, 1988).

107. Donald R. Rothwell, *The Polar Regions and the Development of International Law* (Cambridge: Cambridge University Press, 1996).

108. Ingrid Schild, "The Politics of International Collaboration in Polar Research," PhD dissertation, University of Sussex, 1997; Christopher C. Joyner, *Governing the Frozen Commons: The Antarctic Regime and Environmental Protection* (Columbia: University of South Carolina Press, 1998); A. Elzinga "Antarctica: The Construction of a Continent by and for Science," in Elisabeth Crawford et al., eds., *Denationalizing Science* (Dordrecht, The Netherlands: Kluwer Academic, 1993), 73–106; R. Fifield, *International Research in the Antarctic* (New York: Oxford University Press, 1987); Christy Collis and Quentin Stevens, "Modern Colonialism in Antarctica: The Coldest Battle of the Cold War," in Gunter Lehman and David Nichols, eds., *Proceedings 7th Australasian Urban History/Planning History Conference* (Geelong, Australia: Deakin University, 2004), 72–95; James Spiller, "Re-imagining United States Antarctic Research," *Public Understanding of Science* 13 (2004): 31–53; Colin Summerhayes, "International Collaboration in Antarctica: The International Polar Years, the International Geophysical Year, and the Scientific Committee on Antarctic Research," *Polar Record* 44 (2008): 321–334; Katrina Dean, Simon Naylor, Simone

Turchetti, and Martin Siegert, "Data in Antarctic Science and Politics," *Social Studies of Science* 38/4 (August 2008): 571–604; Quentin Stevens and Christy Collis, "Living in the Cold Light of Reason: Colonial Settlements in Antarctica," in Maryam Gusheh and Naomi Stead, eds., *Proceedings Progress: The 20th Annual Conference of the Society of Architectural Historians, Australia and New Zealand* (2003): 291–297; Rip Bulkeley, "Aspects of the Soviet IGY," *Russian Journal of the Earth Sciences* 10 (2008).

109. George J. Dufek, *Operation Deepfreeze* (New York: Harcourt Brace, 1957); Dian Olsen Belanger, *Deep Freeze: The United States, the International Geophysical Year, and the Origins of Antarctica's Age of Science* (Boulder: University of Colorado Press, 2006).

110. A few examples of this are John C. Behrendt, *Innocents on the Ice: A Memoir of Antarctic Exploration, 1957* (Niwot: University Press of Colorado, 1998); John C. Behrendt, *The Ninth Circle: A Memoir of Life and Death in Antarctica, 1960–62* (Albuquerque: University of New Mexico Press, 2005); and Charles Swithinbank, *Forty Years on Ice: A Lifetime of Exploration and Research in the Polar Regions* (Lewes, Sussex, United Kingdom: Book Guild, 1998).

111. Solid overviews of the history of space exploration include William E. Burrows, *This New Ocean: The Story of the First Space Age* (New York: Random House, 1998); Howard E. McCurdy, *Space and the American Imagination* (Washington, D.C.: Smithsonian Institution Press, 1997); Roger D. Launius, *Frontiers of Space Exploration* (Westport, Connecticut: Greenwood Press, 1998). All of them discuss the IGY and the satellite program that resulted. Specifically on *Sputnik 1*, see these important works: Robert A. Divine, *The Sputnik Challenge: Eisenhower's Response to the Soviet Satellite* (New York: Oxford University Press, 1993); Rip Bulkeley, *The Sputniks Crisis and Early United States Space Policy: A Critique of the Historiography of Space* (Bloomington: Indiana University Press, 1991); Roger D. Launius, John M. Logsdon, and Robert W. Smith, eds., *Reconsidering Sputnik: Forty Years Since the Soviet Satellite* (Amsterdam: Harwood Academic Publishers, 2000); Matt Bille and Erika Lishock, *The First Space Race: Launching the World's First Satellites* (College Station: Texas A&M University Press, 2004); Paul L. Dickson, *Sputnik: The Shock of the Century* (New York: Walker and Co., 2001); Matthew Brzezinski, *Red Moon Rising: Sputnik and the Hidden Rivalries that Ignited the Space Age* (New York: Times Books, 2007). Other aspects of this are recounted in Martin Collins, ed., *After Sputnik: 50 Years of the Space Age* (New York: Collins, 2007); Asif A. Siddiqi, *Sputnik and the Soviet Space Challenge* (Gainesville: University Press of Florida, 2003); James J. Harford, *Korolev: How One Man Masterminded the Soviet Drive to Beat America to the Moon* (New York: John Wiley & Sons, 1997); Philip Nash, *The Other Missiles of October: Eisenhower, Kennedy, and the Jupiters, 1957–1963* (Chapel Hill: University of North Carolina Press, 1997); Peter J. Roman, *Eisenhower and the Missile Gap* (Ithaca, New York: Cornell University Press, 1995); Kenneth Osgood, *Total Cold War: Eisenhower's Secret Propaganda Battle at Home and Abroad* (Lawrence: University Press of Kansas, 2006); Homer Hickam, *Rocket Boys* (New York: Delacorte, 1999); Constance M. Green and Milton Lomask, *Vanguard: A History* (Washington, D.C.: Smithsonian Institution Press, 1971). Legislative responses are discussed in Enid Curtis Bok Schoettle, "The Establishment of NASA," in Sanford A. Lakoff, ed., *Knowledge and Power* (New York: Free Press, 1966), 162–270; Alison Griffith, *The National Aeronautics and Space Act: A Study of the Development of Public Policy* (Washington, D.C.: Public Affairs Press, 1962); Barbara Barksdale Clowse, *Brainpower for the Cold War: The Sputnik Crisis and the National Defense Education Act of 1958* (Westport, Connecticut: Greenwood Press, 1981).

112. Walter A. McDougall, *The Heavens and the Earth: A Political History of the Space Age* (New York: Basic Books, 1985; reprint Baltimore, Maryland: Johns Hopkins University Press, 1997).

113. McDougall, *The Heavens and the Earth*, 5.
114. McDougall, *The Heavens and the Earth*, 6–7.
115. Richard Sorrenson, "The Ship as a Scientific Instrument in the Eighteenth Century," *Osiris*, 2nd series 11 (1996): 221–236.
116. David H. DeVorkin, *Science with a Vengeance: How the Military Created the U.S. Space Sciences after World War II* (New York: Springer-Verlag, 1992), 66–68.

# Part Two

# National Case Studies

## Chapter 5

# Going Global in Polar Exploration: Nineteenth-century American and British Nationalism and Peacetime Science

*Christopher Carter*

For much of the nineteenth century, national competition defined attempts to investigate the poles. International scientific cooperation depended on a variety of particular social and political factors. Such mutual collaboration could best occur during periods of relative diplomatic tranquility. Just as the decades between the Congress of Vienna (1815) and the outbreak of the Crimean War (1854) had provided such a space for scientific teamwork, so the political situation after the Congress of Berlin (1878) opened the possibility of another era of international scientific partnership. Indeed, rather than being novel, the efforts leading up to the coordinated study of the Arctic during the first International Polar Year (IPY) (1882–83) represented the return to a model of international action first seen earlier in the century. These synchronized attempts to solve the mysteries of the Arctic momentarily blunted nationalistic bravado in favor of a cosmopolitan study of geoscience but also gave domestic politics a more significant role in shaping those scientific ventures. Such cooperative scientific pursuits seemed well suited to the peacetime situations of the nineteenth century and serve as a model of analysis for history. By comparing parallel efforts by Great Britain and the United States to launch an Antarctic expedition in the early nineteenth century, it is possible to see the effect that domestic partisan realities could have on the supposedly impartial study of science in this period. These influences continued to play a role during the planning of the IPY later in the century. In the end, politics could have as much of an impact on the success of a scientific venture as the theories and techniques utilized during the venture.

Due to improved modes of navigation and communications, science had become global by the nineteenth century. While earlier efforts to study worldwide phenomena had been made (such as Halley's survey of terrestrial magnetism in the seventeenth century and the observations of the transits of

Venus in the eighteenth century) these had been restricted to one-time voy-ages that produced limited results. By the nineteenth century, the conditions existed to make a sustained, continuous study of global phenomena. The steamship and the telegraph allowed for faster transmission of data, while the expansion of empires opened fresh spaces to scientific inquiry.[1] This new capability to study nature on a systematic, worldwide scale promoted efforts in the geosciences, especially terrestrial magnetism and meteorology. The success of these new endeavors required both international cooperation and a relatively stable political world.

Peace and war could have important effects on science. In the eighteenth century, the Seven Years War (1756–63) had spanned a large part of the planet and created logistical problems for observation of the 1761 transit of Venus. Similarly in the nineteenth century, the Crimean War interrupted a period of mid-century scientific cooperation that started after the Congress of Vienna. The 1830s had been an auspicious time for international science. The global approach taken during this time included both active coopera-tion, in the form of coordinated observations carried out by various states, and passive cooperation, in which one power did not directly interfere with a rival's projects. In 1838 British astronomer Sir John Herschel, returning from a survey of the Southern Hemisphere skies at the Cape of Good Hope, wrote to his colleague Francis Baily that "if peace continues and the world goes on for another quarter century as it is doing now, we shall know something!"[2] This period of calm also witnessed various state-sponsored scientific and geographic expeditions directed toward the poles. Examining the methods used to gain state sanction for these projects demonstrates the influence that internal political conditions came to have on peacetime science.

Various factors, both foreign and domestic, played an important role in the success of scientific projects undertaken after 1815. While the return of peace freed resources for scientific ventures and allowed the possibil-ity of international cooperation, cessation of hostilities also redirected the competitive incentive for nationalist science. As a result, when making new proposals, scientists had to emphasize the practical benefits that might be achieved, rather than relying on an appetite for nationalistic achievement at the expense of the state's rivals. Such practical concerns could either bol-ster or weaken scientific projects. Supporters might portray the projects as opportunities to strengthen the state, while opponents might accuse scien-tists of taking resources needed elsewhere. In the absence of direct interna-tional competition, proponents often had to utilize other incentives to gain state support for scientific ventures. This situation meant that internal politi-cal factors became more important, because science had to satisfy domestic needs and appeal to the current government. Thus politics acquired a greater role in shaping science.

Launching any scientific project required a political balancing act. Idealistic appeals for international cooperation had to be poised against the nationalist incentives that often drove scientific ventures in the first place. Additionally, in different states, diverse political climates existed in which science could either flourish or stagnate. Proposals that started out as similar plans could transmute in different states into largely different projects, influenced as much by the politicians who approved the funding as the scientists who crafted the theories. In Great Britain, a more aristocratic form of government allowed for the development and implementation of projects by scientists working within the halls of government without the need to appeal to Parliament or public opinion.[3] The American system, which included a larger degree of popular participation, often subordinated pure science to those practical results that appealed to both the public and elected officials.[4] These contrasting systems had a noticeable effect on both the type and the extent of scientific ventures carried out in the nineteenth century.

## Magnetic Crusade versus Exploring Expedition

One example of the impact that politics could have on science involved Anglo-American efforts to launch an Antarctic expedition in the 1830s. This was an example of passive cooperation in which friendly rivalry, rather than military goals, drove two powers that had been at war as recently as 1814. Both sides used the possibility of discoveries by the other as an incentive for their own goals. In 1828 U. S. Secretary of the Navy Samuel Southard claimed, "We now navigate the ocean, and acquire our knowledge of the globe, its divisions and properties, almost entirely from the contributions of others."[5] British scientists backing a southern expedition also invoked the fear that Americans would beat them to the Antarctic as part of their lobbying efforts.[6] The differences in the final form of the two projects reflect to some extent the different domestic situations in the two nations that conceived of them. Great Britain, with its many institutions of learning, had more established men of science than the young republic, and the British aristocratic democracy provided them with a path to appeal directly to the upper layers of the state. Personal connections among men of science and the government or military played a large role in the success of the British effort, in no small part because all the major institutions of science and state were located in London.[7] Joseph Henry had noted something of this sympathy when he commented that the British Association for the Advancement of Science "is quite as aristocratical as the government of the nation."[8] The United States, by contrast, had a smaller number of scientists spread over a greater area and a system of frontier democracy, which sometimes forced advocates to win popular support for major new proposals that could not be funded out of existing appropriations.[9]

Terrestrial magnetism emerged as a major field of study in the nineteenth century. Variations in the compass had been noted for centuries, but no theory had yet been able to successfully explain or predict these changes. Maritime states like Great Britain were especially interested in developing a theory of Earth's magnetic field that would aid in navigation. Alexander von Humboldt, who, in 1829, had founded an observatory at Berlin to measure magnetic variation, put the study of geomagnetism on a new footing in the early nineteenth century. With the help of other scientists, Humboldt created a system of continental magnetic observatories, primarily in Russia and the German states.[10] Additionally, the work of Carl Friedrich Gauss in the 1830s suggested that a general theory of terrestrial magnetism was possible. His 1832 paper presented the hypothesis of a single terrestrial magnetic axis with two poles, in contrast to Halley's earlier two-axes/four-pole theory.[11] Although Halley's theory retained some supporters, S. Hunter Christie helped to popularize Gauss's theories in Great Britain through Christie's 1833 article for the British Association for the Advancement of Science.[12] Interest in geomagnetism continued to grow in Great Britain, encouraged both by magnetic surveys of the British Isles and Gauss's 1838 paper, "Allgemaine Theorie des Erdmagnetismus." In the late 1830s, the British turned their attention south, attracted by the Antarctic magnetic pole(s) and the prospect of a global theory of magnetism.[13]

John Herschel became a prominent figure in the British Antarctic expedition, although the voyage itself was not his primary concern. Herschel's time observing at the Cape had convinced him of the valuable role such colonial stations could play in the study of phenomena such as meteorology and terrestrial magnetism. He saw the possibility of establishing a network of observatories to conduct geophysical observations, not just at the Cape, "but in India, Australia, Mauritius, and in short at as many stations as possible in the English Colonial possessions."[14] On returning to Europe, he began to consider ways of building such a system, consulting with German scientists like Gauss and Wilhelm Weber who were already conducting geomagnetic observations on the Continent.[15] As a result, Herschel developed a plan to establish geophysical observatories in British colonies to make coordinated observations of Earth's magnetic field "over the whole surface of the globe" that would cooperate with "those in Europe on the plan recommended and practiced by Gauss and Weber."[16] Such an expansive scheme of observatories required government backing in order to secure both funding and posts for observing. Herschel approached Captain Francis Beaufort at the Admiralty with his plan but was unable to gain support from an institution that already funded numerous geographical surveys around the empire.

Then in the summer of 1838, a new possibility presented itself. Humphrey Lloyd and Edward Sabine, fellow members of the British Association for the Advancement of Science, recruited Herschel to their effort to win government backing for an Antarctic expedition to determine the location of the south magnetic pole(s). Their original plan was for a single expedition to the Antarctic that would only make temporary magnetic observations on the voyage. As recently as that spring, Lloyd had expressed his reluctance to join in the continental plan of continuous observations.[17] Now Herschel convinced Lloyd and Sabine to add his colonial observatories to their Antarctic plan. The three scientists and their naval allies presented the entire venture to the Association, which, at its meeting in Newcastle, passed a series of resolutions urging the government to take advantage of such a propitious moment for science.

Herschel emphasized the importance of the peacetime setting, while highlighting the domestic advantages of the project. "Great physical theories, with their trains of practical consequences, are preeminently national objects, whether for glory or for utility," he declared in his report on the Newcastle resolutions. "The peace which now happily subsists may not continue many years longer, and in the turmoil of war such objects are little likely to engage attention."[18] Following the Newcastle meeting, Herschel, Lloyd, and Sabine formed a lobby to try to convince the British government to back their Antarctic ambitions while peace still provided an opportunity. They coordinated their plans with Weber and Gauss, linking the new British system of observations into the existing German union (the Göttinger Magnetische Verein) of observatories, which studied terrestrial magnetism. This international project came to be known as the Magnetic Crusade.[19]

Herschel, a baronet, used his aristocratic connections to further the cause of the crusade. He also enlisted allies in the Royal Society, to whom he assumed the government would turn for advice.[20] Within the first weeks of the lobby, Herschel set up meetings with Prime Minister William Lamb, Lord Melbourne, and officials at the Admiralty.[21] In October, he dined with Melbourne and Queen Victoria, using the opportunity to discuss plans for the upcoming expedition.[22] A week after meeting with the Queen, Herschel wrote to Lord Minto at the Admiralty about the plan, "which cannot but prove productive to the most momentous results, especially to a great maritime nation."[23] Meanwhile Sabine, a colonel in the Royal Artillery, was able to use his own connections. He corresponded with the chancellor of the Exchequer, Thomas Spring-Rice, attempting to arrange funding for plans that he stressed "cannot be obtained by private means," only state sanction.[24] Herschel and his supporters arranged for an additional meeting with Melbourne in November.[25] Because of the structure of the British government, advocates of the crusade could go directly to the top, meeting with government officials who were themselves well positioned to put the plans into action.

The lobby for the crusade displayed an interesting mixture of appeals to nationalism and internationalism. British scientists were eager to catch up with the work already being done by their French, German, and Russian counterparts in the geosciences.[26] Herschel appealed to national pride, citing Great Britain's maritime tradition and Halley's earlier geomagnetic voyage.[27] The appeals helped sell the crusade to the British government. At the same time, however, Herschel desired international cooperation. He hoped that the crusade would lead other European nations, and the United States, to follow Great Britain's example and establish observatories of their own, further extending the system.[28]

The greatest challenge to the lobby for the Magnetic Crusade came from internal divisions. Sabine and James Ross, the more naval-minded members of the lobby, favored the Antarctic expedition over Herschel's observatories and appeared ready to abandon the colonial outposts if they threatened the success of the voyage. Sabine worried that the government would be wary of any additional scientific expenditures, a fear that Minto had confirmed when he pointed out to Herschel that the Admiralty already spent £60,000 a year on scientific research. Minto suggested that Herschel find some way to establish his observatories before the expedition departed, as he could not expect "any very immediate measures would be derived from the Admiralty."[29] Sabine and Ross expressed their preference for an expedition over the stations, while Lloyd sided with Herschel in favor of the observatories, arguing that they must "not yield to any attempt... to have this thing done by halves."[30] Herschel appealed for unity among the lobby, insisting that "a perfect harmony of ideas should subsist between men of science whenever application to Government is in question."[31] By December, Herschel feared that the expedition would not depart by the spring, although he remained hopeful "that something *will* be done worthy of this country."[32] Sabine and Ross, however, made clear their intention to pursue the expedition even without the observatories. "If the expedition is not appointed for next spring," Sabine warned Herschel, "The labours of yourself, and of your worthy colleagues, in the deputation, will... be chiefly lost. I will not reurge that all might be ready for the Spring, both for the expedition & the observatories... but no one can question that *the branch of research for which the expedition is designed needs no delay.*"[33]

As a result, Herschel called in support from Spencer Joshua Alwyne Compton—Lord Northampton and president of the Royal Society—to secure the observatories' position in the overall plan. The Royal Society sent its own delegation to meet with the prime minister in January 1839, but time was important because the expedition still hoped for a spring launch.[34] Ross in particular worried about any further delay. He complained to his wife "the proper season for active operation has surely been frittered away by audience after audience, deputations and petitions and still nothing is decided!"[35] The main problem was the fate of the physical observatories, which the Admiralty did not want to finance. Lord Minto still insisted that Herschel find the means to establish the observatories first, before any expedition departed.[36] Herschel convinced the other members of the lobby, including Ross and Sabine, to link the observatories to the expedition for practical reasons, so that the expedition could drop off the instruments and observers as it visited the colonial stations on its way to the Antarctic.[37]

In March 1839, the Admiralty ordered the Antarctic expedition to leave by June.[38] Minto still refused to have anything to do with the observatories though, and "threw the fixed observations overboard."[39] Melbourne solved this problem by transferring their funding to the Chancellor of the Exchequer.[40] "The two halves of our magnetic prayer are granted," Herschel wrote to Whewell.[41] By April, the Royal Society received notice that the government would establish stations for geophysical observations in Canada, St. Helena, and the Cape.[42] By using personal connections and working through the aristocratic British system of government, the lobby for the crusade accomplished their goals: both an Antarctic expedition and a series of geophysical observations. While the lobby had faced difficulties and long delays, in the end the project remained essentially the same one Herschel had proposed to Sabine and Lloyd the previous summer. Partisan considerations were minimal in the British system, which allowed scientists to work closely with the governmental and military figures who provided the financial and logistical support for their plans. The scientists' geographic and scientific aims remained intact and the project was international. The observatories added to an existing series of observations on the European continent, headquartered at the University of Göttingen in Germany.[43] They not only extended the existing system but also inspired the establishment of new observatories in India and northern Africa, "so that certain of them shall be always on duty."[44] Eventually this observing network spanned dozens of countries and more than 30 stations.[45] Geoscience now had a tool to study the world on a global scale.

By contrast, the American Antarctic expedition faced multiple hurdles, forced to win broad support in both houses of Congress in order to avoid the perception that it was too narrowly focused to benefit "visionary" science or whaling interests. The scientific content added to the expedition did not find favor with all members of Congress and often threatened to become a liability to the proposal. Approval, in 1838, of the Exploring Expedition came only after earlier attempts had floundered. To win this political victory, supporters of the expedition had been forced to recast it, downplaying sectional benefits for New England and scientific objectives. Only by convincing Congress of its national importance, economic benefits, and navigational prospects did supporters succeed in gathering the support needed to counter the resistance to new state expenditure. In the process, though, the expedition lost much of its scientific content.

Selling scientific projects to a skeptical Congress required a great deal of political skill. In early nineteenth-century America, exploration into Louisiana and the West held a greater attraction than polar voyages. Interest in Arctic and Antarctic exploration had been stimulated in the 1820s by the geographical theories of John Symmes, who posited that Earth was composed of a series of concentric hollow spheres open at the poles.[46] Symmes held that the polar openings created a more temperate climate in the polar regions, an idea that survived through the nineteenth century in a different form with the belief in an open polar sea.[47] Apparent evidence for this theory came from Ross's 1818 voyage, which found Arctic inhabitants at surprisingly high latitudes.[48] In 1823, Symmes's supporters petitioned Congress for an expedition to the North Pole to locate an entrance to the inner world, both for the sake of geographical discovery and to open up trade with any interior inhabitants! Although the House of Representatives rejected the petition on a vote of 56 to 46, Symmes and his student Jeremiah Reynolds continued their crusade, raising interest in the idea of Arctic and Antarctic exploration.[49] Reynolds and Symmes pushed for an expedition to the polar regions, lecturing on the hollow Earth theory across the country and gaining some interest, even if their theory found few supporters.[50] "The question to be determined by enlightened minds is not so much whether Symmes' theory is correct," declared the *New-York Mirror*, "as are there not important discoveries to be made, and shall we not make them?"[51] "Even error often leads to important results," opined the *Saturday Evening Post*. "Columbus wanted to go to the East Indies, westerly.... Symmes in searching for an inner world might reach the poles."[52]

While many agreed with Joseph Henry, who denounced the hollow Earth theory at his inaugural lecture at Albany in 1826, it did incite some popular interest in polar exploration.[53] Reynolds, who by this time was downplaying his earlier role in popularizing the hollow Earth theory, continued to promote an Antarctic expedition for geographical and scientific gain into the 1830s. An initial attempt to convince Congress to fund such a voyage had failed in 1828, defeated by the political machinations of congressional Democrats eager to embarrass the outgoing Whig administration of John Quincy Adams, which had backed the expedition.[54] Reynolds continued to advocate his cause, finally convincing Congress to reconsider the expedition. In April 1836, he addressed the House of Representatives concerning the proposed expedition in an appeal that featured the themes of nationalism, honor, and safety for commerce. Reynolds's address marked the culmination of the latest push for the Exploring Expedition. Soon afterward, the House approved the expedition on a vote of 92 to 68, with Senate approval coming later the same day.[55]

A close analysis of the House vote is of interest, because party, location, and sectional affiliations all influenced whether representatives voted for the expedition. One immediate factor in deciding whether to vote in favor was party affiliation. The expedition enjoyed the overwhelming support of Whigs (71 percent in favor), while Democrats narrowly split against it (52 percent opposed).[56] Local conditions were also a major influence. Representatives from coastal districts and New England (83 percent in favor) tended to support the expedition over inland districts. Finally, sectional differences also contributed to the result. Seventy-eight percent of Southerners voted against the measure—including all the Whigs who opposed it— although local factors could still win out over party and section. For example, southern Whigs (and northern Democrats) from coastal districts tended to favor the expedition. In the end, though, party loyalty was the most important factor. Here the Whigs' traditional strength in the West helped immensely. The deciding votes for the expedition came from a group of inland Whigs, representing the states of Ohio, Indiana, Kentucky, and Tennessee. Isolated from the coast, these representatives had no immediate reason to support the expedition, as it offered no direct benefit to their constituents. Yet the majority of these Whigs (12 out of 21) voted in favor of the expedition, providing the margin of victory.

Despite the solid backing obtained from Congress, the expedition faced problems in the implementation stage. While the outgoing Democratic president, Andrew Jackson, had taken "a lively interest in the exploring expedition directed by Congress," under the administration of his successor, Martin van Buren, the expedition's fate became uncertain.[57] The Panic of 1837, the worst economic depression the young nation had yet seen, tightened resources and made it difficult to justify the expenditure laid out by Congress the previous year.[58] Additionally, the secretary of the navy, Mahlon Dickerson, was the brother of Philemon Dickerson (Democrat, New Jersey) who had vocally opposed the expedition in the House the previous year.[59] Despite Jeremiah Reynolds's plea that "no professional pique or petty jealousies should be allowed to defeat the objects of the expedition," Secretary Dickerson showed as little enthusiasm as his brother for the Exploring Expedition.[60]

As if administrative resistance was not enough, the delays in implementation allowed time for new challenges in Congress. Opponents again attacked the expedition, placing American interests above international ones and practical naval needs above speculative science. Henry Wise (Whig, Virginia) moved a resolution to convert the entire Exploring Expedition into a squadron for surveying the coasts and protecting American vessels.[61] William Montgomery (Democrat, North Carolina) also favored keeping the ships in American waters rather than sending an expedition to "Symmes' cylindrical [sic] world"![62] The expedition still had its supporters though. John Reed (Whig, Massachusetts) denounced this attempt to destroy the Exploring Expedition, while Charles Mercer (Whig, Virginia) argued that it was the duty of the United States, being the second commercial country in the world, to add its quota to the cause of science.[63] While the Whigs managed to beat back Wise's attempt to convert the Exploring Expedition into a coast guard, the greater challenge still loomed.[64] On April 10, 1838, Churchill Cambreleng (Democrat, New York) brought a motion to the floor of the House to strike out the appropriation altogether, thus discontinuing the expedition. This motion brought the House to a direct vote on whether to continue supporting the Exploring Expedition.

Science threatened to become a liability to the proposal because the scientific content of the expedition did not find favor with all members of the House. Supporters had to compromise on scientific aims in order to pick up necessary votes from representatives reluctant to continue funding a scientific voyage during a period of economic uncertainty. Isaac Crary (Democrat, Michigan) represented a group of Democrats who were ready to continue the expedition without its explicitly scientific content, accepting the need for a geographical survey, while opposing the scientific components.[65] Crary blamed Commodore Thomas ap Catesby Jones, the commander of the expedition, for its delay, charging that by converting the expedition from an exploring to a scientific voyage, Jones had changed the whole design and intention of Congress.[66] "He must have a bodyguard of scientific men from all of the most celebrated institutions of the country to catch birds and flies, toads and fishes. This was all very pretty.... It was very well to take advantage of so favorable an opportunity to augment the stores of science, but that object should have been only incidental and subordinate to the great purpose of the squadron."[67] Crary and his supporters felt that Congress had no reason to feel confident in the success of the expedition "as it was [then] organized; and he was glad that an opportunity was afforded of placing it upon a proper footing."[68] Crary's message was clear: the Exploring Expedition could survive, but only as a surveying mission, stripped of the bulk of its scientific character.

On one final vote lay the fate of the Exploring Expedition. With the help of Crary and other Democrats, supporters of the expedition held together a coalition in its favor, defeating the motion to derail it 57 to 91.[69] This time the deciding votes came from Democrats such as Crary, who supported the effort now that its focus was no longer on science, while Whigs were more lukewarm toward the revised goals. This change in backing was evident in the vote: for the first time in all the debates about the Exploring Expedition, a larger proportion of Democrats backed the expedition (61 percent in favor) than did Whigs (58 percent in favor). The shift was also apparent when comparing the original 1836 vote on final passage with the 1838 vote to continue funding. Most of the representatives voting both times kept their position for or against the mission. Eleven representatives present for both votes switched sides, from opposition in 1836 to support in 1838. Seventy-three percent of these votes were Democrats. To appeal to representatives, such as Crary, and to pick up conservative Democratic support, the scientific aims of the expedition had been curtailed. The Exploring Expedition survived its last trial, but in the process, a radical shift in its focus had taken place.

The partisan bickering about science changed the goals of the endeavor, allowing a new commander, Charles Wilkes, to continue weakening its scientific components. At the time, there was some suspicion that Wilkes had been selected not because of his "pre-eminent scientific attainments, or his particular fitness, but because he had pledged himself to dismiss a part of the scientific corps."[70] Indeed one of Wilkes's first moves was to cut down on the number of scientists included in the expedition.[71] Wilkes initially believed that naval officers and the ship's medical corps could handle all of the major scientific objects of the effort.[72] Although this reassignment proved impractical, he still reduced the number of scientific personnel from 23 to 9.[73] The final plan of the Exploring Expedition reflected congressional reservations about its scientific content. The instructions for the commander stated that the principle objects of the mission were the promotion of the "great interests of commerce and navigation." Only on occasions when not incompatible with those greater goals would it "extend the bounds of science, and promote the acquisition of knowledge."[74]

This comparison between British and American efforts reveals a number of interesting points about the interaction between science and state in this period. British scientists could rely on aristocratic contacts both within the Admiralty and government to push through their agenda without having to appeal to the House of Commons. The American lobby had to make its way through the channels of Congress under a republican system. Forced to appeal to a popularly elected body like the House of Representatives, supporters of the Exploring Expedition had to emphasize the national, economic, and navigational benefits of the expedition before they won approval. Even after passage, science remained a problem for the expedition. Although some saw science as an integral part of the mission, they ran up against strong resistance in the bureaucracy. In the end Captain Wilkes felt free to dispense with much of the scientific content.

As a result the scientific community received less than they had hoped. Botanist Asa Gray resigned from the project, explaining that he preferred not to accompany a voyage that was "so essentially different from the original."[75] As a result of the political difficulties it had faced, the American expedition became more a geographical mission than a scientific voyage; a survey of the south seas that conducted only temporary observations along the way and collected specimens for later study. Titian Ramsey Peale, one of the few scientists finally attached to the expedition, later lamented the lost opportunity of what he called one of "the expensive blunders of the past."[76] He charged that in order to save money after the return of the expedition, the scientific corps had been dismissed without even a chance to write up the results of their studies. Peale also claimed that many of the specimens brought back by the explorers had been dispersed, taken in "a general scramble for curiosities... in which some 'Honorable' men thoughtlessly took part."[77]

The American expedition also differed in its failure to establish any permanent observing posts on the European model. Attempts had been made to establish American meteorological observations in the 1820s and 1830s, with some local success in New York and Pennsylvania.[78] Later attempts to gain federal support and bring the United States into the international system of geophysical observations ran into the same opposition that had stalled the scientific aspects of the exploring party. In December 1839, Joseph Henry felt confident that scientists could convince the American government to establish a series of geomagnetic and meteorological observatories on a plan similar to those in the British colonies.[79] A committee of the American Philosophical Society petitioned Secretary of War Joel Poinsett in favor of the plan.[80] Unfortunately the American government proved less responsive than the British to the idea.[81] The Democratic-controlled House of Representatives overwhelmingly defeated the proposal by a vote of 34–97 in July 1840, even though former president John Quincy Adams had introduced it.[82] Elias Loomis was gloomy about the outcome and feared that governmental support for science would disappear, leaving scientists to their own designs. "Our own Government seems to feel very little interest in the cause of science," he complained. "The election of a President is a matter of far greater importance than the discovery of the laws of magnetism."[83] Despite the lack of funding, a number of private observations did occur, including those at Alexander Bache's observatory at Girard College in Philadelphia. These efforts corresponded with those established by the British for the crusade but were supported through a private subscription by members of the American Philosophical Society.[84]

More so than in Great Britain, the cause of science was hard to sell in the young republic. Laissez-faire political beliefs and anti-international traditions stood in the way of the successful cooperation of science and state that had been demonstrated by the Magnetic Crusade. In the 1840s, for instance, amid the partisan antebellum atmosphere, funding for a national meteorologist became a political issue, constantly in danger of being cut off by a skeptical Congress that balked at paying a $2,000 annual salary on "visionary projects for the regulation of storms and the making of rain."[85] By contrast, the system of British observatories established by the crusade continued to operate through the 1840s, receiving additional grants of government funds at regular intervals. As the time came for the renewal of the project, the new prime minister, Sir Robert Peel, had no objection to another three-year period of observations.[86]

Even British support for scientific observatories began to wane and could not survive the return of war in the 1850s, though. What politics could not end, war did. As European countries again took up arms, resources for science dried up. The Cape magnetic observatory had already ceased independent operation in 1845, while the St. Helena observatory closed in the spring of 1849.[87] After renewing the funding for the Toronto observatory three times, the British government refused to continue its operation beyond 1851.[88] The Tasmanian observatory also fell victim to budget cuts in 1853, as the British government reduced scientific expenditures in the buildup to the Crimean War. Despite attempts by the local colonial government to keep the observatory running, it closed completely by the end of 1854.[89] Just as the long period of relative European peace in the early nineteenth century gave the crusade a chance to gain state aid, so the return of war signaled the end of its objectives. The United States also soon became preoccupied with its own internal conflict, crippling its ability to conduct scientific observations.[90]

## IPY Buildup

The British Magnetic Crusade had established an example of international cooperation for geoscience in the first half of the nineteenth century. This system had been interrupted by the Crimean War, which both diverted funds from science and distracted the major European powers that later played a role in Arctic exploration. Russo-Turkish conflicts continued to consume resources and diplomatic attention until the Congress of Berlin in 1878 opened a new opportunity for international scientific cooperation. The impetus for this venture came in part from the Austro–Hungarian polar expedition of 1872–74. The voyage, intended to locate a northeast passage to Asia, led to the discovery of Franz Joseph Land in the Arctic Ocean.[91] This naval accomplishment by a relatively landlocked empire renewed interest in polar exploration by the traditional maritime powers.[92] It also led to an attempt to replicate the system of observations that had characterized the crusade 40 years earlier.

The colonial stations set up by the crusade became the template for similar outposts around the world. Additional colonial observatories in Aden, Algiers, and throughout the East India Company holdings, in addition to the American geomagnetic stations proposed by the American Philosophical Society, had used the same design as those original stations.[93] The observatories for the IPY echoed these earlier outposts. As before, the stations used Göttingen mean time to make their observations. British observers again came primarily from the Royal Artillery, following the precedent set when Edward Sabine of the Artillery had been one of the leading figures behind the crusade.[94] In the 1830s, Sabine and Sir John Herschel had used their military and political connections to gain support within the British government for a series of simultaneous observations at colonial outposts. Karl Weyprecht echoed their arguments in 1875 when he called for "a series of synchronous expeditions, whose task it would be to distribute themselves over the Arctic regions, and to obtain one year's series of observations made according to the same method."[95]

As the IPY approached, supporters utilized various arguments to gain state sanction for this ambitious project. Weyprecht argued from the standpoint of international science, urging the nations involved to "agree to lay aside their rivalries, and combine for the common good of mankind." He stressed cooperation over competition, claiming that "as long as Polar Expeditions are looked upon merely as a sort of international steeple-chase, which is primarily to confer honour upon this flag or the other, and their main object is to exceed by a few miles the latitude reached by a predecessor," the mysteries of geomagnetism and meteorology would remain unsolved.[96] Such altruistic

Moonlight in the Arctic Regions.

**Figure 5.1** Moonlight in the Arctic regions; published in *The Voyage of the Fox: A Narrative of the Discovery of the Fate of Sir John Franklin* (1860), by Captain F.L. McClintock. (Courtesy of National Oceanic and Atmospheric Administration, NOAA Library Collection, Washington, D.C.)

appeals helped to win over some reluctant European governments. Even Great Britain, which had already been involved in decades of polar voyages, contributed a station in the Canadian Arctic, operated in conjunction with the Hudson's Bay Company.[97]

The American successor to Herschel and Sabine in the years after the Civil War was Henry Howgate, who in 1880 explicitly invoked the memory of the crusade in support of a new polar expedition.[98] Howgate had been pushing Congress for an Arctic expedition since the 1870s and bills to "authorize and equip an expedition to the arctic seas" had been introduced in the 44th and 45th Congresses without success.[99] Congress proved reluctant to invest money in Arctic exploration regardless of the motivation in this period. In the mid-nineteenth century, numerous expeditions had been sent out to search for Sir John Franklin, who had never returned from an 1845 voyage to the Arctic. The British government alone sent nine rescue missions in the years after Franklin's disappearance.[100] Attempts to encourage American involvement in the search for Franklin and the Northwest Passage, however, had met staunch resistance from Congress.[101] Even when Henry Grinnel offered his own ships to sail under the American flag, some members of Congress balked at the notion of subcontracting out for a "wild-goose chase."[102] Despite clear support from the majority, opponents tried numerous procedural techniques to derail the resolution, which finally required no less than 17 separate votes to pass the closely divided House of Representatives.[103]

Convincing the United States to go along with the IPY also proved difficult. Howgate adopted the plan for the IPY as part of his Arctic expedition, but recognized the problems in selling such a plan to the American government. "It is undeniable that hitherto Congress has been, as a rule, averse to the equipment of expeditions for exploration and strictly scientific purposes" he admitted in 1880.[104] Indeed there was a natural resistance in Congress to expending funds on voyages that often became self-perpetuating, as one ill-fated expedition might require a later rescue mission. "Experience has proven...that the constant warfare against the North Pole which civilized nations are undertaking from time to time is a losing warfare on every occasion," lamented Senator George Edmunds (Republican, Vermont). "The quantum which is gained for science is entirely outweighed by the suffering and loss of life and property by whoever undertakes it."[105] Yet the allure of the pole still attracted explorers northward. "So long as the American people exist and the North Pole remains undiscovered," responded Senator John McPherson (Democrat, New Jersey), "we shall have enterprising and progressive spirits...who will be seeking to find it."[106] Advocates of the Arctic expedition now had to use a mixture of motives to appeal for support from Congress, where bills for two similar polar projects had already perished despite Howgate's offer to provide one of his own ships for the voyage. As before, political realities decided the fate of American participation in the IPY, when the United States ultimately sponsored two expeditions, the Greely expedition to the northeast and a separate Alaskan expedition to Point Barrow.[107]

Complicating the task of gaining approval was the addition of a fixed observing station to the original plan for the Greely expedition. Another factor was the partisan atmosphere in Congress in the years following the U.S. Civil War. While often seen as a period when few major issues separated the two parties, the decades after the Civil War still featured intense political competition. Rarely did the same party control both houses of Congress, much less Congress and the presidency simultaneously. Bills passed by one house could easily fail due to inaction by the other if a different party controlled it. Thus it is unsurprising that while polar bills failed in the 44th and 45th Congresses (with a Republican Senate but a Democratic House), they fared better when the same party controlled both houses, regardless of which party that was. The *Polaris* expedition, proposed in 1870, had passed the Republican 41st Congress with little debate.[108] Similarly Howgate's proposal for an Arctic expedition in 1880 found a warmer reception in the now-Democratic 46th Congress.

Nonetheless supporters of the proposed expedition still recognized that science alone was not enough to sell the project. While some might deny that any "direct commercial advantages [we]re expected from exploring the region to the north," Isaac Hayes, a polar explorer who had led an Arctic expedition shortly before the Civil War, recognized the allure of profit.[109] "A good year, with skilled hunters, might yield $20,000 or $30,000 worth of valuable commodities," he argued. "The fur of the Arctic fox...blubber from the seal, and walrus ivory, and seal skins and eider-down."[110] Even Howgate declared, "the government has a right to provide for our general scientific and commercial interests."[111] Nationalism was also a popular point to make, despite traditional views of science as neutral.[112] Howgate appealed to American pride and duty in the face of a "national apathy" that "let foreign countries survey, and map, and discover, and also reap the benefits."[113] "England has beaten us in our best efforts to reach the northern limit of the earth" he warned, "For our government to make a special effort in the same field will be to arouse the ambition of many brave spirits, and enable the chosen few not only to trace England's footsteps over the ice-fields of the North, but also, under the guidance of a good Providence that from the first destined man to 'dominion,' to stand at the Pole itself."[114]

Debate in Congress was limited but not without objection. The bill contained a provision "to establish a temporary colony at some point north of the 81st degree of north latitude, on or near the shore of Lady Franklin Bay, for the purposes of scientific observation and exploration and to develop or discover new whaling grounds."[115] Amendments made it clear that the station was to be a temporary outpost, not a permanent colony, and authorized the president to accept a ship from Howgate for use in the expedition. The majority Democrats in the House dominated debate. Washington Whitthorne (Democrat, Tennessee) pointed out that the American expedition would participate along with seven other nations in "making the circle of the North Pole." He saw the expedition as a chance to enrich the public. "We have heretofore expended millions of dollars in this country subsidizing

corporations and individuals. Here is a proposition that looks to . . . the elevation and improvement of mankind . . . for the benefit not only of our own
people but of the whole human race." George Converse (Democrat, Ohio)
objected to the participation of private ships in the expedition. "Is it possible
that the Government of the United States is reduced so low in the scale of
nations that it is obliged to borrow the vessel of one rich man?" Roger Mills
(Democrat, Texas) derided the whole proposition, asking "what will you do
with the pole when you find it?" Appeals to scientific and material progress
won out however. In the polar region, declared Whitthorne, lay "the secret
of human life and human progress. It is there that . . . the tides and currents of
the sea emanate; there is the birthplace of the atmosphere of the world."[116]

In the end the resolution easily passed Congress and was signed into law
by President Hayes.[117] The Greely expedition went on to have its own event-
ful history, accomplishing several noteworthy goals. In addition to participating in the observations of the IPY, it spent two consecutive winters in
the Arctic and "reached the highest latitude ever attained in any part of the
world" to date.[118] Despite the nationalist rhetoric that helped to sell the new
expedition it was, from the beginning, an international idea. The European
nations involved were able to put aside their rivalries because of the new
period of peace that followed the Congress of Berlin. Such cooperation would
have been unthinkable just a few years before, when Austria, Great Britain,
Germany, and Russia had nearly gone to war over the Bulgarian question. At
the same time, domestic political considerations remained important. Great
Britain's overextended commitments limited its involvement to a single station, while a fortuitous period of single-party control in the 46th Congress
allowed a more substantial American contribution.[119] The participation of
both nations still depended on the international conditions that fostered
scientific cooperation in this period.

## Conclusion

Scientific proposals had to make their own way through the diplomatic and
political environment of the nineteenth century. In many ways the same structures that helped to extend global science, such as the creation of worldwide
empires and centralized states that could support expeditions and observatories, also provided the greatest obstacles to scientific research, as internal and
external competition threatened to derail scientific ventures before they had
a chance to begin. While moderate amounts of political rivalry might help
to further scientific goals, conflicts among states (or among parties within a
state) could fatally weaken these attempts. Thus international science in the
nineteenth century fared best during periods of peace when friendly rivalry
could encourage new ventures without draining resources from them. The
possibility of joint polar expeditions and observations in the nineteenth century depended on diplomatic conditions that allowed states with Antarctic or
Arctic interests to cooperate. At the same time these scientific proposals had
to negotiate with domestic political realities. Successful projects navigated

through political systems that offered funding either through friendly patronage or popular appeals. A negative reception or failure to adapt to the prevailing domestic political climate could doom a project from the beginning. Even victorious proposals still could not help but bear the imprint of the political conditions that had shaped them, both in their extent and outlook. As science came to increasingly rely more on state sanction, such influence was unavoidable.

## Notes

1. See Bruce Hunt, "Doing Science in a Global Empire: Cable Telegraphy and Electrical Physics in Victorian Britain," in Bernard Lightman, ed., *Victorian Science in Context* (Chicago: University of Chicago Press, 1997), 312–333.
2. Herschel to Baily, April 7, 1838 (Royal Society Herschel Papers 25.8.10).
3. John Gascoigne, *Science in the Service of Empire* (Cambridge: Cambridge University Press, 1998), 34–47.
4. Robert Bruce, *The Launching of Modern American Science: 1846–1876* (New York: Alfred Knopf, 1987), 130–134.
5. Quoted in Jeremiah Reynolds, *The South Sea Surveying and Exploring Expedition* (New York: Harper & Brothers, 1841), 186.
6. "I hope the Yankees will not get the start of us," wrote James Ross about the planned British Antarctic expedition of 1839, "that would really be most provoking." Ross to Sabine, October 10, 1838 (Public Record Office BJ3.16.16).
7. Jack Morrell and Arnold Thackray, *Gentlemen of Science* (Oxford: Clarendon Press, 1981), 340.
8. Henry to Bache, August 9, 1838. Marc Rothenberg, Paul Thompson, Kathleen Doman, John Rumm, and Deborah Jeffries, eds., *The Papers of Joseph Henry* (Washington, D.C.: Smithsonian Institution Press, 1992), 4:103. Hereafter cited as JHP.
9. Although concentrated in New England, American scientists were spread out over the whole nation. In 1846 there were no fewer than 32 different scientific societies in the United States. Robert Bruce, "A Statistical Profile of American Scientists, 1846–1876," in George Daniels ed., *Nineteenth-Century American Science: A Reappraisal* (Evanston, Northwestern University Press, 1972): 63–94.
10. James Gabriel O'Hara, "Gauss and the Royal Society: The Reception of His Ideas on Magnetism in Britain (1832–1842)" *Notes & Records of the Royal Society of London* 38, no.1 (August 1983): 28.
11. O'Hara, "Gauss and the Royal Society," 27.
12. S. Hunter Christie, "Report on the State of Our Knowledge Respecting the Magnetism of the Earth," *BAAS Report* (1833): 105–130; Sabine remained committed to the "superior claim" of the four-pole hypothesis as did Ross. Sabine to Lloyd, March 23, 1838 (Royal Society, Terrestrial Magnetism Archive # 43).
13. Beaufort to Ross, August 24, 1838 (Public Record Office BJ2.3.24).
14. Herschel to Humboldt, July 31, 1838 (Royal Society Herschel Papers 21.255).
15. Herschel to Lloyd, June 11, 1838 (Royal Society Herschel Papers 21.248).
16. Herschel to Beaufort, June 29, 1838 (Royal Society Herschel Papers 21.253).
17. Lloyd to Sabine, April 28, 1838 (Public Record Office BJ3.8.30).
18. John Herschel, "Report on the Subject of a Series of Resolutions Adopted by the British Association at Their Meeting in August 1838," *BAAS Report* (1839): 33, 39.
19. The earliest occurrences of the term "Magnetic Crusade" in the nineteenth century appear to be from America in the 1840s. Farrar and Lovering refer to "the present magnetic crusade" in their 1842 textbook on electromagnetism, while Robert Patterson used the term in his 1843 address to the American Philosophical Society. John Farrar and Joseph Lovering, *Electricity, Magnetism and Electrodynamics* (Boston:

Crocker & Ruggles, 1842), 243; "Celebration of the Hundredth Anniversary, May 25, 1843," *Proceedings of the American Philosophical Society* 3, no. 27 (May 25–30, 1843): 34. I am grateful to Marc Rothenberg for his assistance in finding these citations.

20. Herschel to Whewell, October 22, 1838 (Royal Society Herschel Papers 21.264).
21. Herschel to Melbourne, August 27, 1838 (University of Texas 1054:257.1).
22. Morrell and Thackray, 340.
23. Herschel to Minto, October 20, 1838 (University of Texas 1054:263).
24. Sabine to Lloyd, October 23, 1838 (Royal Society Terrestrial Magnetism Archive # 58).
25. Lloyd to Sabine, October 25, 1838 (Public Record Office BJ3.8.51).
26. John Cawood, "Terrestrial Magnetism and the Development of International Collaboration in the Early Nineteenth Century," *Annals of Science* 34 (1977): 585.
27. Herschel, "Terrestrial Magnetism," *Quarterly Review* 66 (1840): 294.
28. Herschel, "Report," (1839): 42; Herschel, "Terrestrial Magnetism," 293.
29. Minto to Herschel, October 24, 1838 (University of Texas 1087:372).
30. Lloyd to Sabine, October 29, 1838 (Public Record Office BJ3.8.52).
31. Herschel to Lloyd, November 16, 1838 (Royal Society Herschel Papers 21.270).
32. Herschel to Sabine, December 2, 1838 (Public Record Office BJ3.26.13); Herschel to Gauss, December 3, 1838 (Royal Society Herschel Papers 21.271).
33. Sabine to Herschel, December 3, 1838 (Royal Society Herschel Papers 15.25).
34. Sabine to Ross, January 3, 1839 (Public Record Office BJ2.13.7).
35. Ross to Ann Ross, February 6, 1839 (Public Record Office, BJ2.5.2).
36. Whewell to Sabine, n.d. [before February 8, 1839] (Public Record Office, BJ2.13.8).
37. Sabine to Northampton, February 8, 1839 (Public Record Office, BJ2.13.9); Ross to Sabine, February 1, 1839 (Public Record Office, BJ3.16.78).
38. Herschel to Whewell, n.d. [March 1839] (Royal Society, Herschel Papers 22.20).
39. Ross to Sabine, March 14, 1839 (Public Record Office, BJ3.16.82).
40. Sabine to Lloyd, March 16, 1839 (Royal Society Terrestrial Magnetism Archive # 61); Sabine to Herschel, April 4, 1839 (Royal Society Herschel Papers 15.33).
41. Herschel to Whewell, March 31, 1839 (Royal Society, Herschel Papers 22.6).
42. Sabine to Herschel, April 11, 1839 (Royal Society, Herschel Papers 15.36). Ross also founded another observatory in Tasmania.
43. O'Hara, "Gauss," 61.
44. Sabine to Lloyd, July 5, 1839 (Royal Society Terrestrial Magnetism Archive #70).
45. William Whewell, *History of the Inductive Sciences* (London: Parker & Son, 1857), 55.
46. Symmes employed centrifugal force to explain the formation of these spheres. Halley had earlier suggested that Earth might have an interior sphere to explain the phenomena of magnetic variation. T. J. Matthews, "Lecture on Symmes' Theory of Concentric Spheres," *The Cincinnati Literary Gazette* 1, no. 12 (March 20, 1824); no. 14 (April 3, 1824).
47. The idea of an open polar sea had existed at least since the fourteenth century. Kirsten Seaver, *The Frozen Echo* (Stanford: Stanford University Press, 1996), 262. Both Howgate and Hayes continued to believe in the existence of such a sea that could be traversed with little difficulty. Henry Howgate, "Arctic Meeting at Chickering Hall," *Journal of the American Geographical Society of New York* 10 (1878): 280, 293.
48. "Capt. Symmes' Theory," *Niles' Weekly Register* 16, no. 416 (August 21, 1819).
49. *Annals of Congress*, 17th Congress, 2nd session, January 1823, 698–699. Symmes himself offered to lead an Antarctic expedition to the interior to "establish our claim by right of discovery, and extend our commerce and fisheries, in immeasurable extent." John Symmes, "The Inside of the World," *Zion's Herald* 3, no. 19 (May 11, 1825).
50. "Capt. Symmes' Theory," *Masonic Mirror and Mechanics' Intelligencer* 1, no. 50 (December 3, 1825).
51. "Symmes' Theory," *The New-York Mirror* 3, no. 43 (May 20, 1826).

52. "Capt. Symmes' Theory of the Earth," *Saturday Evening Post* 6, no. 289 (February 10, 1827).
53. Wise to Henry, May 1, 1849 (JHP 7, 521–522).
54. William Stanton, *The Great United States Exploring Expedition* (Berkeley: University of California Press, 1975), 17–27. "Whig" here refers collectively to the various anti-Jacksonian parties of the period.
55. *House Journal,* 24th Congress, 1st session (May 9 and 10, 1836): 794–795, 805.
56. Goetzmann suggests that "Jackson's forces pushed the exploring expedition bill through Congress," an assertion that is difficult to sustain given the relative lack of support from Democrats (only 48 percent in favor) when compared to the anti-Jacksonian Whigs. William Goetzmann, *New Lands, New Men* (New York: Viking, 1986), 271.
57. Jackson to Dickerson, July 9, 1836. Nathan Reingold, ed., *Science in Nineteenth-Century America: A Documentary History* (Chicago: University of Chicago Press 1964), 110.
58. Nathaniel Philbrick, *Sea of Glory* (New York: Viking, 2003), 35.
59. *Congressional Globe,* 24th Congress, 1st session, 441.
60. Reynolds, *Expedition,* 98.
61. *Congressional Globe,* 25th Congress, 2nd session, 273.
62. *Congressional Globe,* 25th Congress, 2nd session, 295.
63. *Congressional Globe,* 25th Congress, 2nd session, 274.
64. *Congressional Globe,* 25th Congress, 2nd session, 280.
65. Crary certainly had no problem with asking for state funding in aid of navigation: in February 1838, he had petitioned the House for an appropriation to erect several lighthouses along waterways in his home state of Michigan. *House Journal,* 25th Congress, 2nd session (February 19, 1838): 481–482.
66. *Congressional Globe,* 25th Congress, 2nd session, 296.
67. *Congressional Globe,* 25th Congress, 2nd session, Appendix, 263.
68. *Congressional Globe,* 25th Congress, 2nd session, Appendix, 264.
69. *House Journal,* 25th Congress, 2nd session (April 10, 1838): 739.
70. *Congressional Globe,* 25th Congress, 2nd session, 297.
71. Henry to Bache, August 9, 1838, (JHP 4, 98).
72. "Wilkes Memo," Reingold, *Science,* 119. Even Secretary of War Poinsett, while generally approving of Wilkes's actions, felt it "would appear injudicious to dismiss entirely the whole of the Scientific Corps," Reingold, *Science,* 120.
73. Wilkes to Secretary of Navy, July 16, 1842, Reingold, *Science,* 124.
74. Quoted in C. Ian Jackson, "Exploration as Science: Charles Wilkes and the U. S. Exploring Expedition, 1838–42," *American Scientist* 73 (1985): 455.
75. Quoted in William Stanton, *The Great United States Exploring Expedition* (Berkeley: University of California Press, 1975), 68.
76. Peale felt that the collected items were only "of value to 'closet-naturalists,' stay-at-home philosophers, and others who could profit by Congressional appropriations of money, liberally made for the care of the articles." Titian Ramsey Peale, "The South Sea Surveying and Exploring Expedition," *American Historical Record* 3, no. 30 (June 1874): 249. In 1846 Henry had questioned the need to spend thousands of dollars displaying the expedition's specimens, commenting "a collection of curiosities at Washington is a very indirect means of increasing or diffusing knowledge." Henry to Hawley, December 28, 1846 (JHP 4, 612).
77. Peale, "South Sea," 250.
78. James Rodger Fleming, *Meteorology in America* (Baltimore, Maryland: Johns Hopkins University Press, 1990), 56–66.
79. Henry to Henslow, December 2, 1839 (JHP 4, 310).
80. American Philosophical Society to Poinsett, December 20, 1839 (JHP 4, 315).
81. The new 26th Congress, elected in 1838, convened in December 1839.
82. *House Journal,* 26, no. 1, 1314–1315.

83. Loomis to Sabine, October 3, 1840 (Public Record Office BJ3.25.9). Loomis referred to the presidential campaign of 1840 between Democrat Martin van Buren and Whig William Henry Harrison. The result was a Whig victory.

84. Henry to Petty Vaughan, March 30, 1840, (JHP 4, 342); Bache to Sabine, July 21, 1841 (Public Record Office BJ3.25.13).

85. *Congressional Globe*, 30th Congress, 1st session, 826.

86. Sabine to Herschel, July 19, 1844 (Public Record Office BJ3.26.252–3).

87. Bynam to Sabine, January 26, 1849 (Public Record Office BJ3.27.386).

88. In 1853 the Toronto observatory resumed work with funding from the colonial government, Sabine to Cherriman, August 11, 1853 (Public Record Office BJ3.40.164). The structure was disassembled in 1907 and relocated, becoming the home of the Department of Surveying and Geodesy for the University of Toronto. The stone observatory still stands today, the oldest building on campus. I.R. Dalton and G.D. Garland "The Old Observatory's Noble History," *The Cannon*, 3, no.3 (October 1980).

89. The stone observatory buildings of Rossbank still stand on the grounds of the Royal Botanical Gardens and Government House in Hobart, now restored and used as residences. Ann Savours and Anita McConnell, "The History of the Rossbank Observatory, Tasmania," *Annals of Science* 39, no. 6 (November 1982): 541–543, 546.

90. Fleming, *Meteorology*, 146–148.

91. Julius Payer, "The Austro–Hungarian Polar Expedition of 1872–4," *Journal of the Royal Geographical Society of London* 45 (1875): 1–19.

92. Similarly the contributions of Norway ("though a small and poor state") to funding geomagnetic research had helped to inspire the Magnetic Crusade. Herschel, "Report," 39.

93. Sabine to Lloyd, July 5, 1839 (Royal Society Terrestrial Magnetism Archive # 70); Lloyd to Sabine, December 1, 1839 (Public Record Office BJ3.9.72–75).

94. William Barr, "Geographical Aspects of the First International Polar Year," *Annals of the Association of American Geographers* 73, no. 4 (December 1983): 465, 473.

95. Karl Weyprecht, "Scientific Work of the Second Austro–Hungarian Polar Expedition, 1872–4," *Journal of the Royal Geographical Society of London* 45 (1875): 19–33.

96. Weyprecht, "Scientific Work," 32–33.

97. The British government did not officially sanction the expedition until April 1882, leaving no time for special instruments to be made. Those taken on the expedition were borrowed from the Kew Observatory, which had been set up by Sabine and Herschel to join in the series of European observations. *Observations of the International Polar Expeditions, 1882–83*," (London: Eyre & Spottiswoode, 1886), vii–viii.

98. "When, nearly thirty [*sic*] years ago, one man of science proposed that magnetical observations should be extended, it was at once answered by the government by sending out to the Antarctic regions an expedition...which has never yet been eclipsed as to the importance of its results and the luster it shed on the British navy." Howgate, "Congress," 84.

99. Howgate, "Arctic," 284–85.

100. Nancy Fogelson, *Arctic Exploration and International Relations* (Fairbanks, University of Alaska Press, 1992), 10–11.

101. W. Elmer Ekblaw, "The Arctic Voyages and the Discoveries of Dehaven, Kane and Hall," *Proceedings of the American Philosophical Society* 82, no. 5 (June 1940): 879.

102. *Congressional Globe*, 31st Congress, 1st session, 819.

103. No party held a majority in the House during the 31st Congress. As a result it had taken 63 separate votes to elect a speaker in December 1849.

104. Howgate, "Congress," 73.

105. *Congressional Record*, 46th Congress, 3rd session, 1203.

106. *Congressional Record*, 46th Congress, 3rd session, 1202–1203.

107. James Rodger Fleming and Cara Seitchek, "Advancing Polar Research and Communicating Its Wonders: Quests, Questions, and Capabilities of Weather and Climate Studies in International Polar Years," in I. Krupnik, M.A. Lang, and S.E. Miller, eds., *Smithsonian at the Poles: Contributions to International Polar Year Science* (Washington, D.C.: Smithsonian Institution Scholarly Press, 2009), 1–12.

108. *Congressional Globe,* 41st Congress, 2nd session, 1773, 2211, 2807, 3144. For the *Polaris* expedition, see Richard Parry, *Trial by Ice* (New York: Ballantine Books, 2001).

109. Howgate, "Arctic," 288.

110. Howgate, "Arctic," 297.

111. Howgate, "Congress," 82.

112. Lord Dufferin, Governor General of Canada, had claimed that "geographers are brothers the world over. To a geographer these lines of ethnological and political demarcation which divide nations do not exist." Howgate, "Arctic," 291.

113. Howgate, "Congress," 73.

114. Howgate, "Congress," 77, 79.

115. *Congressional Record,* 46th Congress, 2nd session, 2417.

116. *Congressional Record,* 46th Congress, 2nd session, 2418.

117. *Congressional Record,* 46th Congress, 2nd session, 2931. Following the condemnation of the *Gulnare,* the 46th Congress again voted to support the expedition the following year. Alden Todd, *Abandoned* (Fairbanks: University of Alaska Press 2001), 16.

118. "The Greely Expedition," *Christian Union* 30, no. 4 (July 1884): 75.

119. Barr, "Expeditions," 465.

# Chapter 6

# Swedish Polar Policies from the First International Polar Year to the Present

*Lisbeth Lewander*

This essay focuses on Sweden's seemingly official disinterest in polar politics, from the earliest international polar years up to the most recent of 2007–09. While there were, in fact, particular policies connected with each International Polar Year (IPY), in addition to the International Geophysical Year (IGY) of 1957–58, it was not always self-evident who actually formulated or promulgated the somewhat fragmented policies for Sweden. Swedish policies were, in general, noninterventionist but cautious. An array of scientific proto-inquiries provided a widely comprehensible argument for a Swedish presence in both the Arctic and the Antarctic from the late nineteenth century and onward. For several reasons, the scientific quest never lost its prime position as *the* argument for a Swedish presence, later complemented by a frequent and highly visible concern for the global environment. As will be shown, however, over the years there were also other practical concerns, from the strengthening of national manly identity to whaling, the latter remaining a potential but undeveloped Swedish interest until the late 1950s. Sweden's generally tacit foreign-policy and security concerns concerning the Arctic have existed ever since the late nineteenth century.

## First IPY, 1883–84

The first polar year took place at a time when European cultural and scientific elites were busy promoting reason and rationality, while more or less systematically reserving these traits for themselves. This was evident in the rather emotionally based quest for nationalism undertaken by Sweden. The logistics of the Swedish IPY expedition of 1882–83 were only partially supported by the government, while the planning and fundraising were almost exclusively matters for the Royal Swedish Academy of Sciences. In the end, the expedition to Kapp Thordsen was privately financed by Lars Olsson Smith, a businessman and industrialist from Stockholm; even so, the Swedish Royal Navy made a major contribution, offering two gunboats and several men to help with loading and unloading. The scientists on the expedition made meteorological and geomagnetic observations. Parallel to the official IPY expedition to Kapp Thordsen undertaken by Sweden, Otto Torell at the Geological Survey of Sweden planned a small expedition for geological purposes. He had hoped to take advantage of the IPY by using the same transport as the official expedition but eventually he hired a vessel formerly used for sealing purposes. Briefly, both expeditions were constituents of Swedish polar policies at the time. There had been no formal Swedish claim to Arctic land, that is Spitsbergen, since 1870, but there were still interests in mapping and securing access to the area; other states were to be prevented, by all means, from putting forward claims. The two expeditions signaled Sweden's intention of linking new discoveries with Swedish commercial interests, especially in the area of natural resources. Plans for exploitation, however, do not necessarily imply actual conquest of a particular territorial area.

Dr. Marie Jacobsson, legal expert at the Swedish Foreign Ministry, has examined Swedish and Norwegian misconceptions about Spitsbergen as *terra nullius*—land areas available for national occupation, in contrast to *terra communis*, an area open for use by everyone with the potential for resource-exploitation unbound by any rules.[1] Although Sweden did not favor a formal process for claiming Spitsbergen, some scholars point out that the Swedish government consistently used the wrong label, *terra nullius*, when referring to their foreign and security policies. In actuality their plans focused on the future exploitation of resources and on establishing precedents to prevent formal claims to the area from Norway and other countries, especially Russia. Although the general historiography of Swedish polar research before the 1880s claims that scientific concerns were the prime motives for most expeditions, a continuous open-door policy allowing all countries to engage in activities in the region was crucial in the event of a large-scale commercial engagement in the decades to come.[2] The Swedish phosphate-mining expedition of 1872, led by Per Öberg, had attempted to develop a mine at Kapp Thordsen, but that effort did not succeed.[3] As the expedition proceeded it was successful "only" in bringing back a substantial geological collection rather than accomplishing the larger agenda intended for Spitsbergen.

The drive to secure and extend Swedish access to land in polar areas coincided with an urge to heighten the sense of a Swedish identity. The forces behind the notion of strengthening nationhood included: the quest for modernity through science and technology, the spread of industrialism, the growth of cities and markets, claims for democracy, ideological conflicts, and women's movements seeking civil rights. In this context, polar exploration and travel offered valuable arenas for the creation of modern Swedish manliness. Besides adding to current knowledge in geology, paleontology, botany, zoology, and meteorology, polar scientists made systematic contributions to the image of Swedish nature and to the role model for a Swedish manly temperament. Being brave, courageous, and rational in enduring the harsh climate became important elements of what ought to constitute a "real man," an image of lasting importance in media and school books. As an example, one of the most telling national projects, which also had gender-formation aspects, was Adolf Erik Nordenskjöld's voyage through the Northeast Passage in 1878–80. This journey led to a form of hero-worship based on polar research that continued to be a latent, though diminishing, feature of Swedish society until the 1950s. The next event shaping national sentiment was Salomon Andrée's balloon expedition in 1897, aimed at drifting across the North Pole. As is well-known, the flight did not go as planned; the expedition vanished, and its remains were not found until 1930 on White Island, northeast of Spitsbergen. This did not dissuade the Swedes from presenting themselves in public as children of the Ice Age. In fact they celebrated the northern region as Sweden's true national landscape and Swedish men as particularly suitable and acclimated for polar research.

Thus polar research from the time of the first IPY and the two successive decades also helped to alleviate the crisis of maleness and manliness in this period of general instability in gender roles and class struggles. The right of women to vote, to inherit property, and to become employed by the state were issues for fierce debates parallel to frequent strikes among male workers for better living conditions.[4] So besides serving Swedish foreign policy, polar research had important domestic political functions in strengthening cultural roles in times of social upheaval, such as strikes and mass demonstrations. The public representation of polar research had nothing to do with territorial claims, resource exploration or identity issues: the one and only theme was brave Swedish men doing science. At the time, there was congruence among manly heroic themes at the individual, institutional, and symbolic levels. The individual male scientists and their support teams certainly took great risks that were recognized by the sponsoring establishments, such as the Academy of Science and the Royal Navy, with the media following suit. Individual heroes in the service of science provided timely symbols for progress and civilization and therefore served the project of nation building well.[5] In Norway we find Fridtjof Nansen, in Sweden Adolf Nordenskjöld, but also Gerhard de Geer and Alfred Gabriel Nathorst. Snow, ice, and successful men became the symbols for a Sweden in need of unifying national sentiments. This congruence gradually decreased, however, since modern nation-states required other kind of heroes or an altogether different set of symbols of success.

In the decade after the first IPY, Swedish polar research proceeded aggressively and generally effectively, with several Arctic expeditions, including the Swedish-Russian Arc of Meridian expeditions during five summers and one winter in 1898–1902. Meanwhile several countries, assisted by mining companies, made headway over Spitsbergen. Norway acquired mining concessions that were subsequently sold to British and American companies. Another major event, directly inspired by the IPY, was the Swedish South Pole expedition of 1901–03. Despite its name, the expedition, led by Otto Nordenskjöld, had no intention of reaching the pole. I see this endeavor as a (rather unsuccessful) attempt to obtain funds from sponsors for both current and future polar research. Nordenskjöld, who did not achieve the status of a Swedish hero, represented a blend of old and new ways of producing science. In contrast to most of his biographers, who stress his focused scientific approach, my research has shown that when he was planning the expedition, Nordenskjöld had only a vague idea of what to investigate. As things turned out, however, some of his research on tertiary plants and vertebrates contributed knowledge to understandings of plate tectonics and continental drift.[6]

Further, Captain Carl Anton Larsen, who commanded the expedition's vessel, was instrumental in opening the Antarctic to large-scale whaling, which was his main purpose. Nordenskjöld also had certain obligations in this respect, because some of his sponsors expected reports on the whaling

**Figure 6.1**  Spitsbergen has been a longstanding base for Swedish operations in the Arctic. This satellite image from Terra of portions of Spitsbergen was taken on July 12, 2003. It shows Longyearbyen, the administrative center of Svalbard, located on Spitsbergen, the largest island of the Svalbard archipelago, and part of the Kingdom of Norway. It is the world's northernmost town, with more than 1,000 people. The settlement was founded in 1906 by John Longyear, owner of the Arctic Coal Company. Until the early 1990s the coal mining industry was the major employer at Longyearbyen. (Courtesy of NASA Goddard Space Flight Center, Greenbelt, Maryland)

issue. Of particular interest are the oft-cited lack of state funding and the outright hostility of the Swedish Academy of Science to his plans. While Nordenskjöld struggled to do science and satisfy his sponsors, the Academy's president accused him of deliberately initiating a mass slaughter of whales in southern waters. Before his departure Nordenskjöld had prepared contacts with the Swedish consulates along his route to Antarctica. His relationship with Argentina, buying coal and visiting ports, involved a great deal of diplomatic correspondence, and the Swedish government closely monitored the expedition's progress. The archives testify to the involvement of some ten ambassadors and the foreign ministry in Stockholm. So although there was no formal state funding, there was considerable moral and practical state support. However, the whaling efforts on behalf of the Swedes were in vain, although hunting of fur and elephant seals resulted in long periods of inactivity for the party aboard the *Antarctic* while the seal blubber was boiling. Diplomatic activity was essential as Nordenskjöld and Larsen tried in 1904 to recover the incomes from the seal hunting. However, as is well-known, the whaling fully became a Norwegian matter rather than a Swedish one.[7]

Later authors have asked why, after the 1901–03 expedition to Antarctica, Sweden did not use the opportunity to establish claims in the Southern Hemisphere, but that was simply not an issue at the time. Legal adviser Marie Jacobsson asserts that the Swedish administration was hesitant to act in polar areas in general, but in Antarctica in particular. The main reason, apart from legal hesitations with regard to Spitsbergen, presumably was the passivity and lack of knowledge of international law on the part of the foreign office administration.[8]

It was not until shortly after World War II that Sweden made its next trip to Antarctica. Instead, long-term political, cultural, scientific, and economic relations were established with Argentina and Chile, while nonintervention was the standard approach to Antarctica and also gradually to the Arctic. As with the IPY expeditions, the Swedish South Pole expedition reiterated the rhetoric of aiming for pure science, even though resource-exploration was an important component. It is noteworthy that in 1911 the Swedish government granted substantial funding for an Anglo-Swedish project largely related to whaling in the Antarctic. British hesitation led to the cancellation of this project shortly before World War I.[9] This bilateral project would have contributed to the strengthening of Sweden as a true polar nation in the sense that the United Kingdom (alongside the newly independent Norway) was without any doubt the leading polar nation. A common project with British researchers would have lent some glory and credibility to Sweden, a nation slowly beginning to lose pace in the league of polar nations. In the North, an expedition was undertaken in 1910 to explore the feasibility of coal mining at Isfjorden, Spitsbergen, and Swedish operations were initiated in Sveagruvan (Svea mine) during World War I and continuing until 1925. These operations gave Sweden a visible and substantial presence in the Arctic. The Swedish policy of nonintervention, in the sense of abstaining from territorial claims or intense diplomatic activity directed at polar areas, was a fact. The decline of Swedish polar research was visible for a period of some two decades.[10] Meanwhile the heroic reputation of Arctic explorers remained as an important ingredient in the Swedish self-image. Schoolbooks, general mass-media, museum exhibitions, and world exhibitions were forceful channels of communication.

## Second IPY, 1932–33

In the period between the first and second IPYs, Swedes concentrated on further expeditions in the Arctic. These included a venture in 1932–33, again under the auspices of the Royal Academy of Sciences and without substantial state funding. Two of the three sites were close to the mining villages Svea Gruvan and Longyearbyen, and the third was at a high altitude on Nordenskiöldfjället. Extensive observations were made north of Sweden. The two stations on Spitsbergen were used to collect meteorological and geological data. This event did not attract much attention in the Swedish media or from Swedish polar historians.

Public rhetoric continued to depict Swedish policy as a matter of pure science, although issues to do with civilian and military communications were in fact being added to the research agenda. One important feature of polar research was the increased use of aviation, which modernized research and raised the issue of firm and reliable governmental support, but also affected male ideals. The modern hero was now a man of reason and intellectual capability rather than merely a hero of physical ability.[11] But active female scientists in polar areas were still unthinkable in most cultures. Louise Arner Boyd (1887–1972) is considered the first female explorer to the Arctic, even though research shows that some 500 women traveled in the Canadian northwest territories between 1867 and 1939.[12] Instead of being tied to "trawling/fishing" in the endless seas of knowledge without any guiding hypothesis, scientists carefully planned activities in the domains of meteorology, geomagnetism, and auroral studies. As for the public representation of polar scientists, there still was coherence between the manly ideals on the individual, institutional, and symbolic levels, although on a new technological level. The commemoration of Salomon Andrée when the remains of his lost balloon expedition were found in 1930 was immense as to media coverage. This permitted the revitalization of one old national hero, while new airborne heroes were on their way. Meanwhile, political concerns were ambiguous and still noninterventionist but, regardless, the Swedish government did provide some funds for this polar year without any debate.

By now Swedish polar research had entered the "Ahlmann epoch," with glaciology and studies of climate change.[13] Hans W. Ahlmann was initially interested in economic geography and thereafter in natural geography and glaciology. He established the geophysical and geological elements of glaciology, and his research presaged the modern research of global climate change. In the 1930s several Swedish-Norwegian expeditions journeyed to the Arctic, and the United Kingdom was still the model polar nation.

**Figure 6.2** Norwegian explorer Fridtjof Nansen who undertook several expeditions into the Arctic, including a 1893 expedition to drift to the North Pole in the ship, *Fram*. (Courtesy of National Oceanic and Atmospheric Administration, NOAA Library Collection, Washington, D.C.)

Although some right-wing military army officers had openly declared their admiration for Germany (and German had long been the first foreign language in the Swedish school system), Swedish foreign policy after World War II was totally reoriented toward dealing with the cold war rivalry between the United States and the Soviet Union. The cold war also became evident in polar politics. Small countries with territorial ambitions in polar areas revived their claims.[14] In addition, both the United States and the Soviet Union increased their activity, the latter by sending a large whaling fleet to the Antarctic in 1946.[15] The United States contemplated claims on large portions of the Antarctic but in late 1947 a policy change produced two proposals for resolving sovereignty disputes in the southern polar area. These proposals were based on a semi-international solution rather than a strict and openly declared demand for a national claim on behalf of the United States. Conflicting territorial claims would be set aside, and the area would be accessible for research by the parties to proposed agreements, for instance the Escudero plan from 1947. (This plan was partly a prototype for the Antarctic Treaty of 1959.) With the prevalence of these increasing tensions, the trilateral venture for research in Antarctica, involving Great Britain, Norway, and Sweden, occurred in a politically turbulent period.

## The Norwegian–British–Swedish Expedition, 1949–52

Although it was not directly related to any of the preceding IPYs, apart from the collection of meteorological data, in most accounts of the IGY 1957–58, the Norwegian-British-Swedish Expedition (NBSX) of 1949–52 is said to have set an example with regard to the form and degree of international cooperation.[16] Lately, some people, including myself, have questioned whether NBSX was a textbook case of science's internationalization.[17] Rather, there were scientists representing three countries and with a distinct division of labor. All in all, in the endeavor the benefits of geophysical research proved out; so, too, did the value of extensive data collection for meteorology and glaciology.

Although the scientific results as such were disappointing if scrutinized vis-à-vis the hypothesis of polar warming in the Antarctic, there were other concerns in the domain of foreign and security policies. Great Britain and Norway nurtured their respective territorial ambitions in the aftermath of German attempts to claim parts of Antarctica, while the Swedish government's motives were less apparent. According to accounts by Ahlmann, the instigator of NBSX, the German aerial mapping of parts of Queen Maud Land was of considerable scientific interest for climate research. In addition Sweden saw an opportunity to strengthen its ties with Norway and establish discrete connections with North Atlantic Treaty Organization (NATO) forces by engaging the Swedish Air Force. Sweden was mainly keen on depicting itself as an actor on the Arctic stage, while at that time operations in the opposite polar area, Antarctica, were useful not only for contacting other relevant Arctic actors but also for testing cold equipment. The NBSX was launched in a period when the negotiation of the Scandinavian Defense Alliance recently had failed and NATO was emerging. Further, the buildup of international cooperation in meteorology was of major importance for both civilian and military purposes in Norway and Sweden, in addition to Great Britain and the United States. Representations in Swedish media were all about science, and politics was an absolute nonissue.

## The IGY, 1957–58

The NBSX served as something of a template, at least rhetorically, for the IGY of 1957–58, the scientific event that constituted the third IPY and paved the way for the Antarctic Treaty System. In the late 1940s and early 1950s territorial disputes about Antarctica were growing in number.[18] Meanwhile negotiations to settle the sovereignty issue had already started, as had preparations for the IGY. In October 1955 the American embassy asked Sweden about its interest in the IGY in Antarctica and its general political interest in the continent.[19] The inquiry also concerned the Swedish attitude toward the notion of a conference on conflicting claims and toward Soviet interests in the area. Although Sweden had by now adopted the view that Antarctica was a continent rich in natural resources, the official reply was that at present there were no plans for a Swedish expedition. The political issues were declared nonissues and, therefore, Sweden had nothing to say about its position. But in the event of the area being internationalized, Sweden would require an opportunity to present its interest in whaling rights.

This answer must have come as a great surprise to the governments involved, such as the United States, Great Britain, and Norway (with regard to whaling). Sweden was not known as a whaling nation, but in Stockholm there was strong backing to develop an autonomous whaling industry. The Swedish Kooperative förbundet, an influential organization in Swedish food and light consumer industry, did not want to remain dependent on Norwegian supplies of animal-oil–based margarine.[20] When asked by the U.S. embassy about the possibilities of international cooperation, Sweden somewhat passively answered that before making a decision it would be preferable to know the purpose of the cooperation. A similar strategy was used in spring 1956 when India's proposal for the entrance of Antarctic issues onto the provisional agenda of the Eleventh General Assembly of the United Nations (UN) was passed on by Ambassador Gunnar Jarring from the UN to the foreign office in Stockholm.[21] There was a moment of confusion because it initially was unclear whether India was advocating some kind of internationalization of Antarctica. In the end India's proposal was withdrawn, and Sweden once again was absolved from expressing an opinion about the status of Antarctica, avoiding difficult diplomatic issues with several countries on this subject. Shortly after the IGY, the signing of the Antarctic Treaty at the Washington conference in 1959 froze all claims, in addition to the Swedish position with regard to Antarctica.

In the Arctic, Swedish policy was somewhat more distinct. Meteorology, geophysics, and oceanography were the disciplines for which Sweden continuously asserted its neutrality and freedom of alliance. This doctrine lasted until the early 1990s, albeit with several flagrant exceptions, such as during the NBSX and IGY. Not only did logistical operations include close cooperation with foreign military services, but there were also instances of intelligence operations, unspoken of until 2004. However, there had been long-lasting, close cooperation on the Oslo-Stockholm-Washington axis, and much polar research was done on aerial and naval (including submarine) warfare. There was a 1958 Swedish-Finnish-Swiss expedition to Nordlandet, Spitsbergen, led by Gösta Liljequist (also a member of the NBSX). Meteorology, geophysics, cosmic radiation, magnetism, atmospheric electricity, and the aurora were all on the program. Valter Schytt (also on the NBSX and Ahlmann's student) went on a glaciological expedition in 1957–58 to Kinnvika, along the same route as Ahlmann had taken in 1931, and obtained data to complement his mentor's research.[22]

The expedition led by Schytt was not an official contribution to the IGY because the Swedish contribution to the IGY did not include glaciology. A constellation of neutral states interested in the Arctic, including Finland, is noteworthy, while the exclusion of Norway presumably signaled Sweden's desire not to arouse Soviet suspicions. By avoiding open cooperation with NATO countries, Sweden maintained its freedom of alliance within the auspices of the IGY. For various reasons Sweden did not send its own expedition to Antarctica but did have a serious engagement in the IGY through Schytt's participation in setting up the Scientific Committee for Antarctic Research (SCAR), which preceded the Antarctic Treaty. During the planning of the IGY, Schytt made several visits and expeditions to the Arctic: in 1954 with a Canadian expedition and, in 1956, a visit to North Greenland on the Soviet icebreaker *Ob*.

Expeditions to Spitsbergen continued during the 1960s, now led by Gunnar Hoppe, but their contemporary relevance for Swedish foreign policy has not been studied. These activities in the 1950s and 1960s fit well into what I would like to label the rhetoric of Swedish polar research, namely the message that all activities in polar areas have been about science and have been entirely civilian. But considering the degree of participation from the navy and air force during the IGY, it is hard not to conclude that the government made a deliberate decision in granting this support, despite the cautious choice of partners for cooperation. Moreover, Schytt was granted other substantial funding outside the IGY. The policy was to use science as a tool for advancing foreign affairs and security interests in a sensitive area, but without making long-lasting formal commitments to any political structure. Whereas the former policy of noninterventionism rested traditionally on a lukewarm interest. or even incompetence. from 1945 onward, the same policy turned out to be of immense value in times of perceived crisis.[23] In post-IGY times the heroic dimension no longer automatically accrued to the individual male scientists of various nationalities. Evidently, there were still certain persistent heroic connotations to enterprises, such as the Trans-Antarctic Expedition

**Figure 6.3**  Sir Edmund Hillary after accompanying the first plane to land at the American station at Marble Point during the International Geophysical Year. (Photograph by William McTigue. Courtesy of National Oceanic and Atmospheric Administration, NOAA Library Collection, Washington, D.C.)

with Sir Edmund Hillary, but in general the immense American and Soviet undertakings during the IGY signified a new kind of institutionalized "Big Science" in which old-time heroism had been substituted by high-tech polar research among competing states. On the symbolic level the state of the art remained remarkably unchanged. Polar research was still a matter for men, while the few female scientists and crew members aboard the Soviet vessels were made fun of in Western newspaper magazines.[24] Apart from an attempt to diminish the prime enemy at the time, this may also be seen as an expression of the still-unisex domain of polar science.[25]

## A New Era for Swedish Polar Research

In the 1970s some individuals in Swedish defense circles tried to raise the issue of potential conflicts within the Western alliance with regard to

resources in the Arctic. They advocated the production of a major study
of actual resource exploitation, living and non-living resources (fish, seals,
whales vs mining, oil and gas) , but were turned down by the foreign ministry. Only a few years later, due to the implications for the Arctic of the
Third Convention of the Law of the Sea, there was a change of attitude and,
in 1980, the icebreaker *Ymer* departed for a celebration of the centenary of
the *Vega's* voyage through the Northeast Passage. The official organizers
were the Swedish Society of Anthropology and Geography and the Royal
Swedish Academy of Sciences, with ample assistance from the National
Administration of Shipping and Navigation and the Royal Swedish Navy.
A joint Swedish-Soviet expedition had been proposed, but Soviet participation was not forthcoming. More than a hundred scientists from nine countries took part.[26] On the Swedish side only one person actually spoke up,
a few years later, about the real reason for this trip—to manifest a Swedish
and Norwegian presence and an international interest in keeping the Arctic
Ocean open as a free space, available for free international passage, and
opposing any attempt at enclosure by any state.[27] Even so, in 1982 journalists were asking whether or not Sweden had a policy for polar areas. After
that things happened rather quickly—polar institutions were created and
the strategy emerged for entering the Antarctic Treaty System (ATS) to
govern the continent. The Swedish Polar Research Secretariat was set up
under the Ministry of Education in 1984. However, as Sweden joined the
ATS in that year as a nonconsultative member and then acquired consultative status in 1988, besides becoming a full member of SCAR, Sweden
wanted not only to demonstrate support for a politically open Antarctica
but above all to be visible as a relevant actor in polar affairs, also with
regard to the Arctic. The cost at the time was the construction of a permanent research station in Antarctica. The same dual interest has continued
throughout modern Swedish polar research, although cooperation with
the Nordic countries on the issue of Antarctic expeditions has gradually
been extended to Germany, New Zealand, South Africa, and more recently
to the United States.

The policy toward the Arctic that involved setting up the International
Arctic Science Committee was another indication of the sensitive relations
with close and distant partners. Membership and the location of the secretariat were issues for debate. Behind these policies were Swedish companies, such as shipbuilders, and business interests in transport and natural
resources, such as oil and gas. The Soviet Union no longer exists, but the
attitude to Russian intentions in the Arctic remains cautious and somewhat
vigilant. Another feature of modern polar policies is the aspiration for a say
in the growing market for cold-tech products of use in polar logistics worldwide, from clothing, weasels, and ice-breakers. At the same time Sweden
guards its international reputation as a small state with a time-honored and
far-reaching concern for the environment. This image rests on earlier foreign-
policy achievements emanating from the UN system. Sweden was one of the
initiators of the International Environment Conference held in Stockholm

in 1972. Research and environment are the key concepts for modern polar research, in times of growing political unrest in the Arctic.

## Lessons for Today

Today there is a continuous Swedish presence in both polar regions—and in my view we still adhere to rhetoric that harks back to the late nineteenth century: assertion of a political disinterest that is not as apolitical as it may sound. The phenomena I have described as "polar policies" feature in public debate as issues that boil down to research policy or logistics. At the same time the yearbooks of the Swedish Polar Research Secretariat indicate that political developments—and, not least, the issues of environment and global warming—have become of increasing importance, not only for economic and political developments in polar areas but as matters of global concern. With an increasing amount of actors interested in resource-exploitation in environmentally vulnerable areas, areas also of vital importance for research of global needs, the propensity for conflicting interests has become a growing concern on a worldwide basis. One might have expected this to be reflected in novel Swedish postures within international political affairs with regard to polar areas. There is already a comparatively high degree of international cooperation in research and logistics, implying at least some participation of foreign policy makers. It is still an open issue whether cooperation with other nations will become more competitive or not. Although modern Swedish polar research was institutionalized in the early 1980s, with its own favorable budgetary praxis, the growing Swedish engagement in distant polar areas has so far remained relatively unproblematic as a policy-making issue.

Several countries, with or without geographical borders with the Arctic or Antarctic, are now formulating comprehensive policies for the polar areas. On the agenda are issues of security, sovereignty, resources, climate, transports, minority rights, and democracy in the Arctic. The will to maintain open access to the Arctic is still a highly important item on the polar-policy agenda. However, the issues have not yet become a matter for genuinely public policy-making in an ordinary sense, with particular parties spelling out the pros and cons of specific policies. It is rather the case that the policies are being driven by interest groups that have had varying success in persuading politicians, regardless of ideology, and civil servants, to act along certain lines. In Sweden, polar policies are still a matter for a small minority, although historically they have attracted substantial funds and ought to become increasingly important as an instrument for foreign, security, and environmental policy in the relatively nearby Arctic, a region where national concern has been superseded by global concern.

What about the ever-changing manly ideals? In my view there is no longer a unity with regard to male heroism on the analytical levels of individuals, institutions, and symbology. Even if we still may find individuals of both sexes setting out on one-person heroic missions, these adventures will no longer be endorsed by societal institutions active in polar research and polar politics for purposes of national glory. Modern society needs successful scientists regardless of their gender. On the symbolic level the no-longer-pristine polar areas have taken on a new meaning in their capacity as threatened wilderness and global laboratory. Although every now and then the media still refer to the "macho" coded work in polar research, climate research is produced by men and women, mostly without heroic aspirations.

## Notes

1. Jacobsson, "Acquisition of Territory at the Time of Otto Nordenskjöld," in Aant Elzinga et al., eds., *Antarctic Challenges* (Göteborg, Sweden: Royal Society of Arts and Sciences, 2004), 301–326.
2. T. Frängsmyr, "Swedish Polar Exploration," *Science in Sweden: The Royal Swedish Academy of Sciences 1939–1989* (Canton, Massachusetts: Science History Publications, 1989), 177–198.
3. C. Holland, *Arctic Exploration and Development c. 500 b.c. to 1915: An Encyclopedia* (New York: Garland Reference Library of the Humanities, vol. 930, 1994), 287.
4. U. Wikander and U. Manns, *Det evigt kvinnliga—en historia av förändring* (Lund, Sweden: Studentlitteratur, 2001).
5. B. Hettne, S. Sörlin, and U. Östergård, *Den globala nationalismen* (Stockholm: SNS förlag 1998), 344–346.
6. L. Lewander, "The Representations of the Swedish Antarctic Expedition, 1901–1903," *Polar Record* 38, no. 205 (2002): 97–114.
7. Lewander, "The Representations."
8. Jacobsson, "Acquisition of Territory," 319.
9. Jacobsson, "Acquisition of Territory," 190.
10. U. Wråkberg, *Vetenskapens vikingatåg: Perspektiv på svensk polarforskning 1860–1930* (Stockholm: Centrum för vetenskapshistoria, Kungliga vetenskapsakademin, 1999).
11. R. Berg, "Gender in Polar Air: Roald Amundsen and His Aeronautics," *Acta Borealia* 23, no. 2 (2006): 130–144.
12. B. Kelcey, *Alone in Science. European Women in the Canadian North before 1940* (Winnipeg: McGill-Queens University Press, 2006) and M. McLeone *Women Explorers in Polar Regions* (New York: Capstone Press, 2004).
13. S. Sörlin, "Hans W:son Ahlmann, Arctic Research and Polar Warming: From a National to an International Scientific Agenda, 1929–59," *Northern Studies* 14 (Umeå, Sweden: Umeå University, 1997), 13.
14. L. Lewander, *Polariseringens politik, studier av nation och kön* (Karlstad, Sweden: Karlstad University Studies, 2004), 51.
15. L. Gould, "Strategy and Politics in the Polar Areas," *The Annals of the American Academy of Political and Social Sciences* (January 1948).
16. C. Swithinbank, *Foothold on Antarctica* (Lewes, United Kingdom: Book Guild, 1999).
17. L. Lewander, "The Norwegian–Swedish–British Expedition (NBSX) to Antarctica, 1949–1952: Science and Security," in C. Lüedecke, ed., *Reports on Polar and Marine Research* 560/2007 (Bremerhaven, Germany: Alfred Wegener Institut, 2007).
18. P. Beck, *The International Politics of Antarctica.* (London: Croom Helm, 1986)
19. UD/Documents of the Swedish Foreign Ministry. PM on Antarctica by Belding/IB/MO-O, October 18, 1955. Archive Box HP 53 P.
20. L. Lewander, "Svunna visioner-svensk valfångst" in *Vision&Verklighet – umanistdagarna 9–19 oktober 2004* (Göteborg, Sweden: Göteborg University, 2004).
21. UD/Documents of the Swedish Foreign Ministry. Letter from Gunnar Jarring to Sverker Åström, May 15, 1956. Archive Box HP 53 P.
22. Frängsmyr, "Swedish Polar Exploration."

23. Jacobsson, "Acquisition of Territory."
24. L. Lewander, "Snälla flickor åker inte till Sydpolen. Kvinnliga polarfaraares reseskildringar 1830–2000," in *Moderniteter, text, bild och kön* (Göteborg, Sweden: Makadam, 2008), 208.
25. E. Chipman, *Women on the Ice* (Melbourne: Melbourne University Press, 1986).
26. Frängsmyr, "Swedish Polar Exploration," 195.
27. B. Johnson Theutenberg, *The Evolution of the Law of the Sea: A Study of Resources and Strategy with Special Regard to the Polar Areas* (Dublin: Tycooly International Publishing Limited, 1984).

# Chapter 7

# The Polar Years and Japan

*William R. Stevenson III*

National expeditions aimed at the discovery of uncharted lands, new sea passages, and the elusive North Pole dominate the early history of Arctic exploration. It was in reaction against these costly expeditions that Karl Weyprecht (1838–81) initiated what became the first International Polar Year (IPY). Based on his experiences in the far north, he declared that the primary purpose of exploration should not be geographic conquest, but rather the pursuit of scientific knowledge gained through international cooperation. Despite the strength and resilience of Weyprecht's original vision, national ambitions have challenged the core spirit of international cooperation in the polar years from the very beginning. This tension between nationalism and internationalism is particularly evident in the IPY contribution of Japan. Perhaps more than any other nation in the late nineteenth and early twentieth centuries, the Japanese were obsessed with achieving both national development and international, namely Euro-American, acceptance. The polar years played a significant role in the pursuit of both goals.

At the time of the first IPY of 1882–83, Japan had only recently opened its doors to much of the outside world. For its ruling elite, science meant progress, and the polar year inspired the Japanese to establish their own research facilities. By the second IPY of 1932–33, Japan had already visited the Antarctic. Yet, despite this earlier foray, Japan's participation in the second polar year was limited to recently constructed scientific centers around the empire, reflecting a newfound imperialist ambition to which most of its scientists and explorers were equally committed. Finally, in the wake of defeat in World War II, the International Geophysical Year (IGY) of 1957–58 proved the ideal opportunity to promote a new peace-oriented, internationalized Japan, which in turn fostered its own strain of national pride. The following is a historical overview of Japanese involvement in the polar years. It specifically looks at the national trends that have both encouraged and hindered international scientific cooperation, and traces Japan's rise to its current position of leadership in geophysics and the polar sciences.[1]

**Figure 7.1** The crew of the Mt. Fuji observatory during the second International Polar Year, August 28, 1932. (Courtesy of Japanese National Institute of Polar Research, Tokyo)

## The First IPY, 1882–83

By July 1882, the scientific expeditions of the first IPY had long departed for distant stations. The expeditions—all European and American—would establish a total of 14 research stations, the majority of which lay in the Arctic. Observations were to begin in August, and by midsummer most of the parties had arrived at their posts and were busy preparing the facilities and equipment that would sustain them for the coming year.[2]

The Japanese, meanwhile, seem to have been unaware of the polar year. Just over a decade earlier, in the Meiji Restoration of 1868, their leaders had overthrown the former Tokugawa government in a revolution that ended centuries of self-imposed national isolation. The overriding goal of the new leadership was to protect Japan from European and American imperialists, and they recognized that embracing the technologies and institutions used by the imperialists would be an effective means to that end. Military technology topped the list, but Western sciences were not far behind. Leading intellectuals such as Fukuzawa Yukichi (1834–1901) declared the natural sciences to be the foundation of civilization. Through simple reasoning, he implored his countrymen to abandon the impractical tradition of studying Chinese classics, and instead urged them to learn the pragmatic laws of modern science. The Japanese—particularly those in power—believed that for their country to survive foreign encroachment and develop into a strong nation, they would need to promote science and benefit from its application. The IPY was one such opportunity.[3]

Nevertheless, in 1882, the Japanese had yet to take an active role in the international scientific community, and appear to have been oblivious to the polar year until just days before it began. They learned of the plan in late July, when Henry Becquerel (1852–1908), a French physicist who would eventually receive the Nobel Prize, walked into the Japanese legation in Paris and requested the participation of the young nation.[4] During his visit to the legation, Becquerel explained the purpose of the polar year and asked that the Japanese make observations from the central latitudes. Dispatched to Tokyo, the request went first to the Ministry of Industry, but because of the complete absence of meteorological facilities or instruments, they soon turned it over to the Imperial Navy. Here, at last, the proposal found an enthusiastic supporter in Naval Minister Kawamura Sumiyoshi (1836–1904). Recognizing that much could be gained from the exercise, Kawamura ordered the Navy Hydrographic Bureau to begin geomagnetic observations. From November 1882, the bureau began to measure, on a bimonthly basis, magnetic declination, dip, and horizontal force at its Tokyo observatory. In the end, however, the national benefits of the project overshadowed the larger international effort, and it seems they never sent the results of their survey to the IPY commission.[5]

News of Becquerel's request also reached Arai Ikunosuke (1836–1909) of the Ministry of Home Affairs. Famed for his earlier opposition to the Meiji Restoration—he had been naval minister of the short-lived Republic of Ezo—Arai became one of the few former samurai to hold high office in both the old Tokugawa regime and the new Meiji government. His talents were many, for in addition to his knowledge of military technologies, Arai was also a student of the natural sciences. Of note, in 1887 he successfully led a team in taking the first photographs of a solar eclipse on the Japanese islands, and in 1890 he became the first chief of the Central Meteorological Observatory, an institution that would be at the center of activity in later polar years.[6]

On the occasion of the first polar year, Arai responded to Becquerel's request by organizing a committee out of the bureaus of geography and telegraphic communications. Deliberations were soon under way and resulted in the construction of a geomagnetic observatory in Tokyo's Akasaka district. Operations commenced on March 15, 1883. The polar year was already more than halfway through, but the Japanese scientists began their work in earnest, making round-the-clock observations at the top of each hour.[7] Oddly, they did not send their findings to Paris until early 1884, and posted them to Becquerel rather than to the polar year committee. As a result, the data failed to make it to the final April meeting of the IPY commission. Instead, Becquerel's father, Edmond Becquerel (1820–91), presented the findings to the Academy of Sciences in July, and the younger Becquerel reported that the French were much impressed.[8]

In addition to the Naval Observatory and the Akasaka Observatory, there was also a third response to the polar year. It did not, however, originate with Becquerel's invitation. Rather it was the individual initiative of a German scientist living in Japan that led to the most ambitious project of the year. The German, Edmund Naumann (1854–1927), was one of about 3,000 foreigners, or *oyatoi gaikokujin*, hired by the Japanese government during the early years of the Meiji era. The government chose them for their expertise, paid them handsomely for their knowledge, and then replaced them as quickly as possible with their Japanese protégés. The goal was to learn from the Europeans and Americans while avoiding long-term dependence. *Oyatoi* were particularly active in higher education. When the Meiji government established the University of Tokyo in 1877, 12 of the 15 professors of science were foreign hires. Naumann held the chair of geology. Despite his youth—he was only 23—the Japanese recognized his brilliance and valued his initiative. When Naumann suggested the government establish an office of geology that would act as a "mediator between science and the economical requirements of the nation," it created a geological division within the Ministry of Home Affairs. The following year, 1879, he proposed a geological survey of all Japan that was similarly approved. The ministry sent him to Germany to purchase the necessary instruments for the survey, and when he returned in June 1880 they took him out of the university and placed him under their own jurisdiction.[9]

Naumann ambitiously planned to survey Japan, excluding the northern island of Hokkaido, on a 1/200,000 scale within 12 years. At the time, most of inland Japan had never been surveyed, and due to poor infrastructure, traversing its rugged terrain was a daunting task. Remarkably, Naumann himself made a preliminary inland survey of 5,000 kilometers (2,700 miles), traveling at a marathoner's pace with his poor assistants in tow. It should be mentioned that he, like many *oyatoi*, imagined himself to be a harbinger of civilization, and in this case he did not hesitate to physically beat the backwardness out of his trainees. For this and other reasons, he was more respected than liked, and his contract was terminated in 1885. The survey remained incomplete until 1919.[10]

Nevertheless, from the beginning of the project, Naumann believed that the many clefts and folds of Japan's unsettled terrain would correspond to magnetic irregularities. With this conviction he directed two assistants, Sekino Shūzō (1852–1929) and Kōtari Katsuki (1854–1937), to make a geomagnetic survey of Japan, which he timed to coincide with the first IPY. Sekino and Kōtari conducted a two-part operation. They began with bimonthly geomagnetic observations at a permanent location in Tokyo, and then over the course of the year took measurements at 198 locations throughout the interior, including the southern end of Hokkaido. This was the first geomagnetic survey of inland Japan, and it transformed what had been a blank sheet into a broad picture of tectonic and magnetic activity. Yet, like the Akasaka facility, the data collected by Naumann and his assistants never made it back to the IPY commission. Instead, the tabulated results from the stationary observatory were presented at the 1884 International Congress of Electricians, held in Paris, and the findings of the geomagnetic survey of the interior were mapped and displayed at the 1885 Third International Geographical Congress, which convened in Berlin . In both cases the research was then donated to the French Academy of Sciences.[11]

In the end, there is no evidence that Japanese scientific observations made it to the IPY commission. While each program began in response to the polar year, all failed to become part of the greater endeavor. Rather, they remained largely independent projects, partly inspired by the spirit of internationalism, but unable to be fully incorporated into the polar year itself.[12]

While Japanese data never made it to the IPY commission, the international scientific community did not fail to recognize the work. In the summer of 1884, Henry Becquerel sent a letter to Tokyo, explaining–as mentioned–that his father presented the Japanese meteorological and magnetic records to the French Academy of Sciences, and that the great scientific progress of Japan was now known throughout France. He ended the letter with a request that the Japanese government continue to send geomagnetic reports on an annual basis. The French Academy posted a similar letter, asking that Japan also contribute annual data on atmospheric electricity and telluric currents. The Japanese accepted both proposals and, as a result, activities continued at the Akasaka facility, marking the beginning of continuous geomagnetic observations in Japan.[13]

Over the following decades, Japan took an increasingly active role in world affairs. Within a generation of the first IPY, the nation freed itself from unequal treaties, began to expand into neighboring lands, and readied itself to face the West on even terms. One of the defining moments came with a stunning victory over the Russians in the Russo-Japanese War of 1904–05. With this success, the Japanese buried the anxieties of the past and marched forward under the deceptive security of military might. By the second IPY of 1932–33, Japan was a world power with an empire that reached deep into the Asian continent and stretched from Sakhalin in the north to Micronesia in the south.

The Japanese were also venturing far beyond the confines of their empire. Of particular importance to the IGY of 1957–58 was Japan's first push to the Antarctic. It began in 1910 with the international race for the South Pole. The prime mover and sole leader of the effort was Shirase Nobu (1861–1946). Shirase was a professional soldier, a hardened colonist of the Kuril Islands, and an adventurer of mythic proportions; but he was neither a scholar nor a scientist. With few exceptions, universities and governmental ministries did not look favorably on his plan, declaring it to be too spontaneous, too risky, and too focused on adventure at the expense of real science. In an odd twist, they declared the South Pole expedition to be too provincial for their cosmopolitan senses. It was only through the support of the Japanese public and a few key patrons that the expedition proceeded. The *Asahi*, the largest newspaper of the day, played a crucial role in rallying the public, soliciting donations from across the Japanese Empire and beyond. This, too, set a precedent that would be followed with the geophysical year. Shirase and his men sailed for the Antarctic in November 1910. They vowed to reach the pole or perish in the attempt, but failing to do either, they instead made a short dash across an untried stretch of the Ross Ice Field, named all they could see "*Yamato Yukihara*" (Yamato Snowfield), and then returned home as heroes extraordinaire.[14]

## The Second IPY, 1932–33

In September 1929, Sekiguchi Koikichi (1886–1951) traveled to the International Meteorological Organization conference in Copenhagen as a representative of the Japanese Central Meteorological Observatory. Decisions made in Copenhagen soon would have him wintering at the summit of Japan's highest peak, Mt. Fuji (3,776 m or 12388.5 feet). Specifically, in response to a petition presented by a team of scientists, Sekiguchi and the others unanimously approved the call for a second IPY.[15]

The directors of the International Meteorological Organization decided that to ensure the success of the polar year, they would invite all governments of the world to participate. They also chose to entrust the details of the plan to an international commission, which held its first meeting in Leningrad in August of the same year, 1929. Geophysicist Tanakadate Aikitsu (1856–1952) happened to be traveling in Europe, and attended the meeting despite not being a member of the group. Tanakadate was the best known Japanese geomagnetist of his generation. He earned his degree in physics from Tokyo Imperial University in 1882, became the first Japanese to be nominated for a Nobel Prize in physics in 1910, and then went on to become the inaugural president of the International Association of Geomagnetism and Aeronomy (IAGA) in 1919. Following the Leningrad meeting, Tanakadate returned home to begin planning his nation's contribution, and by the following spring he was president of Japan's national polar year committee.[16]

With Shirase's Antarctic Expedition of 1910–12, the Japanese proved themselves capable of polar exploration., But rather than make a second voyage to Antarctica, the Bureau of Fisheries began to dispatch periodic research expeditions to the far north under the leadership of Taketomi Eichi (1886–1955). The bureau provided a small (but state-of-the-art) ship and commissioned Taketomi to study any and all phenomena that might bolster the nation's fishing industry. This included oceanography and meteorology, and Taketomi interpreted it to also include polar sciences. He crossed into the Arctic Ocean on his first voyage in 1923 and would return many times over the following two decades. Such expeditions were without precedent in Japan, so for guidance Taketomi turned to foreign sources. When he learned of international plans for a second polar year, he petitioned the bureau for funds to lead an Arctic expedition. But such plans quickly disintegrated with the Manchurian Incident of 1931. The conflict diverted the attention of the nation, isolated the Japanese from the international community, and propelled the Rising Sun toward disaster, culminating with defeat in World War II.[17]

▲ 第1次 越冬隊 11名

**Figure 7.2** Members of the Japanese Syowa Station Wintering Party aboard *Sōya* in 1958 after the ship rescued Nishibori and the others from Syowa Base. (Courtesy of Japanese National Institute of Polar Research, Tokyo.)

Despite growing international tensions, the Japanese did participate in the polar year from within their imperial borders. Under the leadership of Tanakadate, they developed a plan that called for the construction, operation, and overall coordination of observatories across their expanding empire. They orchestrated most of these activities through the government-operated Central Meteorological Observatory and chose to focus in particular on geomagnetism.[18]

Sakhalin was Japan's northernmost colony, and Sakhalin's Toyohara Provisional Observatory (46°96'N) became her northernmost geomagnetic facility. The Central Meteorological Observatory built it specifically for the polar year, choosing the location in part because of its northern latitude. The Japanese controlled only half of Sakhalin, having acquired all lands below the fiftieth parallel following the Russo-Japanese War of 1904–05. Over the succeeding decades, colonists and industrialists arrived in droves to clear lands and extract natural resources, and the government followed by building schools, museums, temples, and shrines. The observatory was part of this larger trend of territorial development. But besides being Japan's northernmost territory and a colony ripe for development, Sakhalin was also a logical choice for its distance from existing stations. These included the old geomagnetic station at Kakioka, near Tokyo, and a third station that conducted similar observations in Qingdao, China. Several university-run facilities also cooperated in making geomagnetic observations.[19]

In addition to the basic observations conducted by all polar-year participants, the Central Meteorological Observatory suggested the Japanese make
an intensive study of "sudden commencement," or the geomagnetic irregularities that precede magnetic storms. Technological advancements made
sudden commencement an appealing field of research. Specifically, developments in radio communications and the introduction of quick-run magnetic
recorders created possibilities that were never an option during the first polar
year. Results were optimal. For example, in studying an April 1933 magnetic
storm, the Japanese noticed that sudden commencement began in Kakioka
only eight seconds before striking Toyohara. Even allowing for the greatest
possible margin of error, the evidence confirmed that magnetic storms were
not simultaneous; and yet jumping 1,200 kilometers (745 miles) in mere
seconds meant they traveled at a prodigious speed. Impressed by the data,
Tanakadate asked all foreign polar-year observatories with quick-run recorders to submit their observations, and he discovered the storm traveled most
of Earth within a fraction of a minute. The finding did not lead to an immediate explanation, but it opened the door to various hypotheses and proved
once again the value of international collaboration. It also demonstrated
Japanese initiative. The secretary general (later president) of IAGA, Viggo
Laursen, used the Japanese discovery in a speech before the International
Council of Scientific Unions as his sole example of how international cooperation leads to scientific breakthrough.[20]

Despite the meteorological significance of sudden commencement, the
work of the scientists went unnoticed by the Japanese public. Geomagnetism
could not compete with the excitement of invading Manchuria, not to mention the assassination of a prime minister, a failed coup d'état by officers of
the Imperial Navy, and a show trial that would all but exonerate the perpe-
trators. In fact, Japanese participation in the polar year never made it beyond
a few buried newspaper articles. It seems neither scientists nor journalists saw
any benefit in publicizing their polar-year activities.[21]

Observations made on Mt. Fuji were the one exception, for during the
second polar year, a group of meteorologists determined to become the first
persons to winter at the summit. Plans for a year-round observatory on the
mountain had begun nearly 40 years earlier. In 1894, a young meteorologist
named Nonaka Itaru (1867–1955) proposed building a permanent station at
the summit that he alone would staff. In preparation, Nonaka made the first
known winter ascent of Fuji in February 1895. He then built a small hut at
the peak over the following summer and began observations that September.
In what became a widely circulated tale of loyalty and adventure, his wife
determined to join him and, after surviving a perilous climb of her own, they
together braved the elements until weakness, sickness, and cold forced a late
December descent.[22]

In the late 1920s, meteorologists led by Satō Junichi (1867–1939) revived
Nonaka's dream. In 1926, Satō built a hut next to Nonaka's and began a
series of summer meteorological observations. However, like Nonaka, Satō
was not content to limit his work to the warmer months, and in January

1930 he attempted to prove the feasibility of a year-round observatory by making a mid-winter ascent. He returned a month later, declaring that it was a "little difficult but not dangerous" and then spent six months in a hospital recovering from frostbite, beriberi, and injuries sustained during a fall. Despite these battle wounds, Satō urged the Central Meteorological Observatory to petition the government for funds to build a polar-year station at Fuji's summit. Construction of a larger station began in the summer of 1931, and by July 1932 observations on Mt. Fuji were in full swing.[23] Surviving a winter at the peak of Japan's highest and most beloved mountain had far more appeal than other aspects of Japan's polar-year program, and the public followed the developments on Fuji with interest. Under the leadership of Sekiguchi—the same meteorologist who attended the Copenhagen conference in 1929—the Japanese made round-the-clock observations for the duration of 18 months.

At Tanakadate's request, the Imperial Navy and Army also participated in the second polar year. Much like a half century earlier, the Navy Hydrographic Department recognized the polar year could be of inestimable value in improving its geomagnetic maps and participated as part of the third general survey of the Japanese Empire. Over the course of the year, ships took measurements at 172 stations in a network spanning from the Kurils to Micronesia. Similarly, the Army Air Service and the Navy Aviation Bureau made meteorological observations by plane.[24]

By the early 1930s, the Imperial Army had clearly chosen to disregard internationalism in favor of wanton self-interest. The Manchurian Incident set the precedent, and the government supported the move. When, in early 1933, the League of Nations protested the invasion of Manchuria, the Japanese literally picked up their papers and walked out of the general assembly. They would no longer subject themselves to the opinions of the international community and were prepared to defy any and all international agreements that threatened their imperial ambitions.

This gesture of defiance occurred in the final months of the polar year, and while Tanakadate and others continued to work with their counterparts in Europe and the Americas, the future for scientific cooperation could not have been bleaker. The leadership of Japan had chosen to turn their back on the international community, and over the coming years the rest of the nation, including scientists, would follow. With few exceptions, the Japanese willingly abandoned the internationalism that had taken decades to be build, choosing instead a Japan-centered regionalism—or Pan Asianism—that they would attempt to forge through 15 years of war.

## The IGY, 1957–58

In 1950, scientists began to talk about a third polar year. Less than five years had passed since the end of World War II, and the victorious Allied Powers still occupied the Japanese islands. By the time the Japanese regained sovereignty in the spring of 1952, the International Council of Scientific

Unions had already approved a third polar year and, expanding the program, renamed it the IGY. That summer, the council dispatched letters of invitation to various nations, to which Japan responded by creating a national committee under the jurisdiction of the Science Council of Japan.[25] Hasegawa Mankichi (1894–1970) chaired the Japanese committee, and Nagata Takeshi (1913–91) became secretary. Both scientists had worked under the direction of Tanakadate who, only weeks prior to forming the committee, died at the age of 95. Hasegawa had participated in the second polar year as a geophysicist at Kyoto Imperial University's Mt. Aso facility. Nagata, a young Tokyo University professor of geomagnetism, was in many ways Tanakadate's heir apparent. Like their mentor, both Hasegawa and Nagata were committed to scientific internationalism.[26]

In the summer of 1953, Nagata attended, as one of 12 observers, the first meeting of the Special Committee for the IGY (Comité Spéciale de l'Année Géophysique Internationale, CSAGI). The committee reconvened the following year, at which point it decided that in addition to the "relatively inaccessible regions of the earth," such as the polar regions, participants would also focus their observations along three meridians, one of which ran directly through Japan at longitude 140° east. The committee selected Nagata as the leader of the meridian group, which included Japan, Australia, Siberia, and a few Pacific islands.[27]

The Japanese initially focused their program on geomagnetic observations, much as they had in both the first and second polar years. But Japan's involvement quickly spread far beyond geomagnetism; it organized and operated six geomagnetic stations, thirteen ionospheric stations, eighteen meteorological stations, and facilities to study solar activity, seismology, nuclear radiation, oceanography, and other disciplines. These operations spanned from the northern tip of Hokkaido to the southern island of Amami Ōshima and also included several ships that conducted surveys across the Pacific.[28] Despite all these developments, news of Japanese geophysical-year activities rarely made it beyond scientific circles. In general, the public had little interest in what researchers were concocting in their remote observatories, and the scientists made little effort to appeal to the general population. At no point were there any plans to venture abroad, to send sounding rockets or satellites into space, or to conduct any of the more sensational projects undertaken by the Europeans and Americans. Such plans would have resonated with the Japanese public, but they were costly, and in postwar Japan the scientific community assumed that neither funding nor resources were available. In 1953, Nagata informed the international committee that Japan would consider sending a geomagnetic expedition to equatorial islands, but nothing came of this offer, and for the time being Japan remained focused on domestic operations.[29]

This began to change in the autumn of 1954 with the introduction of rocketry. No such program existed in Japan, but Tokyo University's Institute of Industrial Science expressed interest in studying rockets, and Hasegawa's committee agreed to support it. The university established a base on Honshu's northwest coast (Akita), and in 1957 launched the first Kappa rocket. Over the course of the year, they fired nine rockets, laying the foundation for Japan's postwar space program.[30]

The most significant development in Japan's geophysical year program came in 1955, arriving almost as an afterthought. Rather than originate with any member of the scientific community, it began with a high jumper named Yada Kimio (1913–90) who first leapt to fame by placing fifth in the 1936 Berlin Olympics. By the early 1950s, Yada was an investigative journalist for the *Asahi* newspaper. It was in this capacity that he learned of European and American plans to coordinate research operations in the Antarctic, and he decided that Japan should join them by contributing an expedition of its own. Knowing such an unprecedented event would generate newspaper sales, Yada proposed the *Asahi* simply charter a ship, load it with scientists, and then deliver them to their polar laboratory. He reasoned that the Japanese already had the scientific capability for polar research and the only challenge would be to get them near a pole.[31]

The *Asahi* management was also enthusiastic. Thinking in even larger terms than Yada, they intuitively saw the expedition as a declaration of national rebirth. The war had left Japan in a state of brokenness, poverty, and isolation. But, as the *Asahi*'s senior director explained, the Japanese were prepared to emerge from the darkness, and through the expedition once again join the nations under the blue skies of international cooperation.[32] At the same time, the newspaper management recognized they could not proceed alone. In May 1955, they turned for assistance to Kaya Seiji (1898–1988), president of the Science Council of Japan. What followed was a three-pronged effort to gain support. First, Kaya led the movement by quickly winning over the scientific community. Then, in the face of initial bureaucratic opposition, the ministers of education and finance persuaded the government to provide financial backing. Finally, the *Asahi* ran a publicity campaign that soon had the nation dreaming of icebergs and penguins.

The campaign began in September 1955. In an article subtitled "A Determined Postwar Japan," the *Asahi* announced the Japanese would make a "scientific expedition" to the Antarctic, and that it would be "a beacon of new life, showing that our nation has recovered from the despair of defeat, and that the Japanese people intend to contribute to world peace beginning in the area of science Prime Minister Hatoyama Ichirō declared that, in the wake of the war, the plan showed a "new type of courage," giving hope to future generations. The president of the Japan Chamber of Commerce and Industry wrote that Japanese participation was "not simply scientific," but an opportunity to "display the nation's strength before the world, and in the postwar years give a new boldness to the people who have suffered more than enough grief." Similarly, the president of the National Women's Association affirmed that "for years the bloodshed and uncertainty of war mercilessly poured into our homes. But now, we are greeted by splendid news of an Antarctic scientific expedition and it is being discussed in kitchens everywhere, causing much celebration."[33] The reaction elsewhere was similar, for everyone seemed to agree that the expedition embodied a new postwar spirit of peace and development.

Over the course of the following year, the Japanese showed their support by providing financial and material backing. The *Asahi* led the way. Donating a helicopter, the newspaper then ran a national campaign to choose a name for the aircraft. Forty-thousand suggestions later it chose *Penguin*, the Antarctic's only non-flying bird. The newspaper also announced it would give ¥100 million to ensure the scientists could "successfully perform their duties and attain results that were equal to or greater than that of the Europeans or Americans." The public quickly followed suit and provided an additional ¥42 million.[34]

Not everyone in the international community shared their enthusiasm. At the Antarctic Conference in July 1955, the Japanese announced their wish to establish a polar station, naming the Knox Coast as their preferred location. In response, both Australia and New Zealand balked at the notion of seeing Japan in the Antarctic, and in their turn the Americans and Soviets denied them the Knox Coast, instead suggesting they go to Peter I Island, a pile of rocks separated from the Antarctic by 280 miles of sea and ice. The Japanese wanted nothing to do with the island. Regrouping, they sent Nagata as their representative to the second Antarctic Conference in September 1955. He requested that the Japanese be allowed to build a station on Prince Harald Coast, a desolate part of Queen Maud Land that the geophysical year committee had stated was in need of an observatory long before the Japanese began to think of going to the Antarctic. The delegates consented, and the Japanese began to make concrete plans for an expedition.[35]

Throughout their planning, the organizers were conscious of the symbolic importance of the expedition. After a decade of isolation, the journey to the Antarctic would mark Japan's return to international activity, and with wartime memories still vivid at home and abroad, such an enterprise called for careful deliberation. Above all, the Japanese committee stressed that the Antarctic expedition would be strictly scientific. To avoid any hint of past militarism, they declined to work with the nation's new "Self-Defense Force" and instead requested that the Maritime Safety Agency provide a vessel. The agency complied by refitting the 2,207-ton *Sōya* for polar activity.[36]

To further emphasize the centrality of science, they decided a scientist should lead the expedition and, turning to one of their own, they chose Nagata. While Nagata and his colleagues were leaders in their specific areas of research, they had no experience in dealing with harsh climates and knew nothing of surviving a polar winter. For logistical and practical expertise, the scientists thus turned to the Japanese Alpine Club, and chose Nishibori Eizaburō (1903–89) for their second-in-command. With Nishibori onboard, plans again expanded. He insisted the Japanese dispatch a wintering party a year earlier than initially proposed, arguing that to send a group of scientists with no polar experience and expect them to begin full-scale observations would be to court disaster. The committee agreed, and it was decided that from early 1957 Nishibori and a small team of scientists and specialists would winter on the ice to prepare the base for a larger operation the following year. As it turned out, they would be the only Japanese to participate in the geophysical year from the Antarctic.[37]

On January 29, 1957, the Japanese climbed onto the Antarctic coast. They were technically standing on West Ongul Island, which they had already visited by dogsled and snow tractor three days earlier. But it was this second arrival that marked the official moment of glory. They came in an orchestrated sequence, with Nagata and his scientists lumbering across the ice by tractor and representatives from the ship following minutes later by air. Together the men formed two lines and, with a salute, the Rising Sun flag was unfurled above their heads. After the national anthem and a round of banzai, Nagata declared the land and its surroundings to be Japan's new Syowa Base. This rather sensational landing—vividly depicted in the press— was the culmination of two years of preparation and, despite coming six months before the beginning of the IGY, it was in many ways the climactic moment of Japanese participation.[38]

The *Sōya* began her voyage home on February 15. Ice along the Harald Coast temporarily seized the ship, but with the assistance of the Soviet icebreaker *Ob* they managed to turn north. Nishibori and ten others remained on the continent. Over the following year they finished constructing Syowa Base and then began to make preliminary observations in the fields of meteorology, geology, and aurora and cosmic-ray studies. According to the initial plan, *Sōya* was to return in January 1958 with fresh supplies and a team of 20 scientists to begin full-scale observations. The pack ice, however, was even less cordial than it had been the year before, and it prevented the ship from getting within 100 kilometers of the station. Once again the Japanese turned to a foreign icebreaker for help—the USS *Burton Island*—but foul weather forced their retreat. With time slipping away, they scaled back their program from 20 scientists, to 16, to 11, and finally to a mere 7. Yet, in the end, the elements had their way and the Japanese abandoned all hope for a second expedition. Their focus now went to rescuing Nishibori and the others and, flying them out with a newborn litter of puppies, they turned north and headed home.

The most talked-about episode of the year came as a result of the quick evacuation. Not realizing that they were on the last flight out, the Japanese left most of the dogs tethered with only 10 days of provisions. Their abandonment caused outrage across the nation. But when the Japanese returned to Syowa Base in 1959 to resume operations, they discovered that two of the dogs were alive and healthy, having mysteriously survived a year on the ice. The dogs, Taro and Jiro, became instant celebrities and today remain the best-known of Japan's polar explorers.[39]

The failure to land the second party in 1958 was a disappointing end to Japan's Antarctic geophysical year program. Nevertheless, the preliminary team had made basic observations for the first half of the geophysical year, and scientists working from the deck of *Sōya* also conducted studies in such fields as meteorology, polar oceanography, and ionospheric physics. In the

**Figure 7.3** Raising of the Japanese flag on June 3, 1958, along with the pets of the men at the Syowa Station. (Courtesy of Japanese National Institute of Polar Research, Tokyo.)

end, however, the expedition was as important to the public as it was to the scientists involved. Through the geophysical year, and in the name of international cooperation and scientific progress, the Japanese were once again able to stand with pride before the international community.

## Conclusion

Over the past one and a quarter centuries, the polar years have served as a catalyst for the advancement of Japanese geophysics. Many of Japan's scientific stations and observatories began with the polar years, and data gathered from these facilities have laid the foundation for much of today's geophysical research. Yet, science often serves nonscientific aims, and from the beginning decision makers within the Japanese government responded first and foremost to the national benefits they saw inherent in the polar years. These included practical benefits, such as in meteorology, and also less tangible benefits, such as the opportunity to parade national developments before the international community.

But more than just their service to national interests, the polar years are also testament to the larger and long-term influence of scientific development. With regard to scientific progress, each polar year has stood on a foundation laid in part by its predecessor and, as a result, each event has exponentially increased knowledge of the geophysical world. In addition to scientific progress, the shared objective of geophysical research has continually led to opportunities for international cooperation that would not have otherwise existed. With each ensuing polar year, Japanese ties to the international scientific community have strengthened, and in this way the polar years and the accompanying internationalization of science have contributed to the emergence of a more internationalized Japan.

## Notes

1. I am indebted to two persons in particular for leading me to key sources: Kusunoki Kō, oceanographer for the 1956–57 Japan Antarctic Research Expedition and retired chief of the Japan Polar Research Association Polar Information Office, and Nagano Hiroshi of Asahi University. All Japanese names appear in the accepted order of last name first.

2. See James Rodger Fleming and Cara Seitchek, "Advancing Polar Research and Communicating Its Wonders: Quests, Questions, and Capabilities of Weather and Climate Studies in International Polar Years," in I. Krupnik, M. A. Lang, and S. E. Miller, eds., *Smithsonian at the Poles: Contributions to International Polar Year Science* (Washington, D.C.: Smithsonian Institution Scholarly Press, 2009), 1–12. There is slight variation in the official number of expeditions, but most sources cite 11 nations operating 14 stations.

3. Fukuzawa's 1872 work *Kummō kyūri zukai* (*Illustration of Natural Science*) is particularly concerned with this theme. Yajima Suketoshi, "The European Influence on Physical Sciences in Japan," *Monumenta Nipponica* 19, no. 3/4 (1964): 350; Carmen Blacker, *The Japanese Enlightenment* (Cambridge: Cambridge University Press, 1964), 10.

4. Kishō-chō, *Kishō hyakunen shi*, 2 vols. (Tokyo: Nihon kishō gakkai, 1975): vol.1, 95.

5. The minister to France was Ida Yuzuru (1838–1889). Chijiki kansokujo, ed., *Chijiki kansoku hyakunen shi* (Ibaraki: Chijiki kansokujo, 1983), 2.

6. Nemoto Junkichi, "Ārai Ikunosuke–jiseki," *Kishō* no. 471 (July 1996): 40–45.

7. The Akasaka facility, called Jikishikenjo, was built on land that is today used for U.S. embassy housing. Fukushima Naoshi, "Kokusai chikyuu kansoku no rekishi to hyakunen kinen jigyou," in Nagata Takeshi and Fukushima Naoshi, eds., *Chikyuu kansoku hyaku-nen* (Tokyo: Tokyo University, 1983), 5; *Chijiki kansoku hyakunen shi*, 2.

8. Alexandre-Edmond Becquerel was the head of the Meteorological Society of France until 1890. *Chijiki kansoku hyakunen shi*, 2; *Kishō hyakunen shi* vol. 1, 95.

9. There are 2,050 documented cases of foreign employees hired by the Japanese government, but some estimates place the number at more than 5,000. Figures are imprecise because of missing records. There were also many thousands of privately hired foreign teachers and experts. Of significance to later polar years, an *oyatoi* named Henry Batson Joyner had already proposed the construction of Japan's first meteorological observatory in 1873. Ardath W. Burks, "The West's Inreach: The Oyatoi Gaikokujin," in Ardath W. Burks, ed., *The Modernizers: Overseas Students, Foreign Employees, and Meiji Japan* (Boulder: Westview Books, 1985), 194. In addition to 15 full professors of science, there were also four Japanese assistant professors. Kenkichiro Koizumi, "The Emergence of Japan's First Physicists: 1868–1900," in Russel McCormmach, ed., *Historical Studies in the Physical Sciences* 6 (Princeton: University Press: 1975), 103. For Naumann, see Edmund Naumann, "The Physical Geography of Japan, with Remarks on the People," *Proceedings of the Royal Geographical Society and Monthly Record of Geography* 9, no. 2 (February 1887): 102.

10. Heinrich Edmund Naumann was in Japan for a total of 10 years: 1875–85. It is believed that his assistant-turned-rival, Harada Toyokichi (1860–94), persuaded the government to terminate his contract. Tanimoto Tsutomu, "Nauman no Nihon guntō kōzō ron," *Shizen* 3 (March 1983): 82–83; H.J. Jones, *Live Machines: Hired Foreigners and Meiji Japan* (Vancouver: University of British Columbia Press, 1980), 86.

11. Satō Hiroyuki "Chishitu chōsajo shoki no chijiki kansoku: hyakunen shi no ichi koma." *Chishitsu nyu-su* no. 371 (July 1985): 11; *Chijiki kansoku hyakunen shi*, 5. It seems that Naumann learned of the polar year through private correspondence. The government transferred Naumann's survey office to the new Ministry of Agriculture and Commerce (established 1881), and the office was renamed the Geological Survey of Japan (established 1882). It is today part of the National Institute of Advanced Industrial Science and Technology.

12. There are several reasons for the limited scale of Japanese involvement in the first polar year. These include the lateness of the invitation and the absence of equipment and observatories. But a lack of cooperation within the early Meiji government is partly to blame. By March 1883, three distinct branches of the government were conducting the same geomagnetic observations on the same days and in the same city, either completely unaware of the others' activities or consciously operating out of intergovernmental rivalry.

13. Following the polar year, the facility pared back geomagnetic observations from once an hour to eight times a day. It is unclear whether or not the Japanese continued to send data to the academy over the following years. The facility was relocated in 1886. It was again moved in 1913, becoming the present Ibaraki Prefecture Kakioka Magnetic Observatory. *Kishō hyakunen shi*, vol. 1 96; *Chijiki kansoku hyakunen shi*, 6.

14. For the official account of the expedition, see Nankyoku Tanken Kōenkai, *Nankyokki* (Tokyo: Nankyoku Tanken Kōenkai, 1913).

15. J. A. Fleming, "The Proposed Second International Polar Year," *Geographical Review* 22, no. 1 (January 1932): 131; *Kishō hyakunen shi* (1975): vol.1, 193.

16. For Tanakadate's early work, see Cargill G. Knott and Tanakadate Aikitsu, "A Magnetic Survey of All Japan," *The Journal of the College of Science, Imperial University, Japan* 2, no. 2 Tokyo (1888): 163–262. For Nobel Prize nomination, see James R. Bartholomew, "Japanese Nobel Candidates in the First Half of the Twentieth Century," *Osiris* 2nd series, 13 (1998): 240. Also see Kenkichiro, 72–82. For role in IPY, see Okada Takematsu, *Sokkō sadan* (Tokyo: Iwatari, 1937), 306; Ono Suminosuke, "Dai nikai kyokuchi kansoku honpō kansoku no gaiyō hōkoku," *Nihon gakujutsu kyōkai hōkoku* 9, no.4 (October 1934): 148.

17. Kimura Yoshimasa and Taniguchi Zenya, "Taketomi Eiichi senchō to sono senpaku— hokuyō hoppyōyō no aru senzen shi," Nihon Kyokuchi Kenkyūkai, undated manuscript, personal collection; Kimura Yoshimasa, interview by Kusunoki Kō, May 28, 1979, transcribed, personal collection; *Yomiuri Shinbun* (Tokyo), December 17, 1955.

18. Before 1932 the sole geomagnetic facility was the Kakioka Observatory, previously located in Akasaka and built for the first polar year. Ono, "Dai nikai kyokuchi kansoku," 148.

19. For the Toyohara Observatory, see Imamichi Shūichi, "Daini-kai kyokunen kansoku ni tsuite," in Nagata Takeshi and Fukushima Naoshi, eds., *Chikyū kansoku hyakunen* (Tokyo: Tokyo University), 236–242. The proper name of the Qingdao facility was Aoyama Sokkōjo. It originally belonged to the Germans. The Japanese seized it during World War I and, despite returning the city to the Chinese in 1922, they maintained the facility as part of their consulate. Kyoto Imperial University operated the Aso Volcanic Research Facility in Kumamoto, Tōhoku Imperial University maintained an observatory on Sendai's Mukōyama, and Tokyo Bunrika University worked the Shimoda Marine Laboratory in Shizuoka. All three facilities ran multiple programs, including atmospheric electricity, telluric currents, and seismological observations. It seems that most of their programs, if not all, originated independently of the polar year, and that it was only through the invitation of Tanakadate that they cooperated with the Central Meteorological Observatory. Tokyo Bunrika University is today part of Tokyo University of Education. During the polar year, Kyoto Imperial University also studied zodiacal light at their Kwasan Observatory and Kyūshū Imperial University

studied ultraviolet rays. *Kyoto Teikoku Daigaku* (Kyoto: University Press, 1943), 858, 864; Nagano Hiroshi and Sanō Yasuharu, *Hasegawa Mankichi to chikyū denjiki gaku* (Tokyo: Kaisei, 2002), 31–37; *Tōhoku Daigaku gojū-nen shi* vol.1 (Sendai: University Press, 1960), 697; Ono, "Dai nikai kyokuchi kansoku," 148.

20. Okada Takematsu (1874–1956), chief of the Central Meteorological Observatory, presented these findings at the International Union of Geodesy and Geophysics in Lisbon. The April storm was not the largest of the years' magnetic storms. That honor would go to a May 1, 1933, storm. *Chijiki kansoku hyakunen shi*, 12. Ono, "Dai nikai kyokuchi kansoku," 151. Laursen used the example of Japan in a speech before The International Council of Scientific Unions that was then used in the introduction to the *Bibliography for the Second International Polar Year* (Copenhagen, 1951). It finally appears in V. Laursen, "The Second International Polar Year," in *Annals of the International Geophysical Year* 1 (London: Pergamon Press, 1959), 232.

21. For one of only a few examples, see *Asahi Shinbun* (Tokyo), (July 7, 1931): 11. The assassinated prime minister was Inukai Tsuyoshi (1855–1932).

22. This was before the birth of mountaineering in Japan, and Nonaka climbed to the summit without the use of crampons or other such aids. When his father learned of the plan to winter at the summit, he is reported to have said that "I raised him to rise above his peers, but the summit of Fuji is a bit too high." Nonaka Itaru, *Fuji Annai* (Tokyo: Heibonsha, 2006; Shunshōdō, 1901), 244.

23. In the mid 1920s Satō was a temporary employee of the Central Meteorological Observatory. He had been a long-time proponent of mountaintop observatories and had written extensively on the topic. In 1926, a wealthy businessman named Suzuki Seiji commissioned him to build an observatory, which Suzuki then donated to the Central Meteorological Observatory in 1929. This became the *Kikō kansoku jo* of the second polar year. *Kishō hyakunen shi* vol. 1, 437, vol. 2, 371–374; "Monbushō kokuji" no. 178, in *Kanpō* no. 1652 (July 4, 1932), 89.

24. Army Air Service observations took place at Tatekawa (Tokyo). Navy Aviation Bureau observations took place at Kasumigaura (Chiba/Ibaraki) and Ōmura (Nagasaki). Ono, "Dai nikai kyokuchi kansoku," 150. For more on the Hydrographic Department of the Navy, see Kaijō hoanchō suirobu, ed., *Nihon suiro shi: 1871–1971* (Tokyo: Nihon suiro kyōkai, 1971), 248–250. Other Japanese activities during the second IPY include aerological observations at Haneda (Tokyo) and Tsukuba (Ibaraki). Meteorologists at the Tsukuba facility launched pilot balloons on international days from the towns of Honjō (Saitama) and Koga (Ibaraki). For these and other secondary activities, see Ono, "Dai nikai kyokuchi kansoku," 147–151.

25. Harold Spencer Jones, "The Inception and Development of the International Geophysical Year," in *Annals of the International Geophysical Year* 1 (1959), 383–385; *Japanese Contribution to the International Geophysical Year 1957/1958*, 2 vols. (Tokyo: National Committee of the International Geophysical Year, Science Council of Japan, 1958–1960), 89–93. The Japanese national committee is referred to as the "IGY *Tokubetsu Iinkai*."

26. Nagano and Sanō, 31–37; Nagata Takeshi, "Introduction," in Nagata and Fukushima, *Chikyū kansoku hyaku-nen*, iii.

27. The CSAGI met in Brussels in 1952 and in Rome in 1953. Nagata was not selected as the leader of the 140° east meridian group until the 1955 Brussels meeting. In 1956, Japan became a World Data Center for regional observations in several areas, including geomagnetism and nuclear radiation. Harold Spencer Jones, "The Inception and Development of the International Geophysical Year," in *Annals of the International Geophysical Year* 1 (1959), 384–385, 393, 404, 411–412.

28. For details, see *Japanese Contribution to the International Geophysical Year 1957/1958*, 2 vols. (1958–1960). Amami Ōshima is part of the Ryūkū island chain. However, since the seventeenth century it has belonged to what is today Kagoshima Prefecture. During the geophysical year the Americans occupied and controlled the islands to the south of Amami, namely the Okinawan Islands. These were not returned until 1972.

The vessels belonged to the Japan Fisheries Agency, the Japan Meteorological Agency, and several universities.

29. The United States' own plans for equatorial observations put a quick end to Japanese ambitions. M. Nicolet, "The International Geophysical Year Meetings," in *Annals of the International Geophysical Year* 2A (London: Pergamon Press, 1959), 35; Shibata Tetsuji, "Nankyoku koto hajime: kakenda otoko no tondemo nai hassō," in Ono Nobuo and Shibata Tetsuji, eds., *Nippon nankyoku kansokutai: ningen dorama 50 nen*, (Tokyo: Maruzen, 2006), 10; Monbushō, *Nankyoku roku nen shi: nankyoku chiiki kansoku jigyō hōkoku sho* (Tokyo: Ōkurashō, 1964), 2.

30. While the *Kappa* rockets reached a height of more than 12 miles/60 km, the results were not always satisfactory because of difficulties in telemetry, tracking, and sensing. *Japanese Contribution to the International Geophysical Year 1957/1958* 1 (1958), 89–93; 2 (1960), 146.

31. Shibata, "Nankyoku koto hajime," 1–20.

32. *Asahi shinbun-sha shi* (Tokyo: Asahi shinbun-sha, 1994), 224.

33. *Asahi Shinbun* (Tokyo), (September 27, 1955). President of the Chamber of Commerce and Industry was Fujiyama Aiichirō. President of the National Women's Association was Yamataka Shigeki.

34. For helicopter, see *Asahi shinbun-sha shi* (Tokyo: Asahi shinbun-sha, 1994), 224–226; for quote and donation, see *Asahi Shinbun* (Tokyo), (January 20, 1956); for public donation, see Monbushō, *Nankyoku roku nen shi*, 243.

35. The Japanese did not attend the first Antarctic conference and instead cabled their interest in sending an expedition to the Knox Coast. The Americans and Soviets opposed the request, citing their own intent to establish a station. Yet, despite denying the Japanese access, neither nation followed through with a base of their own. At the second conference, in addition to Prince Harald Coast, the Japanese stated their willingness to go to Cape Adare. Once again, however, several bases were already planned in the vicinity of the cape. The conference thus resolved that the Japanese should go to the Prince Harald Coast as it lay center to the largest gap (more than 1,000 miles) between geophysical year stations. M. Nicolet, "The International Geophysical Year Meetings," in *Annals of the International Geophysical Year* 2B (London: Pergamon Press, 1959), 397, 424–425, 430; Monbushō, *Nankyoku roku nen shi*, 261; Taniguchi Zenya, personal correspondence (August 5, 2007).

36. Following alterations it was 2,497 tons.

37. Nishibori Eizaburō, *Jinsei ha tanken nari*, Nishibori Eizaburō senshū, vol. 1 (Tokyo: 1991), 176. Nishibori was popular among peers but lacked much in the way of diplomatic finesse. When critics suggested that a preliminary expedition would be too costly and dangerous, Nishibori responded by going to the press and declaring that anyone who disagreed with him was *aho*, or "an idiot." Kuwabara Takeo, "Nishibori nankyoku ettō taichō," in Nishibori, *Jinsei ha tanken nari*, 319.

38. Two days later the Japanese decided to construct the new station on the northeast coast of East Ongul Island. Nagata chose the name "Syowa" on his own, and apparently did not tell others in advance. He explained his choice, saying that he wanted the base to be a symbol of the Shōwa Era (1926–89), a period that before the geophysical year had been most widely associated with militaristic expansionism. The first flag-raising was followed by a second on the continent. It occurred on the same day with Nagata, *Sōya* Captain Matsumoto Matsuji, and a few others in attendance. *Asahi Shinbun* evening edition (Tokyo), (January 30, 1957); *Asahi Shinbun* (Tokyo), (January 31, 1957). Monbushō, *Nankyoku roku nen shi*, 31–32.

39. For an overview of the expedition, see Monbushō, *Nankyoku roku nen shi* (1964), and the first dozen issues of the Monbushō journal *Nankyoku Shiryō* or *Antarctic Record*. The journal began publication in December 1957. A monument to the dogs stands at the base of Tokyo Tower. The film *Nankyoku Monogatari* (1983) is based on this episode, as is the American film, *Eight Below* (2006). Jiro (1955–1960) died in the Antarctic. Taro (1955–1970) lived a long and full life, retiring to Hokkaido. Both

dogs are today embalmed and on display for their many fans: Jiro in the National Museum of Japan (Tokyo) and Taro at the Hokkaido University Museum. See Inukai Tetsuya, "Nankyoku no Taro ha jyūni sai," *Ushio* (March 1968): 318–323; Kitamura Taiichi, "Taro Jiro, watashi to eiga *nankyoku monogatari* no shinjitsu," in Ono and Shibata, *Nippon nankyoku kansokutai*, 21–40.

# Chapter 8

# China and the International Geophysical Year[1]

*Zuoyue Wang and Jiuchen Zhang*

In the history of science during the cold war, the International Geophysical Year (IGY) 1957–58 often has been viewed as a great success story of global scientific collaboration across the Iron Curtain, with the only exception being the withdrawal of the People's Republic of China (PRC) from the endeavor when the IGY organizers admitted Taiwan in 1957. Thanks to research by Ronald Doel, we now know that the U.S. State Department played a central role in the controversy by prompting Taiwan to apply for IGY membership, but little is known about the mainland Chinese side of the story.[2] In this essay we propose to examine Chinese sources to reconstruct the considerations that led China to join the IGY in the first place and the reactions to the Taiwanese issue that eventually led to its withdrawal. We will also examine the impact of the IGY on Chinese geophysical research even after its formal withdrawal from the collaboration.

## China and IPY

Although it did not send polar expeditions, China was involved in both the first and second International Polar Years (IPY), 1882–83 and 1932–33, that were sponsored by the International Meteorological Organization (now the World Meteorological Organization) and were precursors to the IGY. During the first IPY, the French-run geomagnetic station at Sheshan near Shanghai provided data for the international collaboration.[3] The second IPY coincided with a great buildup of Chinese scientific effort under the new Nationalist government. As a result China was able to expand considerably its participation in the project: Besides measurements of geomagnetism and solar radiation at Sheshan, observatories at Shanghai and Qingdao carried out measurements related to the determination of longitudes and latitudes, and, perhaps most notably, two stations were established at the mountaintops at Taishan in Shandong and Emei in Sichuan to carry out meteorological measurements. Coordinating most of these activities was the leading Chinese meteorologist and geographer Zhu Kezhen, who had received his PhD from Harvard in 1918 and who at the time headed the newly established Meteorological Institute of the Academia Sinica. Working under difficult conditions and without the resources to participate in polar explorations, Chinese scientists played only a marginal role in the second IPY, but they were nevertheless delighted to make a contribution to the global endeavor.[4]

The international political environment changed dramatically for Chinese scientists when the third IPY, renamed the International Geophysical Year to indicate a broadened scope, was proposed in the early 1950s for implementation in 1957–58. In 1949 the Communist forces under the leadership of Mao Zedong had won the civil war against the U.S.-backed Nationalist government under Jiang Jieshi (Chiang Kai-shek), which had fled to Taiwan. Most Chinese scientists were unfamiliar with the Communists but nevertheless decided in the end to stay on the mainland, primarily due to their loss of confidence in the corruption-ridden Nationalists and with the hope that the Communists would provide a stable political government for the reconstruction of the country and the support of scientific research. For example, Zhu Kezhen declined, at great personal risk, Jiang's special invitation to retreat with him to Taiwan. Instead, Zhu responded to the new government's call to Beijing to become a vice president of the Chinese Academy of Sciences (CAS). The CAS was established in those institutes of the Academia Sinica that had remained on the mainland; the Academia Sinica moved its headquarters and a few institutes to Taiwan, where it continued operations.

## Joining IGY

Thus by 1951–52, when the International Council of Scientific Unions (ICSU) approved the proposal for the IGY—a proposal that had originated in 1950 at the famous Washington dinner party attended by, among others, Sydney Chapman of Oxford, James van Allen of Johns Hopkins's Applied Physics Laboratory, and Lloyd Berkner of the Carnegie Institution—China became both an important link in the international project and a sensitive political issue. Any comprehensive investigation of the geophysics of Earth would be amiss if it did not include the vast Chinese mainland, but China, even more than the Soviet Union, posed a problem for the IGY organizers: The recent revolution had placed the country's representations in international scientific associations in dispute as both sides of the Taiwan strait—the PRC under the Communists and the Republic of China under the Nationalists—vied for China's seat. Furthermore, the United States, in armed conflict with China in Korea, continued to back the Nationalists in the United Nations and other international bodies, such as ICSU.

The IGY organizers were determined to find a way to get China involved in the IGY. In 1952, vowing its non-political nature, the ICSU Special Committee (Comité Spéciale de l'Année Géophysique Internationale,

[CSAGI]) for the IGY reportedly issued invitations to both the Chinese Academy of Sciences in Beijing, which at the time also used the Latin name Academia Sinica, and the Nationalist Academia Sinica in Taipei.[5] Clearly, from the point of view of the IGY organizers, the best scenario was that both sides would agree to participate, which would contribute not only to the gathering of geophysical data but also to the perception that science indeed could bring political enemies together.[6]

In December, the CAS in Beijing received the CSAGI invitation, dated November 28, 1952, and signed by its interim secretary, Ernest Herbays, a Belgian radio scientist. The General Office (*bangongting*) of the CAS promptly translated the letter into Chinese and sent it to the bureau in charge of foreign cultural affairs in the Commission on Culture and Education (CCE) of the Administrative Council (the cabinet) and the foreign ministry for instructions. The foreign ministry responded on January 23, asking the CAS to consider three questions: (1) What were the benefits and drawbacks of participation? (2) What preparations would be required and what difficulties would be encountered, and in what name would China participate? (3) How were the two earlier IPYs carried out, and how were earlier Chinese governments involved in them? The ministry also advised that the CAS consult with the People's Revolutionary Military Commission (PRMC).[7] This step was necessary not only because geophysical research would have obvious military implications but also because, until August 1953, the military controlled the People Republic of China's meteorological services.[8] Meanwhile the CAS received a second letter from Herbays in March 1953 requesting that it notify CSAGI of its plans and suggestions for the IGY by May 15.[9]

As a result of consultation with not only the PRMC but also relevant institutes within the academy—especially its premier Purple Mountain Observatory in Nanjing, the Institute of Geophysics, and the Institute of Geography—the CAS leadership came to the consensus that China's participation in the IGY would depend on that of the Soviet Union. On May 9, 1953, while waiting for an answer from the Soviet Union on an inquiry along this line, the CAS proposed to the CCE a draft reply to CSAGI stating, "The question raised is under consideration; once a decision is made you will be notified." But the CCE apparently did not approve of the message, and it was never sent.[10] Clearly Chinese scientists understood that for the Chinese government, participation in the IGY was more a political than a scientific question. Given the hostility that existed between China and the United States, due in large part to the Korean War, and given the fact that the Soviet Union had not yet decided to participate in the IGY, it was perhaps not surprising that silence remained the best response of the CAS to the repeated CSAGI requests for participation in this period.

A turning point in the Chinese attitude toward the IGY came in 1955, when several developments converged: The death of Stalin and the end of the Korean War in 1953 had marked a relaxation of cold-war tension and, perhaps most decisively, news of the Soviet decision to participate in the IGY paved the way for a decision by the Chinese Academy of Sciences to do likewise. In March 1955 Ivan Bardin, vice president of the Soviet Academy of Sciences and president of the Soviet IGY National Committee, wrote to Guo Moro, Chinese writer, archaeologist, and president of the CAS, announcing the Soviet decision to participate in the IGY and expressed the wish that China do the same.[11] The letter was followed by a visit to China, from April 26 to June 23, of a high-level delegation of the Soviet Academy of Sciences as part of increased contact between the two academies.[12] The delegation included Bardin and Vladimir Belousov, a seismologist who was by then not only a member of the Soviet IGY committee but also a Soviet nominee as a member of the CSAGI bureau that made executive decisions on the IGY.[13] During his trip Belousov did his best, in the words of Zhu, who accompanied him on an extensive tour in southwestern China, to "abet" the CAS to join the IGY.[14] Chinese scientists were aware of the benefits—for both Chinese science and for the success of the IGY—of their participation in the IGY, but it would have been difficult for them to make the case to join the IGY if the Soviets had decided to sit out, given China's push for closer Sino-Soviet scientific collaboration, and its concern about Western resistance to its return to international science. Perhaps equally important for Chinese scientists was the expectation that Soviet participation would help convince the Chinese government not only to give the CAS the green light on IGY, but also to allocate more resources to geophysical research in general.

Soviet influence proved decisive. On June 18, 1955, the CAS made the decision to participate in the IGY and to organize a national IGY committee under Zhu Kezhen, presumably with the blessing of the central government.[15] Two weeks later, the CAS announced the formation of the Chinese IGY committee: Besides Zhu as chair, it included Zhao Jiuzhang (Jeou-jang Jaw), director of the Geophysical Institute, as vice chair; Tu Changwang, director of the Chinese Meteorological Bureau; Chu Shenglin, chair of the Physics Department of Beijing University; and CAS scientists Wang Ganchang, Zhang Yuzhe, Wu Heng, Chen Zongqi, Ye Duzheng, Shi Yafeng, and Zhu Gangkun (no relation to Zhu Kezhen). Chen Zongqi and Zhu Gangkun served as secretary and deputy secretary, respectively.[16]On September 3, 1955, five days before a major CSAGI meeting was to take place in Brussels, Zhu informed Marcel Nicolet, the CSAGI secretary, of China's decision to participate in the IGY and the formation of its national IGY committee, with the condition: "If the IGY includes Taiwan, it will be difficult for us to participate in it." CSAGI quickly responded, on September 8, that it welcomed all scientists, including those from the Chinese Academy of Sciences and those from Taiwan, to participate in the IGY. Two days later, CSAGI telegrammed Zhu that Taiwan had not formed its own IGY committee, with the implication that Taiwan would not be an issue in the IGY.[17]

Initially the CAS determined that China would participate formally only in meteorology, geomagnetism, solar radiation, and longitude and latitude measurements, with atmospheric physics, ionospheric studies, and tectonics as possibilities, but not oceanography or glaciology.[18] By the time the Chinese IGY committee held its first meeting two weeks later, however, it was decided to add cosmic rays and atmospheric physics to the list.[19] Ionospheric studies and seismology were added in early 1956.[20] Zhu soon got a glimpse of the benefits of participation in the IGY: In early 1956, he spent two days studying IGY country reports sent by Belousov to get a sense of recent geoscientific developments in the world. With alarm, he noted that "India and Japan are far ahead of us."[21]

Taking advantage of China's official participation in the IGY and the expected Soviet support, Chinese scientists proposed an ambitious geophysical program that promised to greatly advance research in the field and in weather forecasting, but which would also entail a huge increase in government funding and the importation of key scientific instruments from the Soviet Union and Europe. In September 1956, for example, the Chinese Meteorological Bureau, under the direction of Tu Changwang, gained the approval of the State Council for the expansion of solar radiation stations from a handful to more than 20 by the time the IGY started.[22] In late 1956 Zhu Kezhen secured the purchase of cosmic-ray detectors from the Soviets in addition to the participation of two Chinese scientists on Soviet oceanographic vessels for training purpose.[23]

Such rapid expansion met with various obstacles, such as the lack of qualified scientists and technicians to operate the new instruments and the costs associated with the dependence on the Soviets as almost the sole source of the latter. In January 1957, for example, Zhu learned to his dismay that the price for a Soviet radar wind system had increased within a few months from 50,000 to 300,000 rubles. Thus China had to cut back its order from eight sets to three.[24] Already in July 1956, Premier Zhou Enlai had cautioned the CAS that it should not pursue too many IGY projects but should focus on a few that it could do well in order to showcase Chinese achievements on the international stage. "We should be realistic, be good at hiding our weaknesses, and impress the world with our best effort (*yiming jingren*)," as he told Zhang Jingfu, CAS vice president and Communist party leader.[25]

The CAS was torn between Zhou Enlai's instruction for consolidation, on the one hand, and pressures for expansion, on the other hand, from the international scientific community. In August and September 1956, the Chinese IGY committee sent three scientists, Chen Zongqi, Zhu Gangkun, and Lü Baowei, to three IGY conferences held in Moscow, Brussels, and Barcelona. As a result of suggestions made at these conferences, the CAS proposed an expansion of China's IGY projects and received the approval of the State Council: The number of meteorological stations participating in the IGY was increased from five to 90, and measurements taken at those stations extended beyond surface and high-altitude observations to include radiosonde observation and solar radiation measurement, the installation of a magnetic fluxmeter at the Guangzhou Geomagnetic Station, aurora observations at meteorological stations in northern China, and the establishment of a station south of Beijing for the measurement of longitudes.[26] Two new stations in Lhasa, Tibet, one on geomagnetism and another on seismology, were also constructed specifically for the IGY. According to Zhu Kezhen, the foreign currency needed for the purchase of instruments to equip these stations more than doubled the total annual national funds for scientific research allocated by the Nationalists before 1949.[27]

Chinese enthusiasm for the IGY reached a height in February 1957 when Zhu Kezhen published an article in *The People's Daily*, the most influential official newspaper in the country, titled, "The Organization of the International Geophysical Year and International Scientific Collaboration." In it he called the IGY "not only a great movement in the development of geophysics, but also a new approach to international scientific collaboration." International collaboration was necessary for all sciences, but especially for geophysics, he noted. He mentioned the expected benefits of the establishment of a global network of high-altitude meteorological stations for the forecasting of weather, and of polar research for the understanding of ocean currents. Remarkably, he also pointed out that "the climate of the northern temperate zone and the northern frigid zone has undergone a warming trend during the 20th century," citing, among other pieces of evidence, Chinese observations of the upward movement of the snow line in Xinjiang, northwestern China, by as much as 70 meters in the last 40–50 years. There was a lack of evidence to determine whether a warming trend was also happening in the tropical regions and in Antarctica, and was therefore global, but he hoped the IGY would help establish enough stations to obtain the necessary data for this purpose.[28]

Zhu then went on to describe in detail the various IGY projects, including Antarctic exploration, atmospheric and Earth sciences, and the launching of satellites by the United States and the Soviet Union. In connection with the latter, he made a point of mentioning the Chinese invention of the rocket in the eleventh century during the Song dynasty, "marking the first step toward the artificial moon." In reviewing the subsequent development of the technology, especially Wernher von Braun's recent proposal for an American military base in space, Zhu made the appeal that "artificial satellites should not become a militaristic weapon of any one country, but a tool for bringing in new knowledge and life for all humankind." He detailed the Chinese IGY program and acknowledged that "our scientific level still lagged quite behind world standards." But he hoped that when the expected next IGY came around in 25 years, China "would have caught up with the scientifically advanced countries."[29]

While Zhu's article primed the public about China's participation in the IGY, the organizers made final preparations by sending a high-level delegation of scientists, under Zhao Jiuzhang, to Tokyo to participate in the IGY regional meeting for the western Pacific from February 25 to March 2, 1957. Such a move also had international political benefits, as Liao Chengzhi, chair of the Communist Party's powerful Central Commission on International Activities, told Zhu Kezhen: "We want to establish friendly relations with Japan."[30] There, Chinese scientists joined colleagues from Australia, Indonesia, Japan, Pakistan, the Philippines, the United States, and the Soviet Union to review IGY programs of participating countries in the region and make suggestions for additional observations and technical cooperation. Calling it "a very successful meeting" in a report published in the CAS *Science Bulletin* on his return, Zhao signaled that China was ready for the IGY.[31]

The finalized Chinese IGY program was published in the March 27, 1957, issue of *Science Bulletin*. It was composed of eight main categories: (1) meteorology, which involved 90 sites, all of which would be engaged in surface measurements; five in radiosonde and wind measurement, and located in Beijing, Shanghai, Hankou, Ganzhou, and Haikou; 23 in radiosonde measurement of high-altitude temperature, pressure, and humidity in addition to wind measurement by theodolites; 23 in solar radiation observation; one, at Wuhan, in high-altitude ozone observation; (2) geomagnetism at four stations in Beijing, Sheshan, Lhasa, and Guangzhou; (3) aurora observation at 23 meteorological stations; (4) ionosphere: six stations at Manzhouli, Beijing, Sheshan, Wuhan, Chongqing, and Guangzhou; (5) solar activities observation at four stations in Beijing, Nanjing, Sheshan, and Kunming; (6) cosmic rays at Beijing and Dongchuan (near Kunming); (7) longitude and latitude measurement at the Xujiahui station in Shanghai and latitude determination at a station near Tianjin; and (8) seismology: seven stations in Beijing, Lanzhou, Nanjing, Sheshan, Lhasa, Kunming, and Guangzhou.[32]

## The Taiwan Issue and Withdrawal

So by March 1957 China was ready for the IGY, but was the IGY ready for China? Quite unbeknown to Zhu and other Chinese scientists, in late 1956 and early 1957 the IGY organizers faced a mounting crisis about the issue of China's representation, which issue they thought they had successfully skirted when Taiwan apparently failed to respond to the IGY invitation even after China joined in 1955. At the Barcelona conference in September 1956 Zhu Gangkun had an informal conversation about Taiwan with Sydney Chapman, his dissertation adviser at Oxford in the late 1940s. When Chapman asked Zhu Gangkun whether China would agree to Taiwan joining the IGY, Zhu Gangkun responded indirectly by asking Chapman how he would react if Scotland, Wales, and England all joined the IGY in the name of Great Britain? Chapman laughed, recalled Zhu Gangkun years later, and said "You are grown up."[33] Having heard nothing from CSAGI about Taiwan after this exchange, Chinese scientists probably breathed a sigh of relief. But the storm was brewing.

Apparently just as Chapman and Zhu Gangkun were talking in Barcelona, the U.S. State Department, alarmed by China's participation in the conference, decided to push the Chinese Nationalist government in Taiwan to apply for participation in the IGY. According to research by Ronald Doel and colleagues, Walter McConaughy, director of the Office of Chinese Affairs at the State Department and a close friend of Jiang Jieshi, first contacted Jiang's embassy staff in Washington, who then prompted Zhu Jiahua (Chu Chia-hua, no relation to Zhu Kezhen), president of Academia Sinica in

Taipei, to write to CSAGI and apply for participation in the IGY. Denying ever receiving an invitation from CSAGI, Zhu Jiahua now requested one for his academy to attend the Tokyo conference and insisted the PRC be deleted from the IGY list of participants.[34] The letter, according to Walter Sullivan, *New York Times* science reporter and author of a history about the IGY, threatened to politicize the international collaboration and result in "the wrecking of the IGY."[35]

Rejecting Zhu Jiahua's demand for the IGY exclusion of the PRC, Chapman and the IGY leadership nevertheless felt that they had to accept the Nationalists, who remained a member of ICSU. Writing on February 12, 1957, Chapman told the CAS about Taiwan's application for participation, CSAGI's intention to issue an invitation to Taiwan for the Tokyo conference if it would send only scientists, not government officials—and would refrain from raising political issues—and his hope that China would find this compromise acceptable. Believing this turn of events was the handiwork of "American imperialists," the Chinese government stood firm in its initial response (sent in Zhu Kezhen's name on February 18), requiring the exclusion of the Nationalists as a condition for the PRC participation in the IGY: "China will join IGY only on condition Taiwan should not be admitted. Otherwise China will withdraw from IGY."[36]

Zhu Kezhen apparently did not see Chapman's letter until he returned from a trip. When he prepared to respond in more detail to CSAGI in early April, Zhu Kezhen, an enthusiastic believer in international scientific collaboration, pushed within the Chinese government for a more flexible stand. In talking to officials at the Foreign Ministry, he agreed that "We certainly would oppose Taiwan's admission into the IGY, but it will be difficult for the IGY to refuse it, for it actually is a member of the ICSU and has the sponsorship of the Americans."[37] He seemed to have succeeded to a certain extent. In his follow-up letters to Nicolet on April 8 and to Chapman on April 10, Zhu Kezhen sounded a more conciliatory tone: China would still insist on the official exclusion of Taiwan and was against the listing of two national committees for China, but would not "object [to] scientists from Taiwan coming to the IGY gatherings" as long as they agreed to acknowledge, "implicitly or explicitly," the PRC as the sole representative of China in international affairs. He also welcomed any qualified Chinese geophysicist, implicitly those in Taiwan, as a member of the Chinese IGY National Committee.[38] Nicolet's initial response to his April 8 letter seemed to give Zhu Kezhen encouragement: "It appears that Nicolet regards this matter very seriously, so CSAGI will likely weigh it carefully when it meets next."[39]

Probably encouraged by the tone of Zhu Kezhen's letters, Chapman and his CSAGI colleagues proposed one last compromise in June 1957: the IGY would quietly admit the Nationalists, but the word "national" would be dropped from the list of all IGY participants, and there would be two consecutive entries on China: "Chinese IGY Committee: Peking" and "Chinese IGY Committee: Taipei."[40]

CSAGI's decision to admit Taiwan intensified the internal debate in Beijing as to how to respond to this unwelcome development. On the one hand, the CAS, especially Zhu Kezhen, took a moderate stand, seeking to protest the decision but stopping short of withdrawing from the IGY. For example, on June 26, the leadership of the Chinese IGY National Committee, including Zhu Kezhen, Zhao Jiuzhang, and Tu Changwang, met with Pei Lisheng, who, as the secretary general of the CAS, was a key link to the central government. They agreed that "we should take a clear stand, firmly opposing CASGI's 'two Chinas' approach, but should not propose to withdraw from the IGY on our own." Specifically they suggested that:

(1) We will protest to CSAGI that we do not accept putting Beijing and Taiwan on equal footing, and ask that it carefully reconsider the matter; otherwise we would not be able to meet our responsibilities (referring to the payment of dues and the supply of our geophysical data); (2) If the problem could not be solved satisfactorily, we will continue our various planned geophysical measurements and observations at home, but we will not carry any responsibilities toward CSAGI and will not exchange scientific data with foreign countries; (3) We naturally can still communicate with the Soviet Union and obtain scientific data from elsewhere in the world through the Soviet Union, thus not suffering very much.[41]

On the other hand, the foreign ministry, especially Qiao Guanhua, assistant to Zhou Enlai in Zhou's capacity as foreign minister, pushed for the hard-line position of withdrawing from the IGY if CSAGI did not reverse itself on the issue of Taiwan.[42] And, as the Communist Party Group of the CAS reported to the party's central committee on June 28, the foreign ministry's position prevailed:

On the 27th, after consulting with relevant departments in the Foreign Ministry, [we] agreed with these basic principles but [we] would like to take a clearer and firmer specific stand, i.e., we should make it clear that if CSAGI does not alter the approach of its executive bureau, we will firmly withdraw from the IGY.[43]

The CAS Party Group did not believe CSAGI's argument that it had to admit Taiwan because it was nonpolitical or Taiwan was a member of ICSU. Rather it saw the move as part of a plot to create "two Chinas" that had been advocated by the United States. Furthermore the Party Group suspected that CSAGI "most likely will force us out" of the IGY if China did not consent to its new proposal.[44]

The next day a cablegram went to CSAGI in Zhu Kezhen's name declaring that Taiwan had been an integral part of China since ancient times, that CSAGI had buckled under the pressure of a conspiracy by a certain international political force to create "two Chinas" when it had admitted Taiwan as an independent country, and that "the People's Republic of China National IGY Committee would withdraw from CSAGI and all its activities until it corrects this decision."[45] Thus the IGY started the next day without the official participation of mainland China, the most populous, and one of the largest, countries in the world.

## Conclusion

After China's formal withdrawal from the IGY, its various IGY activities did, as the CAS Party Group had reported to the Central Commission, continue as planned. The National IGY Committee was changed to the Chinese National Geophysics Committee and went on to hold meetings and coordinate the collection of data.[46] Despite the threat contained in the Party Group's report, some data did find its way into the World Data Center in Moscow and was made available to, among others, Lloyd Berkner, who revealed this fact in a 1959 speech to the Industrial College of the Armed Forces in Washington, D.C. Nevertheless, in the speech he criticized the Chinese government as "so backward that it permitted its political jargon to stand in the way of its active participation."[47]

While the differences in views between Zhu Kezhen and the foreign ministry indicated China's decision to withdraw from the IGY was, indeed, political, not scientific, one could question whether it was not exactly the same politics-driven thinking that motivated the U.S. government to push Taiwan to join the IGY in the first place, with almost certain knowledge that it would lead to China's withdrawal and cause a major interruption in international scientific collaboration. In the end the whole incident probably helped promote geophysical research in Taiwan, but it had serious negative impact on scientific research in China.[48] One of the practical effects of China's withdrawal was the disappearance of urgency in coordinating the collecting, printing, and circulation (even within China) of the vast amount of scientific data produced at the various IGY sites all over the country. By May 1958 only World Day ionospheric observations and sun-spot observations produced monthly reports as planned; monthly meteorological reports lasted only until September 1957, and no reports on geomagnetism and seismology existed at all. Of course, there is no guarantee that such reports would have been forthcoming had China remained in the IGY, but the need to fulfill its international obligations would probably have helped produce the pressure for a much better communications effort.

Perhaps more important, United States intervention resulted in the weakening of the political status of moderate scientists within China, a consequence that was evidently counterproductive to the American goal of promoting "peaceful evolution" in the country. To hard-line party and government leaders, scientists like Zhu Kezhen who had advocated increased Chinese participation in international science appeared to be naïve, at best, in their faith in scientific internationalism and in the possibility of separating science from politics. Coinciding with the launching of the Anti-rightist campaign of renewed ideological control in China, this episode probably contributed to a general tightening of any international scientific opening to the West in this period.[49] Although Zhu

Kezhen survived the Anti-rightist campaign relatively unscathed, the IGY incident certainly did not enhance his political standing. During a "self-criticism" session a year later, he had to mention the IGY as one of those tasks given to him by the CAS that he failed to carry out successfully.[50] Even more painfully, his son Zhu Xiwen, a school teacher in Nanjing, was persecuted as a rightist at this time and died several years later in a labor camp.[51]

Of course the eventual withdrawal of China from the IGY should not obscure the sincere attempts at scientific internationalism on the part of Chapman, Zhu Kezhen, and others in a difficult international political environment. Both Chapman and Zhu promoted such an ideal in the form of the IGY. Chapman did so by trying to persuade China first to join and then, after Taiwan applied, to stay in the project, while Zhu Kezhen did likewise by organizing and publicizing the effort in China and then by making the case, although unsuccessfully, for it to remain in the IGY. What drove their allegiance to scientific internationalism was not just traditional idealism, but also practical needs: Chinese scientists and the IGY needed access to each other's data, and eventually both succeeded, at least to a limited extent. Yet what this examination demonstrated, above all, is that even though scientific internationalism played an important role even at the height of the cold war, ultimately it was the states, on both sides of the Iron Curtain, that determined the conduct of international scientific interactions.

## Notes

1. We express our thanks to the late Dr. Zhu Gangkun of the Chinese Academy of Sciences for an informative interview on his experiences related to China and the IGY during the 1950s, to organizers and participants at the October 2007 "Making Science Global" conference at the Smithsonian Institution for encouragement and constructive feedback, and to Wang Yangzong, Zhang Li, Pan Tao, Ronald Doel, and Rip Bulkeley for helpful discussions and assistance with materials.
2. Ronald E. Doel, Dieter Hoffmann, and Nikolai Krementsov, "National States and International Science: A Comparative History of International Science Congresses in Hitler's Germany, Stalin's Russia, and Cold War United States," *Osiris* 20 (2005): 49–76.
3. "Sheshan dicitai jianjie," http://www.geomag.ac.cn/sitenet/SSH/index.htm.
4. Zhu Kezhen, "Emeishan Taishan guoji jinian guance baogao bianyan" (introduction to *Report on the Measurements for the International Polar Year Carried out at Emei and Tai Mountains*), 1935, reprinted in Zhu Kezhen, *Zhu Kezhen quanji* (the complete works of Coching Chu), vol. 2 (Shanghai: Shanghai Scientific and Technological Education Press, 2004), 277–278. On Zhu, see also Zuoyue Wang, "Saving China through Science: The Science Society of China, Scientific Nationalism, and Civil Society in Republican China," *Osiris* 17 (2002): 291–322.
5. To avoid confusion, we will use "CAS" to refer to the academy in Beijing and "Academia Sinica" for the one in Taiwan unless otherwise noted.
6. Walter Sullivan, *Assault on the Unknown: The International Geophysical Year* (New York: McGraw-Hill, 1961), 36–37.
7. CAS, "Guanyu 'guoji diqiu wuli nian' qingkuang de jieshao he woguo canjia zhe yi gongzuo de jianyi" (background on the IGY and suggestions on our country's participation in this work), June 30, 1956, in Archives of the Chinese Academy of Science, Beijing, folder "1956-4-52."
8. Xue Weimin, ed., *Dangdai zhongguo de qixiang shiye* (meteorological services in contemporary China) (Beijing: Chinese Social Science Press, 1984), 402.

9. CAS, "Guanyu 'guoji diqiu wuli nian,'" June 30, 1956.

10. CAS, "Guanyu 'guoji diqiu wuli nian,'" June 30, 1956.

11. CAS, "Guanyu 'guoji diqiu wuli nian,'" June 30, 1956. On Soviet decision to join the IGY, see Rip Bulkeley, "Aspects of the Soviet IGY," *Russian Journal of Earth Sciences* 10, no. 1 (2008): ES1003, available online at: http://elpub.wdcb.ru/journals/rjes/v10/2007ES000249/2007ES000249.pdf.

12. On the Soviet visit, see "Zhongguo kexueyuan yuanwu changwu huiyi guanyu sulian kexueyuan daibiaotuan de jianyi he zhongsu liangguo kexueyuan zhijian de hezuo de jueyi" (resolution on the suggestions of the delegation of the Soviet Academy of Sciences and on CAS–SAS collaboration made at the regular meeting of the CAS), June 22, 1955, *Kexue tongbao* (Chinese science bulletin), (August 1955). See also Zhu Kezhen diaries for April 27–June 23, 1955, in *Zhu Kezhen riji* (Zhu Kezhen's diaries), vol. 3 (Beijing: Science Press, 1989), 553–572.

13. Bulkeley, "Aspects of Soviet IGY."

14. Zhu diary for June 18, 1955, in *Zhu Kezhen riji*, vol. 3, 571–572.

15. Zhu diary for June 18, 1955, in *Zhu Kezhen riji*, vol. 3, 571–572.

16. Zhu diary for June 30, 1955, in *Zhu Kezhen riji*, vol. 3, 574–575. In January 1957, Tu was promoted to be a vice chair, and several others were added as members. See Wang Zhongjun, ed., *Zhongguo kexueyuan shishi huiyao 1955 nian* (major events in the history of the Chinese Academy of Sciences in 1955) (Beijing: Chinese Academy of Sciences, 1995), 8, 26–27.

17. The exchanges are in the Archives of the Chinese Academy of Science, Beijing, folder "1955–1–30." See also Sullivan, *Assault on the Unknown*, 36.

18. Zhu diary for June 18, 1955.

19. Zhu diary for July 2, 1955.

20. Zhu diary for January 4, 1956, in *Zhu Kezhen riji*, vol. 3, 636.

21. Zhu diary for January 1, 1956, in *Zhu Kezhen riji*, vol. 3, 635.

22. Wen Kegang et al., *Tu Changwang zhuan* (biography of Tu Changwang) (Beijing: Contemporary China Press, 1997), 451.

23. Zhu to Bardin, July 24 and October 27, 1956, in Zhu Kezhen, *Zhu Kezhen quanji* (collected works of Zhu Kezhen), vol. 3 (Shanghai: Shanghai Scientific and Technological Education Press, 2004), 305, 308.

24. Zhu diary for January 14, 1957, in *Zhu Kezhen riji*, vol. 4 (Beijing: Science Press, 1989), 7. See also Wen, *Tu Changwang zhuan*, 460–461.

25. Zhu diary for July 27, 1956, in *Zhu Kezhen riji*, vol. 3, 696–697.

26. CAS to State Council, January 25, 1957, in Xue Pangao and Ji Chuqing, eds., *Zhongguo kexueyuan shishi huiyao 1957 nian* (major events in the history of the Chinese Academy of Sciences in 1957) (Beijing: Chinese Academy of Sciences, 1998), 25–27.

27. Zhu Kezhen, "Guoji diqiu wulinian de zuzhi he guoji hezuo" (the IGY and international scientific collaboration), *People's Daily*, February 19, 1957.

28. Zhu, "Guoji diqiu wulinian de zuzhi he guoji hezuo."

29. Zhu, "Guoji diqiu wulinian de zuzhi he guoji hezuo."

30. Zhu Kezhen diary for February 11, 1957, in *Zhu Kezhen riji*, vol. 4, 15.

31. Zhao Jiuzhang, "Guoji diqiu wulinian dongjing xi taipingyang quyu huiyi" (IGY western Pacific regional conference in Tokyo), *Kexue tongbao* (science bulletin), (October 1957): 312–313. See also Zhu Kezhen diary for April 2, 1957, in *Zhu Kezhen riji*, vol. 4, 39–40.

32. Zhou Hang, "Woguo canjia guoji diqiu wulinian de gongzuo jihua" (our country's working plan for participation in the IGY), *Kexue tongbao* 6 (March 27, 1957): 185–186.

33. Interview with Zhu Gangkun by Jiuchen Zhang and Zuoyue Wang, August 27, 2007, Beijing.

34. Doel and others, "National States and International Science," 69. Sullivan, *Assault on the Unknown*, 37. There is evidence that the IGY was not unknown to the scientific community in Taiwan before Zhu Jiahua made his protest. See, for example, Xue Jixun, "Guoji diqiu wulixue nian" (International Geophysical Year), *Jiaoyu yu wenhua*

(education and culture) 10, no. 1 (December 1955): 17–19. Xue later became a member of Taiwan's IGY committee. See *Final Report [of] Chinese National Committee [for the] International Geophysical Year 1957–58* (Taipei: Chinese National Committee for the IGY, April 1959), 16.

35. Sullivan, *Assault on the Unknown*, 37.

36. Sullivan, *Assault on the Unknown*, 41. On the Chinese belief in the American role, see Communist Party Group of the CAS to Central Commission on International Activities, June 28, 1957, in Xue Pangao and Ji Chuqing, eds., *Zhongguo kexueyuan shishi huiyao 1957 nian*, 182–183.

37. Zhu Kezhen diary for April 1, 1957, in *Zhu Kezhen riji*, vol. 4, 39.

38. Sullivan, *Assault on the Unknown*, 41.

39. Zhu Kezhen diary for April 26, 1957, in *Zhu Kezhen riji*, vol. 4, 48.

40. Sullivan, *Assault on the Unknown*, 42.

41. Communist Party Group of the CAS to Central Commission on International Activities, June 28, 1957, in Xue and Ji, eds., *Zhongguo kexueyuan shishi huiyao 1957 nian*, 182–183.

42. On Qiao's role, see Zhu Kezhen diary for August 12, 1957, in *Zhu Kezhen riji*, vol. 4, 91–92.

43. Communist Party Group of the CAS to Central Commission on International Activities, June 28, 1957.

44. Communist Party Group of the CAS to Central Commission on International Activities, June 28, 1957.

45. Zhu Kezhen to Chapman, June 6, 1957, in *Zhu Kezhen quanji*, vol. 3, 364.

46. See Zhu Kezhen diaries for September 21, 1957; February 1, 1958; and May 21, 1958; in *Zhu Kezhen riji*, vol. 4, 103, 144–145, 179, respectively.

47. Lloyd Berkner, "International Geophysical Year," January 27, 1959, a speech at the Industrial College of the Armed Forces, Washington, DC, available online at: www. ndu.edu/library/ic3/L59–097.pdf.

48. On the Chinese Nationalist effort for the IGY, see *Final Report [of] Chinese National Committee [for the] International Geophysical Year 1957–58*.

49. See, for example, Zhu Kezhen diary of August 24, 1961, in *Zhu Kezhen riji*, vol. 4, 550.

50. "Remarks at 'Jiao Xin' [baring one's thoughts] Rallies of the CAS," June 11 and 13, 1958, in Zhu Kezhen, *Zhu Kezhen quanji*, vol. 3, 439–440.

51. Zhu Kezhen diaries of May 25, 1958, and February 20, 1961, in *Zhu Kezhen riji*, vol. 4, 181 and 509–510, respectively. There is no evidence that Xiwen's persecution had any direct connection with his father's political situation.

# Part Three

# Networked Personalities and Programs

# Chapter 9

# Approaching the Southern Hemisphere: The German Pathway in the Nineteenth Century

*Cornelia Lüdecke*

The German investigation of the Southern Hemisphere connects closely with Georg Balthasar von Neumayer (1826–1909) who, beginning in the middle of the nineteenth century, approached Antarctica step by step. Neumayer was raised during the era of the Magnetic Crusade, when British scientist James Clark Ross (1800–62) found the magnetic pole of the Northern Hemisphere in 1831.[1] In search for the magnetic pole of the Southern Hemisphere, Ross established several magnetic observatories at places like Cape Town, southern Africa, and Hobart, Tasmania. Finally, in 1842 he determined the magnetic pole at 72°35'S 152°30'E.

## Foundation of the Flagstaff Hill Observatory in Melbourne (1857)

In 1845 Neumayer decided to study at the Polytechnische Hochschule or Polytechnical School (today Technical University) and later also at the School of Engineers in Munich.[2] During this time he became familiar with the ideas of American hydrographer Matthew Fontaine Maury (1806–73), who collected meteorological information from ships' logbooks to construct optimal sailing routes. The translation of Ross's travel accounts in 1847 initiated Neumayer's interest in the exploration of Earth magnetism, especially in the still-unexplored regions of the Southern Hemisphere, which became his desired field of work.[3]

Neumayer finished his studies with a theoretical state examination in 1849. Afterward he found work as Karl Joseph Reindl's (1806–53) assistant at the Physical Institute of the Ludwig Maximilians University, where he learned more about physics and electricity.[4] He also practiced with astronomer Johann von Lamont (1805–79) at the observatory in the village Bogenhausen on the Isar River, west of Munich.[5] Lamont, a member

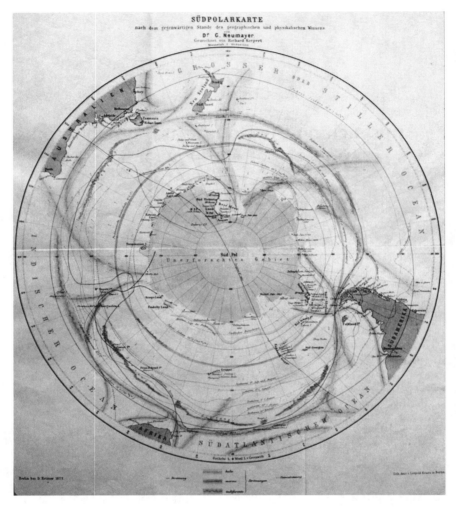

**Figure 9.1**    The Antarctic map of Georg Balthasar von Neumayer in 1872. (Private possession.)

of the former Göttingen Magnetic Association (1836–41), had improved Gauss's magnetic instruments and also invented a travel theodolite, which was used for measurements during expeditions. Under Lamont's guidance Bogenhausen became the most important magnetic observatory in Europe.[6] After Lamont's textbook on Earth magnetism was published, he introduced Neumayer to Gauss's theory of Earth magnetism and the measurement of the Earth magnetic field.[7] In 1850 Neumayer left Munich and traveled to South America aboard the bark *Louise*, intending to learn all about sailing. When he returned to Hamburg he contacted the famous astronomer Charles Rümker (1788–1862), of the Hamburg college for observatory and navigation training. With his knowledge and experience gained aboard the sailing vessel, Neumayer acquired his commission from Rümker within four weeks.

While Austria reestablished its navy, Neumayer applied for an appointment in both Vienna and in Trieste, the Austrian harbor on the Adriatic Sea. Rejected, Neumayer returned to Rümker in Hamburg in January 1852 to teach at the navigation training college, where he introduced Maury's ideas of meteorological navigation.

In early April 1852 Neumayer sailed as an ordinary seaman from Hamburg to Australia, arriving in early 1853 at Port Jackson, the harbor of Sydney, New South Wales. He earned a small profit in the gold mines, explored the outback, and started some magnetic measurements. He also visited Ross's former magnetic station in Hobart. At that stage Neumayer developed a plan to establish an observatory in Melbourne for magnetic and meteorological work in addition to nautical and astronomical studies. It would serve as a base station in the vicinity of the southern magnetic pole in the still-unknown Antarctica.

After his return to Germany in January 1854, Neumayer gave lectures to raise money for his project. Lamont recommended that he practice with the new magnetic instruments built at his workshop in Bogenhausen before taking them to Australia. Following Lamont's advice, Neumayer carried out a detailed magnetic survey of the Bavarian Rhine–Palatinate (Rheinpfalz) and, for nautical purposes, of Hamburg and parts of Schleswig-Holstein.[8]

With the financial support of Bavarian King Maximilian II (1811–64) Neumayer established his magnetic base station for polar regions and central observatory for maritime meteorology on Flagstaff Hill in Melbourne, which he directed from 1857 to 1864.[9] When the observatory was taken over by the colony of Victoria in 1859, Neumayer was appointed "Director of the Magnetic Survey of the Colony of Victoria." After the magnetic survey ended in 1864, Neumayer returned to Germany to continue his plan to investigate the southern high latitudes for the sake of navigation.

### Observations of the Transit of Venus (1874, 1882) from the Southern Hemisphere

Returning to Germany in 1865, Neumayer constantly pleaded for a German South Polar expedition.[10] During the first German Geographer's Day in Frankfurt/Main, July 23–24, 1865, the cartographer and editor August Petermann (1823–78), of *Petermann's Geographische Mitteilungen*, proposed a German north polar expedition, contrary to Neumayer's plan to investigate the meteorological and magnetic conditions of the unknown Antarctic region. Neumayer argued that the upcoming transits of Venus in 1874 and 1882 required preliminary expeditions to the far south in search of good locations to observe this rare astronomical event.[11] Obviously, a German expedition to the closer Arctic seemed much more promising than an Antarctic one. After the rejection of his proposal, Neumayer planned to go south with an Austro-Hungarian expedition, but this could not be

**Figure 9.2** First German balloon ascent on March 29, 1902, in front of the research vessel *Gauss*, trapped in sea ice at Kaiser Wilhelm II Land. (Private possession; Mörder, Feldkirchen Westenham)

realized. Nevertheless, beginning in 1865, Neumayer used every congress or meeting to focus on an Antarctic expedition.[12]

In 1872 Neumayer was appointed hydrographer of the Imperial Admiralty in Berlin.[13] One of his first tasks was to prepare a detailed paper, published by the Berlin Geographical Society, describing the status of the exploration of the south polar region.[14] It included a colored map showing ocean currents and the northernmost expanse of icebergs (figure 9.1), indicating that the area south of Kerguelen at longitude 70° east was most favorable for an approach of the still-unexplored region. Neumayer had also proposed a sailing route aiming south of Kerguelen at longitude 90° east and dividing in an easterly and westerly direction halfway to what he called "unexplored region."

In his new position Neumayer served as an official representative of German interests during international meteorological meetings. At the first International Meteorological Congress in Vienna, September 2–16, 1873, the expansion of the existing meteorological network was an important issue on the agenda. The congress recommended the establishment of meteorological stations in the little-known Arctic, especially in Spitsbergen.[15] After discussions, Neumayer added a footnote in the minutes of the congress saying "and in southern High Latitudes."

When he organized the circumnavigation of the SMS *Gazelle* (1874–76), which would present the scientific power of the new German Reich, established a year before, he could potentially achieve part of his personal goal to send out an expedition to the south. He directed the expedition to Betsy Cove on Kerguelen in the southern Indian Ocean to observe the transit of Venus on December 9, 1874.[16] SMS *Gazelle* stayed at Kerguelen for nearly four months, giving the crew enough time to perform meteorological and other scientific investigations. This data was important for Neumayer to investigate the best route to the far south. From that time on, for him Kerguelen served as the gateway to Antarctica.

Later he expanded his ideas on polar exploration and pleaded for similar and simultaneous measurements over a limited time in both polar regions.[17] Besides magnetic and meteorological observations, he indicated that the transit of Venus on December 6, 1882, should also be included in their observations.

Independent from Neumayer, a German officer of the Austro-Hungarian Navy, Carl Weyprecht (1838–81), also promoted the idea of internationally coordinated measurements in the polar regions.[18] During the 48th meeting of the German Naturalists and Physicians at Graz, Austria, in 1875, he presented six fundamental principles of Arctic exploration, with the aim of establishing a temporary network of stations around the Arctic for at least one year.

1. Arctic exploration is of greatest importance for a knowledge of the laws of nature.
2. Geographical discovery carried out in these regions has only a serious value inasmuch as it prepares the way for scientific exploration as such.
3. Detailed Arctic topography is of secondary importance.
4. For science the Geographical Pole does not have a greater value than any other point situated in high latitudes.
5. If one ignores the latitude the greater the intensity of phenomena to be studied the more favorable the place for an observational station.
6. Isolated series of observations have only relative value.[19]

In preparation for the second International Meteorological Congress in Rome (1877), the Permanent Committee of the Meteorological Congress discussed Weyprecht's proposal in 1876 and recommended the establishment of eight Arctic stations. When the congress had to be postponed for two years due to the Balkan War, Weyprecht's proposal was accepted in 1879. In the deliberations of the congress, Neumayer always pointed to the importance of similar and synchronous investigations on the Southern Hemisphere. In particular he focused on the Kerguelen and Auckland Island stations to continue the previous work of the German stations, where the transit of Venus and meteorological parameters had been observed in 1874. Finally, the congress took over the initiative and set up an International Polar Commission to meet at the German Maritime Observatory (Deutsche Seewarte) in Hamburg, October 1–5, 1879, where Neumayer had acted as director since January 1876.

Nine directors of national meteorological services and institutes attended the meeting, at which Neumayer was elected president of the commission.[20] The main point was to discuss the program of the planned International Polar Year (IPY). Which were the best locations of the stations for the Arctic network? Which instruments and which methods should be used? And what about publication? Additionally, the commission adopted Weyprecht's ideas to receive synchronous meteorological data from the Arctic in order to improve weather and storm forecasting in Europe and the United States. Stations should be also set up on South Georgia, Kerguelen, and Auckland, or on Campbell Island or Balleny Island. The period of 1881–82 seemed to be most favorable for the project, because a maximum of magnetic activities was expected by the scientists.

The next meeting of eight national weather service directors took place in Bern, Switzerland, August 7–9, 1880, just before the meeting of the International Meteorological Committee. Heinrich von Wild (1833–1902), the Swiss director of the Central Observatory at St. Petersburg and president of the International Meteorological Committee, participated as a guest of the International Polar Commission. When only four nations wanted to establish Arctic stations, the start of the measurements was postponed for a year. Besides, Neumayer seemed to be frustrated because the German Reich had not yet decided to finance any German polar expedition to Kingua Fjord (Cumberland Sound on Baffin Island, East Canada) or to the sub-Antarctic island of South Georgia in the southwest Atlantic. Thus, he resigned from the presidency, and Wild, who seemed to be more influential, became his successor.

During the third committee meeting at St. Petersburg, August 1–8, 1881, eight countries assured their participation with Arctic stations, and the final meteorological and magnetic program was scheduled for a period of 13 months starting on August 1, 1882, and ending on August 31, 1883.[21] On special-term days (1st and 15th of each month) magnetic observations should be made each five minutes with the Göttingen Standard Time as reference, which was a relic of the Göttingen Magnetic Association. Magnetic term hours with intensified measurements started at the same term day from midnight to 1 a.m. and continued on the next term day, starting an hour later and lasting from 1 a.m. to 2 a.m., so they would cover each hour of a day during a whole year.

The American stations were planned at Point Barrow (Alaska) and Lady Franklin Bay (Ellesmere Island, northeast Canada); the Austro-Hungarian at Jan Mayen (northern Atlantic); the British at Fort Rae (Great Slave Lake, Canada); the Danish at Godthaab (western Greenland); the Dutch at Dikson (eastern Siberia); the Finnish at Sodankylä (northern Finland); the Norwegian at Bossekop (northern Norway); the Russians at Karmakuly (Novaya Zemlya) and Sagastyr (Lena Delta in Siberia); the Swedish at Cape Thordsen (Spitsbergen); and the French at Cape Horn (Tierra del Fuego) in the Southern Hemisphere.

On the occasion of the inauguration of the new building of the German Maritime Observatory on September 14, 1881, German emperor Wilhelm I

(1797–1888) announced he would finance two polar expeditions.[22] Neumayer had succeeded in finally achieving his goal of sending a second expedition to South Georgia, where many interesting measurements for the investigation of the solar-terrestrial relationship could be made, including of the second transit of Venus on December 6, 1882. Furthermore the German Maritime Observatory equipped six mission stations of the Moravians[23] on the east coast of Labrador with instruments. These would provide crucial data to investigate the Labrador storms, which also influenced the weather in western Europe.

After the first IPY, 20 directors of meteorological services and expedition leaders met in Vienna, April 17–24, 1884, to honor Weyprecht, one of the IPY initiators, who had died of tuberculosis on March 29, 1881. One of the discussions focused mainly on the analysis and publication of the data by each participating country. Later the data and results were included in various products of the German Maritime Observatory, such as magnetic maps and a new cloud atlas.[24] Neumayer's aim to construct daily synoptic weather charts of the southern Atlantic could not be fulfilled because of a lack of data in the far south.[25] The ship data he collected only reflected the main sailing routes across the south Atlantic, the east coast of South America, the coast of Chile around Cape Horn, and north to Valparaiso.

The final meeting of the International Polar Commission took place in Munich on September 9, 1891, in connection with the first conference of the directors of Meteorological Services. Thirteen directors and guests discussed the results and decided that documents and publications should be archived at the Central Physical Observatory in St. Petersburg. The Polar Commission was dissolved at the end of the meeting.

## The German South Polar Expedition (1901–03)

At the end of the nineteenth century, cartographers had yet to map Antarctica, even though commercial whaling and scientific interests had turned south to the terra incognita. Was it a continent, as supposed by the ancient Greeks, or was it a gigantic atoll of ice-covered islands as indicated by the discoveries of the whalers and sealers? This was still a fundamental question waiting to be answered. After the IPY exploration and the resulting data publications, scientists, satisfied by their Arctic investigations, headed south. This time, the geographers, not the physical scientists, such as magneticians and meteorologists, took the lead. During the eleventh German Geographer's Day, under the presidency of Neumayer at Bremen, April 17–20, 1895, the possible future exploration of the South Pole region was presented over the course of three talks.[26] Neumayer opened the session and presented a survey of research needed in Antarctica. He was followed by Erich von Drygalski (1865–1949) and Ernst Vanhöffen (1858–1918), who recently had wintered together on the west coast of Greenland (1892–93). They wanted to expand their glaciological and biological investigations to the far south. Because of the favorable circumstances, a German commission on the South Polar Exploration was founded during the Geographer's Day with Neumayer as chair.

**Figure 9.3** Dr. Georg Neumayer (winter 1872), later director of the *Deutsche Seewarte* (German Maritime Observatory). (Courtesy of Bundesamt für Seeschiffahrt in Hydrography, Hamburg)

At the same time something comparable happened in Great Britain. Clements Markham (1830–1916), president of the Royal Geographical Society, was a similar figure in Great Britain to Neumayer in Germany. For years Markham had tried to send out a British Antarctic research expedition. Finally, the time was ripe. During the fifth International Geographical Congress at Bern (1891), it had been decided to hold the next congress at London to be organized by Markham on July 26–August 3, 1895.[27] Neumayer was invited to support Markham's ideas by giving a talk on southpolar research, which only could be done with international collaboration. After the discussions at the end of the session the congress adopted the following resolution:

### Antarctic Exploration

That the Congress record its opinion that the exploration of the Antarctic Regions is the greatest piece of geographical exploration still to be undertaken. That, in view of the additions to knowledge in almost every branch of science which would result from such a scientific exploration, the Congress recommends that the scientific societies throughout the world should urge, in whatever way seems to them most effective, that this work should be undertaken before the close of the century.[28]

Back home Neumayer could plan the German South Polar Expedition. Nevertheless, the first three meetings of the German Commission of South Polar Exploration at Berlin and a fourth meeting in Jena in 1897 ended without success.[29] The help of a special Action Committee was enlisted to provide financial support for equipping the German expedition. When Drygalski was elected as the leader of the planned expedition during the sixth commission meeting on February 19, 1898, the organization was given over to the young scientist and experienced polar researcher ten days after his 33rd birthday. The polar research vessel *Gauss*, the first of its kind in Germany, was built in Kiel as a three-masted bark. It would serve as a flagship of German science and was fitted with the most sophisticated equipment: a sensitive magnetic compass fixed on a special swinging table to measure Earth's magnetic field; and two winches for oceanographic and marine biological investigations, placed on the additional second bridge behind the main mast. The vessel provided single cabins for five officers and also for five scientists, which was rather unusual at that time. Cabins for two, six, and ten people were used by the crew of twenty-two seamen.

Because of Neumayer's well-known efforts to set up an Antarctic expedition, Germany was chosen to organize the seventh International Geographical Congress (September 28–October 4, 1899) at the Reich capital in Berlin under the presidency of the famous geographer and explorer Ferdinand Freiherr von Richthofen (1833–1905), president of the Berlin Geographical Society. Richthofen had been Drygalski's supervisor when preparing his PhD thesis ("doctor father") and had given him the support needed to continue his investigation of the movements of mountain and inland ice glaciers, such as those found in Greenland. Now Antarctica was the next aim of the international geographical community.

In Berlin an entire session was devoted to Antarctic research. Besides Neumayer and Drygalski, Markham presented his ideas and defined the fields of work of the German and British expeditions.[30] He divided a map of Antarctica into four quadrants, starting at 0° Greenwich meridian. The Weddell and Enderby quadrants, between longitudes 270° west and 90° east, were designated as the working area of the German expeditions, while the Ross and Victoria quadrants, longitudes 90° east to 270° west, were British, in accord with the earlier work of the Ross expedition (1839–43).

Drygalski even proposed collaborating with the British expedition within a broader context.[31] For the investigation of the magnetic field of Earth, more observation stations distributed throughout Antarctica would be needed. Thus, also considered were a German base-station at Kerguelen to continue earlier work done by the German transit of Venus expedition of 1874 and a British base-station at New Zealand. Finally Drygalski proposed the establishment of an international network of similar, and corresponding, meteorological and magnetic observations in and around Antarctica, according to the cooperation of the IPY (1882–83). Magnetic term days and term hours with hourly observations were defined at permanent stations on the first and fifteenth of each month from February 1, 1902, until February 15, 1903, as was the practice during the IPY. Meteorological observations were to be made from October 1, 1901, until March 31, 1903, including all merchant and navy ships sailing on a route south of latitude 30° south.

Scottish surgeon and natural scientist William Speirs Bruce (1867–1921) and Swedish geologist Otto Nordenskjöld (1869–1928), both participants of the Geographical Congress at Berlin, joined the collaboration to explore the still-unknown Antarctic continent and to organize additional expeditions.[32] Further, Drygalski and the British expedition leader, Robert Falcon Scott (1868–1912), started to arrange their measurements and later share them.

When Drygalski presented a plan for his expedition, he followed Neumayer's favored route, using Kerguelen as the German gateway to the south.[33] Older measurements from the circumnavigation of the British SMS *Challenger* had indicated an ocean current leading south, which might allow access to higher latitudes, similar to the area presently known as the Weddell Sea (figure 9.1). Even an ocean current drifting to the South Pole at longitude 90° east had been suggested by the first German oceanographer, Otto Krümmel (1854–1912) from the university at Kiel, in a confidential letter to Drygalski.[34] Krümmel suggested that a ship with a hull like Nansen's famous *Fram* could be frozen-in south of Kerguelen and might drift with the ice and cross the South Pole based on Nansen's model,[35] eventually to emerge in the Weddell Sea. This notion assumed that Victoria Land and the Ross Sea region were connected with the Antarctic Peninsula, forming a comma-shaped Antarctic continent, while the other land sightings were interpreted as islands.

During the thirteenth German Geographer's Day in Breslau (May 28–30, 1901), the German Commission for South Polar Exploration was officially dissolved. With Neumayer's preparatory work finished at age 76, the German expedition to the unknown south would be under way soon. The instructions given by Emperor Wilhelm II were rather general. The expedition was to depart from Kiel in August 1901 and proceed to the Kerguelen Islands, where a meteorological and magnetic station was to be established.[36] Next the expedition should explore the south polar region of the Indian and Atlantic oceans. If land could be reached, a scientific station would be established and maintained for a whole year. The return was to be aimed for 1903 or spring of 1904, according to the decision of the expedition leader.

On August 11, 1901, the expedition left Kiel, passing the new Kaiser Wilhelm Channel to visit Hamburg, and then started its four-and-a-half month journey to Kerguelen, where it arrived at the end of the year. The expedition members helped finish the buildings of the base station, which had already been established with the help of a chartered merchant ship at Royal Bay at the same place, where the British Venus Transit expedition had set up its observation in 1874. A month later *Gauss* set sail for an unknown destination somewhere in the south, while three scientists and two assistants stayed at Kerguelen. Two months later, on March 2, 1902, *Gauss* was beset by ice close to the Antarctic circle at latitude 66°2' south and longitude 89°38' east, some 85 kilometers (45 seamiles) off the Antarctic coast.

Luckily the ice was not drifting, and they could set up a fixed winter station on sea ice close to the ice-covered coast of Kaiser Wilhelm II Land, where they discovered the ice-free extinct volcano "Gaussberg." As they could not build the original observatories prefabricated from wood on the ice, they installed the magnetic instruments for absolute and variation measurements in two caves cut in two icebergs, which functioned well. On March 29, 1902, three ascents with a captive balloon up to some 500 meters were performed (figure 9.2).

During the first ascent, Drygalski measured a temperature inversion of the boundary layer from –15°C (+ 5,9 °F) at the bottom above sea ice and –12.7°C (+ 9,1°F) at 200 meters in height and a decrease of temperature above. Moreover, a device to measure the tides at the high, but frozen, sea was improvised.

After the expedition had been stuck in the polar ice for a year, the ice finally broke up and the *Gauss* came free. When Drygalski failed to approach higher latitudes west of their wintering place in search of the ocean current leading south, he returned to Cape Town to send a cable to Berlin reporting their successful return from the Antarctic. Unfortunately, a second attempt to go south, which Drygalski had wanted, was not possible because the budget was already exhausted. After additional oceanographic measurements, the expedition sailed home, arriving at Kiel on November 25, 1903.

When a French expedition under Jean-Baptiste Charcot (1867–1936) followed in 1903 and established a winter station at L'Île Wandel Island (today Booth Island), international cooperation was expanded until March

**Figure 9.4** Erich von Drygalski, leader of the first German South Polar Expedition (1901–03). (Private possession; Gazert, Garmisch-Partenkirchen)

31, 1904.[37] This information could not be transmitted to the other expeditions, however, because they had no contact with the outer world, unlike the British expedition to the Ross Sea, which was visited each year by a ship bringing new provisions and equipment.

On its return, Drygalski's expedition did not receive the recognition it deserved because Emperor Wilhelm II was disappointed with the outcome. The British led by Scott had beaten the Germans when he reached latitude 82° south, while *Gauss* was trapped in the ice at the polar circle. Of course, due to the political rivalry between historical sea power Great Britain, and an emerging sea power in the German Empire, the world paid more attention to the high latitudes reached by the British than to the scientific observations of the Germans. Geographical achievements were much more valuable than thoroughly measured scientific data, which had to be analyzed and published over decades.[38] During the fifteenth German Geographer's Day in Danzig (1905) preliminary reports were given. When Neumayer made the closing remarks of the polar session, he said: "I am highly surprised, no, I am highly delighted about that, what I have heard today."[39]

After the return of the British, Norwegian, and Swedish expeditions, data was exchanged and synoptic charts plotted for meteorology and magnetism. Earth magnetic data measured by *Discovery* and *Gauss* improved the charts of the southern seas, but the data had been far too scarce to cover large areas.[40] Meteorological data from Antarctica and other southerly stations, in addition to observation logs from ships, was collected in Berlin for the construction of daily weather charts. Among them were 53 German data logs recording time, location, air pressure and temperature, wind direction and speed (in Beaufort), clouds (class, cloudiness, direction of movement), and remarks on the weather.[41] Observations from 350 ships representing different nations were provided by the Hydrographical Office in Washington, D.C. Argentina sent material from 31 stations. Further data from British ships and stations from the British colonies came in exchange with the Meteorological Office at London. The Antarctic expeditions and their base-stations delivered the most interesting material. Additional contributions from 1,000 German ship logs were added, which had to be reduced to Greenwich Time. Altogether about 600,000 single data sets were available in the spring of 1906.

A survey of geographical investigations indicated that Antarctica was a large continent covered by ice. Professor of geography at the university at Göttingen, Wilhelm Meinardus (1867–1952), calculated the mean height of the Antarctic continent using pressure and wind measurements, which resulted in 2,000 ± 200 meters (6560 ft ± 656 ft).[42] The meteorological atlas showed 913 daily synoptic weather charts, 30 charts of monthly means, and several other quarterly, seasonal, and yearly charts.[43] In these charts only South America and the Antarctic Peninsula could be connected with isobars. Neumayer's idea to receive weather charts to improve sailing directions or directions for steamships by plotting charts of the Cape Horn route could not be realized. In the southern region, the meteorological network of stations was not at all dense enough to follow the tracks of polar depressions and provided only descriptive results. Besides, the weather charts were not published before 1915, and at that time were no longer crucial for the Cape Horn route, because the Panama Canal had opened on August 3, 1914.

The expedition leader was commissioned by the Imperial Ministry of the Interior to edit the results. The publication process lasted from 1905 until 1931, almost throughout Drygalski's whole university career as first professor and chair of the Geographical Institute of the Ludwig Maximilians University in Munich. Originally, publication plans included two volumes with geographical and geological results and two volumes each for meteorological and magnetic data and results. One volume was dedicated to bacteriology, chemistry, medical results, and oceanography. Botanical results would be printed in another volume, while the last two volumes were reserved for zoology. These publication plans would not suffice, however, because the biologist Vanhöffen, who had previously accompanied his friend Drygalski to Greenland, had collected 4,030 species, among them 1,470 (!) new species to be described for the first time, which expanded the planned volumes to 12 zoological volumes. Eighty-nine scientists of 105 specialists from Germany (72), Sweden (8), Austria (7), England (4), Denmark (3), Norway (3), Moravia (2), Russia (2), United States (2), France (1), and Switzerland (1) analyzed the huge biological collections, while the other 16 scientists worked on astronomical, geodetic, geographical, geological, magnetic, meteorological, and oceanographic data. A meteorological and a magnetic atlas with various maps and graphs completed the series of results. Despite the political, geographical, and economic changes during World War I, in particular the replacement of the emperor by new governments, Drygalski's relentless efforts to acquire the manuscripts and the finances came to fruition when he finished his commission and published the full material that included the work of the international cooperation.

## Outlook

The German attempt to approach Antarctica was initiated by Georg von Neumayer, who dreamed his entire life of a German South Polar Expedition. First, he became a sailor and traveled to Australia, where no long-term series of magnetic and meteorological data had ever before been taken. Later, he founded the Flagstaff Observatory in Melbourne, which the Colony of Victoria supported and maintained. When he was sure the observatory would continue the measurements, he returned to Germany to prepare the second phase of the expedition to the far south. There were two possibilities in connection with the transit of Venus. Whereas most expeditions traveled by train to the Middle East in 1874, only a few ships were sent to the southern Indian Ocean. Kerguelen at latitude 48° south became Neumayer's fixed point for the 1874 observation, and South Georgia at latitude 54° south for the 1882 observation. While the area around South Georgia was more or less known, the region south of Kerguelen remained unexplored except for one cruise by the SMS *Challenger* in February 1874. For the Germans, exploring this area could yield new information about Earth magnetics and meteorology, which would expand German measurements on Kerguelen from 1874 to an area further south. Having this in mind we can understand that Neumayer chose Kerguelen as Germany's gateway to Antarctica, because it became a matter of prestige to collect new data from a totally unknown region. He was supported by Krümmel's fantastic idea of an ocean current crossing the South Pole. When the International Geographical Congress met in Germany, international cooperation was recommended and organized by the congress secretariat in Berlin.[44] According to the IPY, with its network of 12 observing stations around the frozen Arctic Ocean, a similar idea was approved for the international Antarctic cooperation of 1901–04. But in this case, only four stations were established together with their base-stations some 1,000 kilometers apart. This enterprise revealed that Antarctica was a continent covered by a huge ice mass. A more detailed investigation was not completed until the International Geophysical Year (1957–58).

Germany played an important part in making science global, in sharing crucial meteorological and magnetic data, and in organizing polar networks in the Arctic and Antarctic. Since 1865, Georg Neumayer had been the leading figure. The Bavarian king ennobled him in 1900, which gave him the right to call himself "von Neumayer." When the Federal Republic of Germany established its first Antarctic research station at Atka Bay northeast of the Weddell Sea in 1981, it was called "Georg von Neumayer Station" to honor the German promoter of Antarctic research. Carl Weyprecht's name is closely connected with the first IPY and much attention was given in his birth town, Michelstadt/Odenwald, Germany, to the 125th return of the day of his death in the year 2006. Drygalski's achievements had been remembered during a traveling exhibition, which started in Dresden in 2001 on the occasion of the 100th anniversary of the first German Antarctic expedition (1901–03).[45]

## Notes

1. J. Cawood, "The Magnetic Crusade: Science and Politics in Early Victorian Britain," *Isis* 70, no. 254 (1979): 493–518.
2. H. -J. Kretzer, *Windrose und Südpol: Leben und Werk des großen Pfälzer Wissenschaftlers Georg von Neumayer* (Pollichia: Sonderdruck 4, 2nd add. ed., 1984). H. -J., Kretzer, *Georg von Neumayer (1826–1909)* (Speyer: Pfälzer Lebensbilder, Bd. 4, Pfälzischen Gesellschaft zur Förderung der Wissenschaften, 1987), 205–222.
3. G. v. Neumayer, *Auf zum Südpol! 45 Jahre Wirkens zur Förderung der Erforschung der Südpolarregion 1855–1900* (Berlin: Vita Deutsches Verlagshaus, 1901).
4. Kretzer, *Windrose*.
5. Today Bogenhausen is a part of Munich, close to the city center.
6. R. Häfner, *Die Universitäts-Sternwarte München im Wandel ihrer Geschichte* (München: Universtätssternwarte München, 2003). R. Häfner, *Johann von Lamont 1805–1879: Leben und Werk* (München: Universtätssternwarte München, 2006).
7. J. v. Lamont, *Handbuch des Erdmagnetismus* (Berlin, 1849).
8. G. v. Neumayer, *Eine erdmagnetische Vermessung der bayerischen Rheinpfalz 1855/56* (Dürkheim a.d.H., Schriftenreihe: Mitteilungen der Pollichia, 1905), 62, 21. O. M. Reis, *Georg von Neumayers erdmagnetische Vermessung der bayerischen Pfalz* (Pfälzisches Museum–Pfälzische Heimatkunde, 1926) 5/6, 115–117.

9. R.W. Home and H. J. Kretzer, "The Flagstaff Observatory, Melbourne: New Documents Relating to Its Foundation," *Historical Records of Australian Science* 8, no. 4 (1991): 213–243. C. Stechert, "Georg von Neumayer," *Vierteljahresheft der Astronomischen Gesellschaft* 45, no. 1 (1910): 9–42.

10. Neumayer, *Südpol!*

11. G. Neumayer, *Ein Projekt für die Vorarbeiten betreffs des Venusdurchgangs von 1874* (Wien: Sitzungsberichte der kaiserlichen Akademie der Wissenschaften, 1870), 61, II. Abtlg., 621–646. Neumayer, *Südpol!* 34–35, 40–41, 48–51. The past transit of Venus occurred June 6, 1761, and June 3, 1769, and the recent transit occurred on June 8, 2004. The next: June 5, 2012.

12. During the first International Geographic Congress at Antwerp (August 1871), Neumayer recommended an expedition to explore the Antarctic region referring to geography and physics in connection with the observation of the transit of Venus, see Neumayer, *Südpol!* 62–67.

13. Kretzer, *Windrose.*

14. G. Neumayer, *Die Erforschung des Südpolargebietes* (Zeitschrift der Gesellschaft für Erdkunde zu Berlin, 1872), 6, 120–170.

15. J. Georgi, "Georg von Neumayer (1826–1909) und das 1: Internationale Polarjahr 1882/1883," Deutsche. *Hydrographische Zeitschrift* 17, no. 5 (1964): 249–272. C. Lüdecke, "The First International Polar Year (1882–83): A Big Science Experiment with Small Science Equipment," 21st International Congress of History of Science, Mexico City, *History of Meteorology* 1, no. 1 (2004): 54–63. http://www.meteohistory.org/2004proceedings1.1.

16. A. Auwers, *Die Venusdurchgänge 1874 und 1882,* vols. 1–2 (Bericht über die deutschen Beobachtungen, 1898).

17. Neumayer, *Südpol!*

18. F.W.G. Baker, "The First International Polar Year, 1882–83. *Polar Record* 21, no. 132 (1982): 275–285. G.A. Corby, "The First International Polar Year (1882/83)," *WMO Bulletin* 31, no. 3 (1982): 197–214. Lüdecke, "First IPY."

19. uoted after Baker, "First IPY," 227.

20. Wild, *Mittheilungen der Internationalen Polar-Commission* (St. Petersburg, 1881).

21. Lüdecke, "First IPY."

22. C. Lüdecke, "Das 1. Internationale Polarjahr (1882–1883) und die Gründung der Deutschen Meteorologischen Gesellschaft im Jahr 1883," *Historisch-Meereskundliches Jahrbuch* 9 (2002): 7–24.

23. C. Lüdecke, "East Meets West: Meteorological Observations of the Moravians in Greenland and Labrador in the 18th Century." *Proceedings 22nd International Congress of History of Science,* Beijing, 24–30 July 2005. *History of Meteorology* 2 (2005): 123–132, http://www.meteohistory.org/2005historyofmeteorology2/09luedecke.pdf. The Moravians originally came from Moravia in the Czech Republic. In 1722 they founded a pietistic protestant group at Herrnhut east of Dresden close to the Czech boarder.

24. H. H. Hildebrandsson, W. Köppen, and G. Neumayer, *1890* (Hamburg: Internationaler Wolken-Atlas, 1890). G. Neumayer und C. Börgen (Hrsg.), *Die internationale Polarforschung 1882–1883,* 2 vols. (Berlin: Die Beobachtungs-Ergebnisse der deutschen Stationen, 1886). G. Neumayer, (Hrsg.), *Die internationale Polarforschung 1882–1883,* 2 vols. (Berlin: Die deutschen Expeditionen und ihre Ergebnisse, 1890–1891). G. v. Neumayer, "Atlas des Erdmagnetismus," in H. Berghaus, *Physikalisher Atlas* (Gotha, 1891), Abt. IV.

25. C. Lüdecke, "Beiträge zur maritimen Meteorologie der Südhemisphäre in der Tradition von Heinrich Berghaus (1797–1884) und Matthew Fontaine Maury (1806–1873)," *Historisch-meereskundliches Jahrbuch* 10 (2003/2004): 141–142. W. Schröder and K.H. Wiederkehr, "Über synoptische Wetterkarten des Südatlantik im Ersten Polarjahr und die Beziehungen der Deutschen Seewarte zu Wissenschaftlern in Argentinien," *Meteorologische Zeitschrift* N.F. 3, no. 6 (1994): 337–343.

26. R.A. Krause, "1895, Gründungsjahr der deutschen Südpolarforschung," *Deutsches Schiffahrtsarchiv* 19 (1996): 141–163. C. Lüdecke, "Scientific Collaboration in Antarctica (1901–1903): A Challenge in Times of Political Rivalry," *Polar Record* 39, no. 208 (2003): 25–48.

27. C. Markham, *Antarctic Obsession: A Personal Narrative of the Origins of the British National Antarctic Expedition 1901–1904,* C. Holland, ed. (Bluntisham Books, 1986, posthumously).

28. Lüdecke, "Scientific Collaboration," 37.

29. E. v. Drygalski, *The Southern Ice-Continent: The German South Polar Expedition Aboard the Gauss 1901–1903* (Bluntisham: Bluntisham Books and the Erskine Press, 1989). Translation from E. v. Drygalski, *Zum Kontinent des eisigen Südens* (Berlin: Georg Reimer, 1904).

30. Markham, *Antarctic Obsession,* 10.

31. Lüdecke, "Scientific Collaboration."

32. The four expeditions of the International Cooperation were guided by Drygalski (German South Polar Expedition, 1901–03), Scott (British National Antarctic Expedition, 1901–04), Nordenskjöld (Swedish South Polar Expedition, 1901–03), and Bruce (Scottish National Antarctic Expedition, 1902–04).

33. C. Lüdecke, "Die Routenfestlegung der ersten deutschen Südpolarexpedition durch Georg von Neumayer und ihre Auswirkung," *Polarforschung* 59, no. 3 (1989): 103–111.

34. C. Lüdecke, "Ein Meeresstrom über dem Südpol? Vorstellungen von der Antarktis um die Jahrhundertwende," *Historisch-Meereskundliches Jahrbuch* 3 (1995): 35–50.

35. Fridtjof Nansen (1861–1930) had planned a drift aboard *Fram* for his conquest of the North Pole (1893–96).

36. Drygalski, *Southern Ice-Continent,* 34

37. W. Meinardus and L. Mecking, "Atlas: Meteorologie," in E. v. Drygalski, ed., *Deutsche Südpolar-Expedition 1901–1903 im Auftrage des Reichsamtes des Innern* (Berlin: Georg Reimer, 1911), vol. 1, 12.

38. Drygalski, ed., *Deutsche Südpolar-Expedition 1901–1903,* 20 Bände, 2 Atlasse.

39. E. v. Drygalski, "Auf zum Südpol: Erinnerungen an Georg von Neumayer zu seinem hundertsten Geburtstag," *Pfälzisches Museum, 43–Pfälzische Heimatkunde* 22, nos. 5–6 (1926): 113.

40. E. v. Drygalski, ed., "Erdmagnetismus," *Deutsche Südpolar-Expedition 1901–1903* (Berlin: Georg Reimer, 1925), vol. V, (1). D.E. Yelverton, *Antarctica Unveiled: Scott's First Expedition and the Quest for the Unknown Continent* (Boulder: University Press of Colorado, 2000).

41. Meinardus and Mecking, *Atlas Meteorology,* vol. I, 11.

42. W. Meinardus, "Die mutmaßliche Höhe des antarktischen Kontinents," *Petermanns Geographische Mitteilungen* 9 (1909) part I: 304–309; 12 part II: 355–360.

43. W. Meinardus and L. Mecking, "Atlas: Meteorologie," in: E. v. Drygalski (Hrsg.), *Deutsche Südpolar-Expedition 1901–1903,* vol. I, 2–4.

44. This was similar to the procedures of the first meteorological congresses, which established a Permanent Meteorological Committee to run affairs in the times between the conferences. Later the Permanent Meteorological Committee became the World Meteorological Organization.

45. C. Lüdecke, H. Brogiato, and I. Hönsch, *Universitas Antarctica: 100 Jahre deutsche Südpolar expedition 1901–1903 unter der Leitung Erich von Drygalskis,* exhibition catalogue. (Leipzig: Institut für Länderkunde, 2001). A general recognition of Drygalski's achievements as leader of the last German expedition according to Humboldtian ideas, that is, the investigation of all four elements, is still missing: Earth investigated by the geologist and geographer, water investigated by the oceanographer, air investigated by the meteorologist, and fire as the working area of the geologist, who investigated the extinct volcano Gaussberg. Additionally the biologist investigated the living world. The later German Antarctic expeditions of Wilhelm Filchner (1877–1957) in 1911–12 and the "Schwabenland" expedition of 1938–39 were already specialized and did not perform the complete set of Humboldtian investigations.

# Chapter 10

# Sydney Chapman: Dynamo behind the International Geophysical Year

*Gregory A. Good*

The International Geophysical Year (IGY), a "mass attack" on the problems of global Earth science in the 1950s, is remembered in the early twenty-first century mainly by scholars and the dwindling number of participant-scientists still alive.[1] At the time, though, the IGY commanded the attention of the world. In 1957 and 1958, nearly 70 countries participated in the largest and most ambitious scientific collaboration the world had ever seen. Tens of thousands of scientists, engineers, and support personnel established stations and even small cities everywhere from Arctic ice floes and the South Pole to Pacific islands and—via the first human forays beyond Earth's atmosphere—to artificial satellites *Sputnik* and *Explorer*.

Likewise, Sydney Chapman, a dynamo who drove the International Geophysical Year (IGY) as president of the special committee that organized its activities, was famous at the time but is now largely forgotten. The military metaphor "mass attack" was Chapman's own: an army of scientists, led by a team of scientific generals. Nonetheless, when Chapman died in 1970, his obituary in the *New York Times* proclaimed him "the grand old man of geophysics."[2] The *London Times* called Chapman "a legend" and "an inevitable choice" to head up the IGY.[3] The London obituary continued to describe Chapman as "younger in spirit and pleasure" than much younger researchers and as frankly enjoying both his work and the reputation and rewards that came with it.[4] These strong judgments for a scientist now so little known beg explanation. That explanation requires a closer look at how Sydney Chapman earned his reputation, both as a geophysical researcher and as an organizer of international scientific collaboration on a grand scale. This essay tells how he earned these reputations as a tale of two "migrations," one a migration in scientific interests, the other a physical migration from England to the United States. Chapman migrated from engineering, to mathematics, to geophysics, to space physics, and from the constrained environment of an English university to the free-wheeling, well-funded world of American big science at research

**Figure 10.1** British scientist Sydney Chapman. (Courtesy of AIP Emilio Segrè Visual Archives, Physics Today Collection, American Institute of Physics, College Park, Maryland.)

universities and government institutions, such as the University of Alaska in Fairbanks and the High Altitude Observatory in Boulder, Colorado.

## Chapman's First Migration: From Engineering to Mathematics to Geophysics to Space Physics

Sydney Chapman did not set out to be a geophysicist. Born in 1888 to middle-class parents near Manchester, England, he did well enough in mathematics and science that he entered Manchester University on a competitive scholarship and graduated in engineering with first-class honors in 1907. He stayed and finished a degree in mathematics in 1908. Chapman recalled in a lecture he gave at Boulder, Colorado, in 1965: "By that time I found there was a career in maths & forsook eng'g wh had only been a way to use math'l ideas to earn a living. I went to TCC [Trinity College, Cambridge] on [Horace] Lamb's advice, by means of a schol'p & my father's help."[5]

The lecturer at Manchester, J.E. Littlewood, had impressed Chapman with pure mathematics—"I found his lectures splendidly vital—delta-epsilon work, math'l rigor." Littlewood, just graduated at Cambridge, insisted that if Chapman wished to be a professional mathematician, he needed a Cambridge degree. Horace Lamb, mathematics professor at Manchester, helped him get a scholarship.[6]

Chapman went to Trinity College in 1908. With degrees in engineering and mathematics already in hand, he dove into his studies and sat his mathematics exams after only two years instead of the required three. This earned him a B.A. in mathematics, a typical path for an exceptional English student at the time. Engineering was behind him: "I had by then acquired the hope of being a researcher in maths—my outlook was to become an assistant lecturer—what we'd here [in the United States] be called an instructor—in some provincial univy like Manchester."[7]

Yet Chapman was uncertain what kind of mathematician he wanted to be. While still an undergraduate at Cambridge, he researched nonconvergent series and published five mathematics articles, one with G.H. Hardy. Although Chapman read Karl Weierstrass, Ernesto Cesaro, and other mathematicians on rigor and a pragmatic approach to mathematical operations, he recalled later that he had not known whether to take up pure or applied mathematics. So he sought Professor Joseph Larmor's advice: "But I was not sure whether I wanted to be a pure or applied math'n, so I went to Larmor, whose prof'l lectures I had attended, & asked if he cd suggest a suitable subject of research for me."[8]

Larmor's advice started Chapman on a path toward an applied mathematical approach to physics, with a problem in the kinetic theory of gases. He investigated gas theory, the flow of rarefied gases in tubes, and thermal diffusion. He did not know it, but these theoretical exercises were preparing him for studies of Earth's upper atmosphere decades later.

In 1910, before Chapman had completed his Cambridge degree, the Astronomer Royal, Frank Dyson, came to Cambridge, spoke with Chapman, and chose him as a chief assistant at Greenwich Observatory.[9] Despite Chapman's lack of experience with geomagnetism, Dyson charged him with building a new geomagnetic observatory to replace the one built by George Biddle Airy more than 60 years earlier. He was to modernize the magnetic instruments and observing program. Chapman familiarized himself with up-to-date magnetic instrumentation and participated in magnetic measurement and some astronomical observation. He admitted: "I had never thought of becoming an astronomer and knew very little about it.... I knew even less about the earth's magnetism than I did about astronomy at the time, and had to learn something about it."[10] During his first years at Greenwich, Chapman published on chronometers, photometers, and other observational matters, and he continued to publish on gas theory and pure mathematics.[11] Dyson gave Chapman and the other chief assistant, Arthur Eddington, a start toward scientific careers by encouraging their research and by introducing them to well-known astronomers, such as George E. Hale and Edward C. Pickering.[12] By 1913 Chapman became a fellow of the Royal Astronomical Society, based on his photographic photometry of circumpolar stars, and joined its council that same year.[13]

Chapman and Eddington shared intense interests and commitments. They enjoyed long walks and discussions together. They both loved mathematics and thought in cosmic terms. They were also both pacifists, but because England was at war by 1914, this made them extremely unpopular. Chapman's strict Baptist upbringing and Eddington's faith as a Quaker caused problems for both of them. The Astronomer Royal and other prominent elder scientists rose to their defense. Eddington stayed out of jail (charges had been considered) and Chapman lost his lodgings as a leader with a Boys' Brigade, but he maintained his research position at Greenwich.[14]

The new colleague who most directly impacted Chapman's future career while at Greenwich was the physicist and magnetic researcher Arthur Schuster, who sat on the visiting board of the observatory. Schuster had published several highly influential articles on geomagnetic theory: one in 1889 in which he used the Gaussian method of spherical harmonic analysis to prove that the daily variation of Earth's magnetic field is due to electrical currents in the upper atmosphere. And in 1911 and 1912, just as Chapman began working at Greenwich, Schuster published on the causes of magnetic storms and the main magnetic field of Earth.[15] Chapman reflected later that Schuster's "brilliant sorties" and "striking theoretical conclusions" had inspired him.[16]

Under Schuster's influence, and as chief of magnetic research at Greenwich, Chapman also began to "migrate" toward geomagnetic theory. He said Schuster encouraged his first geomagnetic publications (1913 and 1914) on the influence of the Sun and Moon on Earth's diurnal magnetic variation.[17] Sensing, however, that astronomy would not let him concentrate as much as he wished on gas theory and geomagnetism, Chapman left Greenwich in 1914 to begin an academic career. He lectured at Trinity College, Cambridge, from 1914 to 1916, then returned to conduct his own research at Greenwich until the end of the war.[18]

During this period Chapman concentrated on gas theory and a theory of lunar-induced atmospheric tides and their effects on Earth's magnetism. He based the latter on more than 60 years of routine records compiled at Greenwich since 1854, revealing his simultaneous interest in mathematically elegant theory and thorough empirical evidence.[19] He migrated further into geomagnetism with a theory of magnetic storms in 1918.[20] By 1919 Chapman had published 49 articles. He researched gas theory, the effect of the Moon on the upper atmosphere, and several magnetic phenomena, all with one common denominator: He treated them as problems in applied mathematics. In 1919 the Royal Society named him a fellow at age 31.

The notion that a scientist's research interests "migrate" during a career is central to understanding the history of inherently complex research areas outside the traditional disciplines. Such research areas, and such behavior by scientists, characterize large parts of twentieth-century science. For mid-century, Edge and Mulkay have told this story for radio astronomy, Hufbauer for solar studies, and DeVorkin for upper atmospheric research.[21] Chapman provides an example of such migration several decades earlier.

When Chapman began his geomagnetic research, most activity in this field consisted of routinized, programmatic measurement of Earth's magnetic field elements, tied either to observatories or to expeditions. Exemplary investigations continued long series of measurements at undisturbed locations or increased the spatial density of global mapping. Some researchers, such as Adolf Schmidt in Germany and Louis Bauer in the United States, who held out long-range hopes for advances in theory, in fact had their greatest achievements in empirical work.[22] Schuster provided Chapman with the direct model of how a mathematical physicist, or an applied mathematician, might approach geophysical research more strategically. Chapman attacked geophysical problems through analysis of long-running data sets and by application of physical law to complex phenomena.

With the end of World War I, Chapman returned to university employment. He taught in turn at the University of Manchester (1919–24) and at Imperial College from 1924 until called away by war work in World War II. He replaced Lamb at Manchester as professor of mathematics, and while at Imperial he became the head of the Mathematics Department. Nevertheless, his interest intensified in gases, ionization, and electrical and magnetic properties at high altitude. Chapman applied basic research on gases and on ionization to the upper atmosphere, and in particular he proposed in the 1920s theories of electrical currents in the upper atmosphere.[23] Throughout the 1930s he continued to refine his theoretical treatment of these and other atmospheric and geomagnetic questions. In addition to his theoretical work, Chapman demonstrated his administrative skills at Imperial College, taking a badly divided mathematics group and gradually reforming it into a smoothly functioning unit.[24]

Until the 1920s very little solid theoretical work had been undertaken in geomagnetism and the upper atmosphere, fields characterized more by extensive data-gathering. Chapman earned his research reputation largely by developing mathematical-physical theories in these fields. He outlined his approach to research in a letter to Adolf Schmidt in 1929:

> I have had a busy year since hearing from you, having been working on a field new to me—the sun's magnetism—& having been led to doubt the Stewart-Schuster theory of terrestrial solar-daily variation. I have continued to work on the theory of magnetic storms, but I find that a mathematical error vitiates the last paper I wrote with Mr T T Whitehead, & the large decrease of H.F.

[horizontal geomagnetic force] during storms cannot be explained by induced currents. I am forced back to believe in an extra-terrestrial current such as Stormer [*sic*] proposed, but have not yet found it possible to prove that such a current could exist. I have spent two years in studying the motion of a nearly neutral ionized stream of solar particles, in the earth's field, but the problem is extraordinarily difficult, & I have not made very great progress. Terrestrial magnetic investigation is indeed very difficult.[25]

Chapman imagined physical models of both diurnal magnetic variation and of magnetic storms, elaborated the models using Maxwell's equations, and compared accumulated measurements with his theoretical calculations. When, in his judgment, they did not sufficiently match, or when physical or mathematical errors appeared in the models, Chapman started over with an adjusted model. This letter illustrates how deeply he devoted himself to theoretical geomagnetic research in the 1920s. It also illustrates his migration in search of research problems amenable to the tools he had developed.

Chapman's son, Cecil, recalled in a 1999 interview that his father often sat, engrossed in large tables of data, writing out numbers on special cards with many ruled lines. He also supervised a group of women computers in data analysis.[26] He continually published analyses of observational data in this period, especially regarding the effects of the Moon on the ionosphere. His 17 empirical studies of "atmospheric tides" culminated in his 1939 Presidential Address to the International Association of Meteorology in which he surveyed the theory of such tides.[27] Chapman related that from 1918 to 1930 he and his assistants based their manual calculations on three million observations. After 1930, when he gained access to "a set of Hollerith punching, sorting, and tabulating machines," the number of observations increased by five million.[28] Chapman broadened the number of stations for observing the global atmospheric tide and refined the analytical methods. This project represented a massive effort over two decades, drawing on data reaching back nearly 100 years. He referred to it as equivalent to a search "for a needle in the great haystack of old meteorological records."[29] His motivation was to understand better the effects of the Moon on air in the upper atmosphere, "where the air is ionized and therefore electrically conducting, acts like the motion of the armature in the magnetic field of a dynamo; electric currents are generated, because the air is moving in the field of the great magnet, the Earth itself."[30] He also noted that the lunar air-tide affects the elevation of the E-layer of the ionosphere, which he said was discovered by Edward Appleton's team by bouncing radio waves off the layer. Chapman said he conducted these studies "because of their relation to geomagnetism," not from an interest in meteorology.

This study of lunar air-tides was not Chapman's only research in the 1920s and 1930s requiring long data series, extensive computation, and clear theoretical insight. In 1929 he and V.C.A. Ferraro published on "the electrical state of solar streams of corpuscles" and, in 1932, "a new theory of magnetic storms."[31] These innovative studies of the effects on Earth's near-space environment of ionized streams of particles from the Sun attracted little attention until the 1950s, but Chapman was patient. Likewise, his 15-year-long collaboration, starting in 1925, with the German geophysicist Julius Bartels resulted in the monumental, two-volume, 1,049-page *Geomagnetism* in 1940. This epochal book provided a compendium of the great variety of empirical studies of Earth's magnetism to this time and provided the starting point for many future investigations.[32]

In a notable diversion from his geomagnetic research in this period, Chapman published early ideas on the stratospheric ozone cycle in 1929 and 1930.[33] He continued this work in 1931, adding an influential theory of ionospheric layering.[34] Chapman pulled together results from diverse research programs to provide, for the first time, a unified vision of the structure and properties of the upper atmosphere. He wrote:

> Known facts relating to the upper atmosphere have increased rapidly in recent years, through observations of many kinds—concerning meteors, the absorption and emission of light by the atmosphere, the propagation of radio and sound waves, and the daily and irregular changes in the earth's magnetism. . . . I present a combined picture of many of the phenomena, in the form that now seems to me most probable, though without concealing the uncertainties affecting many points.[35]

These investigations foreshadowed his future focus on "aeronomy," a neologism he introduced in 1945.

Perhaps more important than his considerable productivity in research, however, was Chapman's growing participation in scientific organization on both the national and international levels. His involvement in governance at the Royal Astronomical Society from 1913, and his close association with Schuster, Dyson, and Hale, opened this door. During World War I, these scientists planned toward the reorganization of international scientific bodies at war's end.[36] As part of this process, in 1916 British scientists started considering the organization of a national geodetic institute. By 1917 H.H. Turner, professor of astronomy at Oxford, wrote to Colonel E.H. Hills (geodesist, War Office, retired) that adding a few names to the organizing committee, such as Charles Chree (Kew Observatory) and Chapman, would make this a broadly geophysical proposal.[37] Chapman thus became a founding member in 1917 of what became the British National Committee for the International Union of Geodesy and Geophysics (IUGG).[38] He remained on the National Committee into the 1950s. He sat on Great Britain's subcommittee for terrestrial magnetism in the 1920s with Astronomer Royal Dyson, Director of the Meteorological Office Chree, W.M.H. Greaves (at Greenwich), Schuster, Sir Napier Shaw, and C.T.R. Wilson.[39] He joined an IUGG committee formed in 1927 to produce a photographic atlas of auroral forms, an early step for him into a new research area. From 1924 to 1933 he was president of the Mixed Commission on Solar and Terrestrial Relations of the International Research Council.[40] During the second International

Polar Year (IPY) of 1932–33, Chapman represented the British National Committee in the international discussions of organization. He presided over the international organization related to geomagnetic work within IUGG from 1936 to 1948, and as such was a member of the IUGG executive committee. He was no stranger to the alphabet soup of international scientific organization.

Chapman made especially important visits to the Carnegie Institution's Department of Terrestrial Magnetism (DTM) in Washington, D.C., starting in the early 1930s. During three decades of expeditions at sea and on land, DTM had accumulated the greatest storehouse of data concerning Earth's magnetism.[41] From 1935 to 1942, DTM appointed Chapman a "research associate" and he visited frequently, collaborating on ionospheric and geomagnetic research with A.G. McNish, Harry Vestine (his former PhD student at Imperial College), and Lloyd Berkner. The magnetic, electric, and radio research that DTM conducted at its observatories in Peru and Australia bore directly on Chapman's ideas about the ionosphere.

The 1930s were heady times, but hard times, for Chapman and others who investigated global phenomena. New technologies became available. Berkner perfected his "ionosonde," a tunable shortwave transmitter for echo-sounding ionospheric "layers." Berkner based this ground-based instrument on the one used by Merle Tuve and Gregory Breit in 1925 to determine the height of the ionosphere by the bouncing of radio waves. Chapman and others pursued theoretical directions related to electrical currents in the upper atmosphere, layering of the atmosphere, and even took some tentative steps toward a theory of Earth's main magnetic field. But the promise of international cooperative research was stalled by economic depression, the growth of fascism, and ultimately the outbreak of World War II.[42] The second IPY struggled for lack of sufficient funds, and several decades passed before its magnetic data were incorporated into a useable system of sharing. Chapman tried to encourage collaborations: for example, while at the 1936 IUGG assembly in Edinburgh, he asked Latvian-born scientist L. Slaucitajs to join in a global analysis of the disturbed geomagnetic field, that is, the effect of the Sun on Earth's magnetism.[43]

But more than others, Chapman recognized the storm clouds gathering in Germany, even before Adolph Hitler's rise to full power. In 1932, on a bicycling tour of Germany, a favorite pastime of Chapman's, he was disturbed to hear dark conversations among young German men in the youth hostels.[44] Chapman helped Jewish scientists relocate to England and tried to find them positions and fellowships. Yet he worked with Bartels effectively throughout the 1930s. During the IUGG meeting in Washington, D.C., in September 1939, war broke out in Europe. Chapman and Bartels were finishing their book *Geomagnetism* at the DTM at the time. Bartels soon returned to Germany, where he was president of the Deutsche Geophysikalische Gesellschaft and a member of the National Socialist party. Chapman returned to England, where he took a leave from Imperial College and, for the duration of the war, worked with the War Department on incendiary bombing strategy and operations research—as did his well-known colleagues, Patrick Blackett and Edward (Teddy) Bullard. As deputy scientific advisor to the army council, Chapman was known to stand up to generals and defend the younger scientists.[45] And yet again, although Chapman abandoned his pacifism, he and Bartels continued to exchange page proofs and manuscripts via neutral Switzerland. Like many other scientists during war, Chapman did his best to support his country, but he saw through to publication his scientific collaboration with an enemy national. (It should be noted that Bartels found himself in a nearly symmetrical situation.)

In the 1930s, Dutch colleague and fellow researcher of fluids and plasmas, Johannes M. Burgers, had sat with Chapman on the Committee on Science and its Social Relations. This committee had been established by the International Council of Scientific Unions (ICSU), the major international body overseeing science.[46] Burgers later recalled Chapman's deep concern for social issues, especially world unity and freedom of scientists to communicate internationally. But, said Burgers, Chapman also "understood better what could be achieved and what should be left aside."[47] By 1940, Chapman was renowned not just for his scientific research, but for his tact and dedication in international collaboration.

After the war, Chapman lost no time returning both to his former scientific interests and to international organization. He moved from the Imperial College to the University of Oxford, where he became Sedlian professor of natural philosophy. He became president of the International Association of Terrestrial Magnetism and Electricity (IATME) from 1949 to 1951, and president of IUGG from 1951 to 1954. Chapman was clearly at the peak of his career, still publishing extensively and teaching advanced students, but he was restless at Oxford and sought to start a second career. As he saw it, that new career awaited discovery—or invention—in the United States.

### Chapman's Second Migration: England to America

The traditional starting point of any narrative of the IGY of 1957–58 is a dinner party hosted by James and Abigail Van Allen on April 5, 1950, in their home in Silver Spring, Maryland. Guests at the dinner included Lloyd V. Berkner, S. Fred Singer, E. Harry Vestine, J. Wallace Joyce, and Sydney Chapman.[48] The story holds that, during after-dinner conversation, Berkner suggested the time was right for a new IPY. One polar year had organized observations in the Arctic in the 1880s, another in the 1930s had increased the density of observation stations there and established a few in the Antarctic.[49] Berkner reportedly argued that because of rapid advances since the 1930s, especially in radio, rocketry, and ionospheric research, a 50-year interval was not needed, and it made sense to organize the next IPY for 1957–58. For one thing, this would be a period of maximum solar activity, an advantage for studying terrestrial effects produced by the Sun.

This story has been retold so often it now has the character of a creation myth. Chapman published it in *Nature* in 1953.[50] He repeated the story in 1955 in a fun little publicity piece in *Popular Mechanics*, "Mass Attack on the Earth's Mysteries." He related that this dinner was during his first trip to the United States after World War II. In typically modest Chapman style, he said he was privileged to be in company with a small group of American geophysicists hosted by the Van Allens. And he attributed the idea of a third polar year to Berkner, "one of the pioneers of ionospheric studies." The others present became, in his telling, enthusiastic supporters of the idea.[51]

Chapman neglected to mention that he was the special guest at the dinner, being the elder scientific statesman at 62, among relative youngsters Berkner (45), Vestine (44), Joyce (42), Van Allen (36), and Singer (26). Although he referred to the others as geophysicists, their employers and patrons indicate a definite context. Berkner had just finished a report, "Science and Foreign Relations," for the U.S. Department of State, in which he advocated using international scientific bodies to advance Western (as opposed to Soviet) values.[52] Historian Allan E. Needell argues that "science and national interest were both powerful motivations" for Berkner.[53] As executive secretary of the Joint Research and Development Board, Berkner worked closely with the U.S. army and navy as well as with the Central Intelligence Agency.[54] During the war, Van Allen had played an important role in developing the proximity fuse at the Applied Physics Laboratory (APL) and, from 1947 until the IGY, chaired the "V-2 Panel." He exercised significant sway over rocket-borne ionospheric research. Vestine worked for the Carnegie Institution's analysis of geomagnetic data, had conducted military research during World War II, and later in the 1950s joined the RAND Corporation.[55] Joyce was with the Naval Research Laboratory, and Singer was soon to move from APL to become the scientific liaison officer in the Office of Naval Research bureau in London.[56] Chapman was not meeting with just any group of American geophysicists. He wanted access to rockets and to a network of upper-atmospheric researchers. Berkner needed Chapman to lend his scientific prestige and appearance of disinterested curiosity to this projected international endeavor, with its implications for national security.[57]

This provides the context for understanding the involvement of English scientist Chapman in both a suburban D.C. dinner party and in the organization of an international scientific endeavor as well known for its diplomatic implications and cold-war imagery of *Sputnik* and the space race as it is for its scientific accomplishments. The political context frames Chapman's dedication to massively organized international collaboration and his sincere interest in understanding Earth's magnetism, electricity, and upper atmosphere. To achieve the latter goal, Chapman needed the international collaboration, and this could be achieved only within the political context of the cold war. The two partners who made the IGY possible were Berkner, who handled the politics behind the scenes and kept American interests foremost, and Chapman, whose calm, detached, gentlemanly demeanor always underscored the scientific character of the enterprise. The IGY required both

Berkner and Chapman, one for his invisible activities, the other for his public persona and diplomatic style.

Chapman's persona of the cheerful, inquisitive seeker of deeper understanding, as a person who could overcome personal or political differences to "get the job done," developed over decades. This trait is seen in his accommodation of pacifism and continued activity at Greenwich during World War I, and also in his scientific collaboration with Julius Bartels on the eve of World War II. Chapman, the mediator, was as important to the success of the IGY as was Chapman, the geophysicist.

All the while, Chapman's interests in Earth's magnetism, the upper atmosphere, and the aurora inclined him toward global studies, which benefited from broad international cooperative research. It is easy to see why he was a "person of interest" to Berkner and his colleagues, why they wanted him involved in organizing what became the IGY. The organizational challenges taken on by Chapman, Berkner, and the other organizers of the IGY required several different kinds of individuals.

Chapman's presence at the Van Allen dinner party in 1950 was far from an accident. As noted above, during the 1930s he had known all the principal players in this tale at DTM. The magnetic, electric, and radio research that DTM conducted at its observatories in Peru and Australia bore directly on Chapman's ideas about the ionosphere. Berkner later looked back on that time with Chapman:

> I first remember him in the early 1930s. He was visiting the Carnegie Institution of Washington, Department of Terrestrial Magnetism, where I was working on ionospheric research as a young physicist.... At the time, I was developing the first ionosonde. Sydney Chapman took keen interest in this development and in the first ionograms that it produced, for these were closely related to his great theoretical work on the layered absorption of ultraviolet in the ionosphere. From that time on, we became close friends. Often in the evening, Sydney and Julius [Bartels] would come to our home for dinner and talk—talk that ranged over much of geophysics. This was an incomparable experience for a young physicist.[58]

Chapman, likewise, was attracted to both the youth of American geophysicists and the resources they had at their disposal after the war. Oxford was comfortable, but Chapman knew he could never expect significant support there for the exploration of Earth's near-space environment. Increasingly, he recognized he would have to leave England. In fact, that evening at the Van Allen's house, Chapman was en route to Caltech for a year as a visiting professor. In June 1950 he attended the twenty-fifth meeting of the V-2 Panel in Boulder, along with Van Allen, Homer Newell, and others involved in planning upper-atmospheric research.[59] The panel, working with other groups, was planning an intensely collaborative "Temperature Day," or T-Day, for later in 1950 to measure critical atmospheric data at many elevations using radio, rockets, balloons, airplanes, and ground-based meteor observations. To help plan T-Day, Chapman attended an informal meeting in August and the twenty-sixth meeting of the rocket panel in September. The event came off well in December, and Chapman worked with others on the rocket panel to evaluate the results. The T-Day collaboration provided a dry run for the IGY.

Chapman returned to England after his year at Caltech, but in 1953 was back in the United States at the University of Michigan, then at the University of Iowa in 1954–55 (where Van Allen had moved in 1951), and also at the University of Minnesota. Each of these universities was active in upper-atmospheric research. Chapman's primary duties at each involved lecturing on the upper atmosphere and guiding advanced students in research. In 1953, just before he reached the mandatory retirement age of 65, Chapman resigned his position at Oxford and headed for the United States, to live there more or less permanently. His two longest associations were with the University of Alaska at Fairbanks (from 1951 on) and the High Altitude Observatory at Boulder, Colorado (from 1955 on). Wherever his wanderings took him, Chapman anchored in these ports. He generated a second full career west of the Atlantic Ocean after 1950, at a point when most people contemplate retirement.

For the first ten years of that second career, the IGY dominated Chapman's activity. The American geophysicists at the Van Allen dinner party knew how to get things done in the United States, and Chapman was the international statesman-scientist they needed to carry this scheme to the next stage. Few other European geophysicists had both the research reputation and the organizational ability that Chapman did. More important, he expressed the quiet self-assurance that international collaboration is about science, and that politics needed to be rigorously excluded. Of course politics could not be excluded completely, but Chapman behaved as if he believed it could be.

During the summer of 1950, Chapman and Berkner began working to promote the third IPY. They first went to the Mixed Commission on the Ionosphere meeting in Brussels, Belgium, in July. This commission included scientists from the scientific unions for geophysics (IUGG), astronomy (International Astronomical Union, IAU), and radio research (International Union for Scientific Radio, URSI). All three of these unions adhered to the ICSU. Edward Appleton, with whom Chapman had published several articles, chaired the Mixed Commission; he wrote a resolution supporting the idea of a third IPY and calling for the formation of a group to begin organizing the activities.

Getting this resolution approved was complicated. In September 1950 Chapman and Berkner argued for it at the URSI and IAU general assemblies. In January 1951 they carried the resolution to the ICSU bureau, which decided it would have to go to a higher level within ICSU in October 1951.

Meanwhile, Chapman guided the proposal through the IUGG general assembly in August and September 1951. He saw it as critical that there be a "Polar Year Special Committee" in charge of organizing the activities, and that it answer directly to the highest level of ICSU, not to the individual unions. But he thought it also important that the individual unions and their subdivisions be consulted. As vice president of IUGG and president of the part of IUGG dealing with geomagnetism and the upper atmosphere, Chapman made sure the IPY featured prominently on the agenda at the IUGG general assembly. Consultation, however, opened the door to dissension. At that meeting, Danish scientist J. Egedal strongly argued for equatorial ionospheric research in the "Year" and for a delay until 1958–59 to take better advantage of the solar maximum.[60] In fact, he proposed an "Equatorial Year"! To avoid a crack in the unified support of the IUGG for the IPY, Chapman referred Egedal's proposal to the union's own "Third Polar Year Committee," stressing "that the action did not bind the special representatives in any way and that they should adopt a course of action justified by a careful review of all pertinent factors available to them."[61] Finally, on August 29, the geomagnetic division passed a resolution to be forwarded to the full IUGG and then to ICSU, endorsing the third IPY. Only Egedal and one other delegate expressed regret that their points had not been incorporated.[62] With agreement of IUGG, IAU, and URSI, Chapman and Berkner could now approach final authorization by the parent organization, ICSU.

Each disciplinary union included a wide array of research interests, but in some matters ICSU exercised authority over all the unions. ICSU leaders invited advice and consent from the astronomers, geophysicists, and radio researchers, but ultimate ratification of the special committee was up to ICSU. In October 1951 the leading officers of ICSU met in Washington, D.C., and endorsed the idea of a committee dedicated to organizing and implementing a third IPY.[63] Chapman, attending as the new president of IUGG, emerged successful.[64]

In 1952 the circle of interest broadened beyond astronomers, geophysicists, and radio scientists. The World Meteorological Organization (WMO)—an institution with roots in the nineteenth century, but not affiliated with ICSU—expressed interest in collaboration. This was appropriate, since both the first and second IPYs had involved meteorological organizations. More scientists also chimed in, including Chapman, that upper-atmospheric phenomena the world over should be included in the planned activities. Chapman proposed that this widening global interest required a more general title. Chapman suggested the third IPY be renamed the IGY, or Année Géophysique Internationale (AGI).[65] In October 1952 the ICSU general assembly and executive board made the name-change official. Hence, the Special Committee for the AGI became "CSAGI," referred to as the Special Committee.

In 1952 ICSU appointed Col. E. Herbays as convener of the Special Committee. Herbays was a Belgian engineer prominent in radio research and the Secretary General of URSI. Each union appointed a member of the Special Committee and an alternate. Chapman was not initially on the Special Committee, perhaps because he was president of IUGG. The Special Committee concentrated at first on ionospheric research, but quickly broadened to include latitude-longitude variation and "one or more specific [meteorological] problems which offered promise of early solution."[66]

The Special Committee met informally for the first time in October 1952. Present were a representative of IUGG (J. Coulomb), URSI (Berkner), IAU (Marcel Nicolet), convener Col. Herbays, and ICSU administrative secretary Dr. R. Fraser. The Special Committee began taking specific actions to organize the IGY and motivate scientists to take action in their own countries and research areas. They contacted all the national committees adhering to ICSU and invited interested unions to develop research programs. They planned to contact the United Nations Educational, Scientific, and Cultural Organization (UNESCO) for funding a central office, appointed several provisional officers, and proposed an organizational meeting for 1953. They also discussed how to organize the committee's work.[67]

Chapman and Berkner exchanged views frequently in the lead-up to the first official meeting of the Special Committee in June 1953.[68] Berkner, in New York, suggested Herbay and J.W. Joyce as possible alternates for administrative secretary and stressed that he, Berkner, would be a bad choice. He would draw too much attention to the American role. He suggested that Brussels would be the best location for an office for the secretariat, with possibly a second office in Washington, D.C., if Joyce became secretary. Chapman and Berkner agreed that no matter who was chosen, the person should be "a top-notch man from the scientific point of view." They also agreed that the actual organizing of scientific programs should be delegated out to the respective unions and their sections. They considered the idea that some members of the Special Committee be chosen geographically, specifically, from western Europe, the Soviet Union, India, or Japan. As for the presidency of the Special Committee, Berkner wrote: "Concerning the presidency of the Commission, I believe that in the formal sense this is a matter requiring the action of the [interim committee]. In the practical sense, however, this matter has been settled, since at its provisional meeting [the interim committee] advised the ICSU that if you were named as a member of the Commission you would be elected as its president. Any further action, therefore, is simply a formality."[69] The formality occurred a month later.

During 1953 the Special Committee expanded several times, reaching 14 members, including Chapman and Herbay as representatives of ICSU. At a full CSAGI meeting that summer, the members elected Chapman president, Berkner vice president, and Nicolet secretary-general. Chapman orchestrated this event as a four-day working meeting. After the first, organizational day, the whole committee met along with another dozen scientist-advisors to develop a draft program that could be used to rally interest and inform possible participants. Eleven (later thirteen) working groups developed research programs and drafted resolutions to guide the development of the IGY. Chapman sat on five of these working groups.[70]

From 1953 until the opening of the IGY in July 1957, organizers and researchers met many times, many special committees and working groups formed and dissolved, and gradually a research program and a management system developed. The national committees, set up in each participating country, decided how much of the program and which parts of it they could implement. Most of the detailed work was done in smaller meetings, and meetings of the full Special Committee took place only five times: June–July 1953 (Brussels), October 1954 (Rome), September 1955 (Brussels), September 1956 (Barcelona), and July–August 1958 (Moscow). One dramatic resolution emerged from the Rome CSAGI meeting: the recommendation to launch small satellites carrying scientific instrument payloads.[71] Almost exactly three years later, in October 1957, the Soviet Union launched *Sputnik*.

Chapman had a special hand in organizing the program for auroral investigations. In December 1953, he drafted a ten-point plan that included recommended methods for visual observation, the encouragement and guidance of amateur groups, the recruitment of university physics departments, and the development of an "adequate warning system" of approaching disturbances from the Sun. Chapman suggested: "Propaganda should be used to interest astronomers in making auroral observations in regions where they occur seldom."[72] Typical of Chapman, he made certain that research plans included the processing of historical auroral data. As he noted in a 1955 lecture, historical records—"often buried in local jr'nls"—stand in as proxy data for magnetic storms in periods before researchers have observed them directly.[73] Chapman stayed in contact with other working groups and offered ideas to many of them.

In 1954 Chapman proposed that ICSU create an Advisory Council for the IGY with one representative from each country participating in the IGY to advise the Special Committee on nontechnical matters. He intended this council to promote bilateral cooperation among countries that might otherwise be unable to conduct research in particular topics: loaning of instruments, setting of data archives, and arrangement of regional conferences among national representatives. Chapman also meant to use the Advisory Council as a forum for political discussions, thus insulating the Special Committee and its executive powers somewhat more from politics.[74]

Chapman was as serious a student of the international organization of science as he was of geophysical phenomena. In a lecture to the State University of Iowa Physics Department in 1955 he traced the intensification of international cooperation in science to Carl Gauss's Magnetic Union in the 1830s, Matthew Maury's Maritime Meteorology conference around 1853, and similar efforts, nearer 1900, in weights and measures, geodesy, and other sciences. He pointed to the International Association of Academies, founded in 1899, as "a more general bond of int¹ cooperation" but hindered by "excessive influence of German Acads."[75] The historical sketch, however, merely prefaced Chapman's detailed account of the unions he was coordinating in the buildup to the IGY. He noted the Soviets had not yet joined ICSU, but it had joined the unions for astronomy and chemistry. Chapman noted that although ICSU had the power to accept or reject a country's application to join, in fact it had never rejected any nation.

Underneath Chapman's 18 pages of detailed dissection of the functioning of international scientific unions ran a current of tempered optimism about the possibility of real international cooperation. He was not naïve about the ease of cooperation or of a total divorce of science from politics, but in the rhetoric of internationalism and the institutional leverage of ICSU and its unions, Chapman saw tools that nevertheless could be used to advance scientific research. He claimed that because science is "non-national," international cooperation in science is "natural," and in some sciences is "necessary." Astronomy, geophysics, and radio science, he said, needed international unions, because in them coordinated work was more important.[76] He saw ICSU support of the IGY as essential: "a small sum that catalyses immense cooperative national activities." Moreover, Chapman considered that the unions specifically encouraged cooperation "between E & W" and placed control in the hands of scientists.

Of course Chapman realized politics could not be kept at bay. Tensions between East and West, and especially between the United States and the Soviet Union, necessarily colored all interactions. The Soviet Union did not formally announce its intention to participate in the IGY until October 1954. As Chapman had noted in a letter to Van Allen a few months earlier, he was considering organizing a meeting in "Russia" "if this were found necessary to get Russian cooperation.... It is possible that the question of Russian participation in the IGY may be raised at the meetings of the Foreign Ministers

**Figure 10.2**  Sydney Chapman explaining a key point in science. (Courtesy of AIP Emilio Segrè Visual Archives, Physics Today Collection, American Institute of Physics, College Park, Maryland.)

now being held in Berlin."[77] Given the vast expanse of the Soviet Union, Chapman clearly desired its cooperation, no matter the tensions.

Chapman's more serious problem came once the Soviet Union joined in. Soviet scientists knew that power and prestige rested in having membership on the core group of the Special Committee, called "The Bureau." The Special Committee included representatives of all the unions participating in IGY, while the Bureau was restricted to the main officers of the Special Committee. Chapman addressed the issue at the meeting of the Special Committee at which Vladimir Beloussov announced Soviet participation. Chapman mildly chastised the Soviet Union for not acting earlier when research plans were still being formulated. Overtures made through official channels had produced no result (for example, the Soviet Union did not respond to invitations), and thus the seats that Soviet scientists could have occupied were given to members of other nations. Chapman carefully explained that "members of CSAGI [the Special Committee] have been chosen on a functional, not a national basis—they represent international bodies concerned with particular branches of science, and not the nations of which they are members."[78]

Chapman regretted that the Soviet Union had not presented a national committee report as all the other countries had done, but he said he expected it would be coming along soon enough.

By the 1955 meeting of the Special Committee in Brussels, two Soviet members were appointed to CSAGI: V.V. Beloussov and N.V. Pushkov. In 1957 ICSU appointed Beloussov and French scientist J. Coulomb to the CSAGI Bureau—that is, the small, inner circle—along with Chapman, Berkner, and Nicolet.[79] Chapman took care, however, that they and everyone else understood they were not national representatives, but disciplinary representatives, "a point perhaps not clear" to the Soviet members.[80] Chapman directed as much attention as possible to building structures that at least *seemed* to disregard politics. He emphasized problem-area working groups: "World Days," meteorology, geomagnetism, aurora and airglow, ionosphere, solar activity, cosmic rays, longitudes and latitudes, glaciology, oceanography, rockets and satellites, seismology, and gravimetry. Regional groups were defined as Antarctic, Arctic, Equator, Meridian 10° east, and others. The symbol of the IGY—a sphere with latitude and longitude lines, day and night, the South Pole exposed, and a satellite circling—purposefully omitted any continents or countries: Earth was an abstracted planet.[81]

The fourth and last full meeting of the Special Committee before the launch of the IGY, now much expanded to 39 members—and rather unwieldy—convened in Barcelona, Spain, in September 1956. Alongside this, to complicate matters more, the Advisory Council also met. The most political announcement came from I.P. Bardin, Soviet member of the Special Committee, that the Soviet Union intended to launch satellites during the IGY for studying the atmosphere, cosmic rays, and other phenomena. A thematic side-conference on rockets and satellites devoted several days to the many technical issues raised by this new technology.[82] This became contentious, as some attendees saw the side-conference as an outside issue for all but two countries present. Equally contentious issues related to funding, subsidies of the Special Committee and of poorer countries, and the location of World Data Centers, which would archive data gathered during the IGY. The French delegation to the Special Committee lodged a formal complaint that, "The discussions at Barcelona were often confused, sometimes even agitated." Father Pierre Lejay argued that too many decisions were being made by the small group of Chapman-Berkner-Nicolet, although he couched this in neutral organizational terms. He also argued that basic structural changes were needed in both the Special Committee and the Advisory Council, and that these changes required a formal statute. The French National Committee members, Lejay noted, expressed concern that they could not examine the IGY budget.[83] Also joining in the criticisms was Sir Harold Spencer Jones, general editor of the IGY publications and not an outside, hostile party. Spencer Jones listed as problems the lack of minutes of the previous meeting, insufficient time for the working groups, and lack of time in plenary sessions for discussion. He wrote Berkner: "There was a widespread feeling, which was openly expressed, that the meeting in Barcelona had accomplished very little."[84] Spencer Jones's solution was to "look to the future" and enlarge the central group. Among those rising to the defense of Chapman and Nicolet, at whom most of the criticism was aimed, were Berkner and Joseph Kaplan, of the U.S. National Committee. The IGY, according to Kaplan, was a fast-moving program and required delegation of responsibility: "The [Special Committee] is faced with a major but relatively short-range actual research and operations effort, calling for an organization capable of acting effectively and in time."[85] All in all, Barcelona was a bit of a brawl, and Chapman, Berkner, and Nicolet could not exert much control over events. There were too many new players with little understanding of decisions already made and work already done.

Nicolet, Berkner, and Chapman jointly deliberated many organizational questions. In a May 1957 letter Berkner pushed hard for the widest availability of data generated by the IGY.[86] He distinguished between the level of organization needed to bring about the IGY and the level necessary to promote use of the data afterward: "I have a great fear for the perpetuation of unnecessary organization, and it seems to me that the whole special committee [CSAGI] with its 24 members, etc., is not the sort of organization that will be needed to effectuate the scientific analysis of the IGY data. The cause of Geophysics would best be served by widening the base of responsibility and enthusiasm for the IGY as widely as possible."

To continue the bureaucracy of the Special Committee, he argued, would quell the enthusiasm that had been growing for sciences of Earth. Berkner favored encouraging more responsibility for studying the data among the different scientific unions.

In 1957, in notes for a letter to Col. G. R. Laclavere, secretary general of IUGG, Chapman recorded some of his thoughts on publication of the IGY results: "The Association committees [in the unions] will be dealing with questions of publications, with the study of the general or selected IGY data in their own fields, and with the satisfactory operation & the arrangements for their sections of the World Data Centers.... These are my personal thoughts on the matter; please adopt or refer to them as you think fit in preparing the agenda [for the Toronto CSAGI meeting]."[87]

Chapman elaborated on these organizational questions in a letter to Berkner on May 17, 1957.[88] Just as the IGY neared its opening, the two organizers looked ahead to what would happen after it closed. Chapman agreed with Berkner that CSAGI should wind up quickly, but that its Bureau (Chapman, Berkner, Nicolet, Beloussov, and Coulomb) could continue, directed by the general secretary of IUGG, to coordinate with the unions, the WMO, and "our parent body ICSU."

On the eve of the IGY, Chapman and the Special Committee faced one last political problem, already well-covered here by Wang and Zhang and offered now from Chapman's viewpoint. Throughout 1956 and the first half of 1957 Chapman's most difficult problem involved the participation of Taiwan and the People's Republic of China (PRC) in the IGY. The PRC expressed its interest earlier, had formed a national committee, and was moving ahead with both research plans and coordination with international bodies.[89] Then, in 1956, the Nationalist Chinese government in Taiwan (Formosa) announced its intentions. The government in Beijing objected strongly to Nationalist participation.

Berkner and Chapman were concerned that the Nationalists might attend a western Pacific regional meeting of IGY members in Tokyo, just before the start of IGY Berkner wrote Chapman that they did not attend the Tokyo meeting "and so...no difficulties on that score arose."[90] Berkner spelled out his feelings in this letter. Both Chinas were moving forward, and he knew Taiwan had requested help from the U.S. National Committee "through non-diplomatic channels." Berkner expected the Taiwanese to develop an IGY program. Under this circumstance, he said, the Special Committee would have to "accept their program as part of the IGY work." Berkner seemed to desire participation by both Chinas. He entertained a hope that a change of language might keep the PRC in the IGY. "Perhaps too much emphasis is put on the word 'admission' to the IGY; really the word should be 'adherence.' Perhaps this might help overcome the problems at hand." His conclusion, though, was probably not comforting to Chapman. Berkner thought they should do "everything possible to prevent the Communist Chinese from leaving the IGY," but he clearly would sacrifice PRC participation if Taiwan came in.

For Chapman, who was as usual thinking of the broadest possible participation, the best result was to keep both Chinas in the IGY. He professed not to understand why this was even an issue. The IGY programs, he wrote to the Bureau of the Special Committee on March 15, were "public," and the Special Committee's duty was to promote these programs "in every country" by keeping them informed.[91] Chapman, too, sought a diplomatic way out. They had dodged a bullet in Tokyo only because Taiwan failed to attend. He saw a potential problem with the Advisory Council, which by definition included a member from each participating country. He wrote the Bureau that if the Advisory Council met again "the question of the presence of delegates from China and Taiwan would at once arise." Chapman recommended the council not meet, so that the question of the "admission" of Taiwan would not have to come up. This was an optimistic view.

Chapman's involvement in the "two Chinas question" began at the Barcelona Special Committee meeting in September 1956, where he met Cochin Chu (K. K. Tschu), whose PhD work Chapman had directed at Oxford. Chu, who was assistant secretary of the Chinese National Committee, asked Chapman what China could do to promote geophysics. Chapman suggested China join the geophysics union, and he believed he had arranged with officials to have the union issue an invitation.[92] Chapman learned in February 1957 that the invitation had never, in fact, been issued and reported his disappointment to Berkner. Chapman had asked the union officer, Laclavere, for an explanation, but Leclavere's answer that "there were difficulties, of which I would know the reason" only irritated Chapman. He clearly indicated to Berkner that he wanted China in the geophysical union.

Chapman did not have an easy time exerting leverage on the players, who were spread around the globe. Indeed, in May 1957, as the critical date approached, Chapman was in Fairbanks, Alaska, far removed from any political fulcrum point. He informed his main contacts, Berkner (New York City) and Nicolet (visiting at Penn State University), that it was time to act with regard to the PRC. Chu had been cordial but firm about Taiwan, Chapman said. Chapman encouraged Berkner and Nicolet to take his draft letter, rework it without consulting other members of the CSAGI Bureau (the Soviet and French members), and communicate it directly to Beijing with copies sent to Beloussov and Coulomb. Chapman's idea was to streamline the process in this most delicate negotiation and not let the Soviet or French CSAGI Bureau members complicate matters.[93]

Chapman simultaneously sought assurances from Taiwan. He asked Hugh Odishaw to visit Taiwan to "clarify the role and procedures" of the IGY.[94] Odishaw reported to Chapman that he and Alan Shapley had visited with about 20 geophysicists, assessed their research programs and capabilities, and then turned to the "question of formal participation of Taiwan in the IGY." They explained the "non-Governmental, non-political character" of the IGY and sought assurances that Taiwan would impose no "political conditions" on their participation. Odishaw wrote that he was convinced that Taiwan would not pose such conditions.

Chapman's most recent information from Chinese officials was also reassuring. He wrote Berkner and Nicolet: "I will try to avoid political discussion, as you advise."[95] Berkner was nevertheless uncomfortable with Chapman's effort to negotiate the participation of the PRC in the IGY. He especially wanted to be sure Chapman did not "in any way commit us with respect to the basic principle of permitting adherence of any national group to participation in CSAGI." That is, Berkner did not want Chapman making any promises regarding either the PRC or Taiwan. Berkner continued that this principle had underlain "our whole policy" and that it "must be maintained at all costs."[96] Chapman, on the other hand, thought he was close to an agreement with China. He wrote: "Hence I take it that the Gen. Sec. [Nicolet] will send a suitable letter of acceptance to Peking." Chapman was being hopeful.

Chapman's viewpoint on the PRC differed deeply from Berkner's—or at least his strategy differed. Chapman noted in his May 17, 1957, letter that he was troubled by Berkner's reference to "red" China in a speech he had given to the American Meteorological Society. Chapman thought the PRC was to report its meteorological results to the World Meteorological Organization (WMO) and that the WMO intended to accept them. He advised Berkner: "One may criticize one's own politicians on one's own ground, if conscience compels, but in our position vis-à-vis IGY even that is perhaps inadvisable, and still more so might it be deemed inadvisable to criticize the politicians of other cooperating countries."[97] Chapman had his political principles. But he felt that as chief organizers of the IGY, he and Berkner should avoid discussing political matters critically.

In that same letter Chapman clarified how he differed from Berkner on Taiwan and China, now that a conclusion was imminent.[98] Although he expressed his satisfaction that Taiwan's IGY effort looked good, he did not want formal recognition of Taiwan's IGY participation to derail the PRC involvement. He wrote: "The IGY stands to lose far more than it can gain if Peking withdraws in consequence of formal recognition of Taiwan participation." The choice was clear: Chapman much preferred gathering the results of geophysical research conducted throughout China over making a political point in the cold war. That is to say, Berkner preferred the non-communist Taiwan over the PRC. In the end, the PRC decided at the eleventh hour, just before July 1, 1957, not to participate formally in the IGY.

## Conclusion: The Two Careers of Sydney Chapman

Sydney Chapman combined two careers into one lifetime. This does not refer to a first career in England and a second one in the United States, although that point, too, can be made. Rather, this alludes to his simultaneous and lifelong careers as a research scientist and as a scientific organizer. This career dichotomy did not start for him with the IGY. From his earliest days at Greenwich Observatory, he had involved himself in collaborative work in astronomy and geomagnetism. Especially for a geomagnetic theoretician, the data had to come from around the globe. His first direct work as an organizer of international scientific governance came with the International Research Commission during World War I. From that point forward, Chapman only went deeper into international science. With time, his own research depended more and more on this sort of coordinated effort.

Chapman's activity in organizing the IGY was one step further in this direction. He was the behind-the-scenes organizer of overarching work schemes of the IGY, a complex set of activities involving half a dozen scientific

**Figure 10.3**  Sydney Chapman doing what he did best, sharing his vision of science in an inviting, humorous manner. (Courtesy of AIP Emilio Segrè Visual Archives, Physics Today Collection, American Institute of Physics, College Park, Maryland.)

unions, numerous working groups, and ultimately 66 countries. As president of the Special Committee, Chapman oversaw this all.

Chapman revealed himself to be an adept deployer of the language of international, disinterested science. Unlike Berkner, he did not work directly with national security agencies, such as the CIA, and in his rhetoric did not overtly support either side of the cold war. But he was not naïve. He knew well the potential uses of the ionosphere in warfare, such as the disruption of communication. He understood why both the U.S. and Soviet militaries were interested in radio echoes, radiation belts, and satellites. He also knew, however, that the people who controlled the use of rockets and the funds for near-space research held the key to further understanding of a range of phenomena that had steadfastly resisted all the investigative efforts of scientists up to the mid-twentieth century. Chapman willingly used the rhetoric and institutions of international scientific cooperation to turn these technologies toward achieving his own highest goals. Over the next decade Chapman continued his research of near-space and made important contributions to the explanation of aurora, of magnetic storms, and to other areas of geomagnetism. He died in 1970 in Boulder, Colorado, active until the end.

# Notes

1. Sydney Chapman, "Mass Attack on Earth's Mysteries," *Popular Mechanics*, 1955, 104: 107–112, 260, 262, 264, and 266.

2. "Dr. Sydney Chapman, 82, Dies; Geophysicist Led '57–58 I.G.Y.," *New York Times* (June 20, 1970): 29.

3. "Professor Sydney Chapman: An Outstanding Mathematical Physicist," *London Times* (June 18, 1970): 12.

4. Ibid.

5. Chapman Papers, University of Alaska, Fairbanks, Box 81, folder marked Index Nos. 158–171, "Science & Scientists: Recollections 1904," "NCAR Friday Oct. 8/65, 3.45 for 4–4.45 pm." This lecture is marked "168," an index number provided to these files by Chapman. Three typed sheets. Hereafter referred to as Chapman Papers.

6. Gregory A. Good, "Sydney Chapman," *American National Biography* 4 (Oxford: Oxford University Press, 1999), 716. T. G. Cowling, "Sydney Chapman, 1888–1970, Elected F. R. S. 1919," *Biographical Memoirs of Fellows of the Royal Society* 17 (London: Royal Society, 1971), 54.

7. Chapman Papers. A fuller version, based on an audio recording of the talk, was published as "Talk Given at the National Center for Atmospheric Research, 8 October 1965," in *Sydney Chapman, Eighty: From His Friends,* Syun Akasofu, Benson Fogle, and Bernhard Haurwitz, editors, (Boulder: University of Colorado, 1968), 159–170.

8. Chapman Papers, "Science & Scientists: Recollections 1904," 1965.

9. Chapman, in fact, returned from a long Sunday walk and found Astronomer Royal Dyson sitting in his room at the college. Dyson's visit was unannounced.

10. Chapman Papers, "Talk Given at NCAR," 1968, 168–169.

11. A complete listing of Chapman's publications is provided in Cowling, "Sydney Chapman," 74–89.

12. Chapman Papers, "Science & Scientists: Recollections 1904," 1965. Eddington, six years Chapman's senior, was a pacifist (as was Chapman) and was also a graduate of Manchester, where he too had studied with Schuster and Lamb.

13. Sydney Chapman and P.J. Melotte, "Photographic magnitudes of 262 stars within 25' of the North Pole., Monthly Notices of the Royal Astronomical Society, 1913, 74:40–49.

14. Chapman Papers, "Science & Scientists: Recollections 1904," 1965.

15. Arthur Schuster, "The Diurnal Variation of Terrestrial Magnetism, With an Appendix by H. Lamb. On the Currents Induced in a Spherical Conductor by Variation of an External Magnetic Potential," *Philosophical Transactions* 180A (1889): 467–518; "On the Origin of Magnetic Storms," *Proceedings of the Royal Society of London* 85 (1911): 44–50; and "Critical Examination of the Possible Causes of Terrestrial Magnetism," *Proceedings of the Physical Society of London* 24 (1912): 121–137.

16. Sydney Chapman, "Charles Chree and his work on Geomagnetism," *The Proceedings of the Physical Society* 53 (1941): 629–634.

17. Chapman Papers, "Science & Scientists: Recollections 1904," 1965.

18. Cowling, "Sydney Chapman," 54–55.

19. Sydney Chapman, "The Lunar Atmospheric Tide at Greenwich, 1854–1917," *Quarterly Journal of the Royal Meteorological Society* 44 (1918): 271–280.

20. Sydney Chapman, "An Outline of a Theory of Magnetic Storms," *Proceedings of the Royal Society of London* 95A (1918): 61–83; "The Energy of Magnetic Storms," *Monthly Notices of the Royal Astronomical Society* 79 (1918): 70–83; "Solar and Lunar Diurnal Variations of Terrestrial Magnetism," *Philosophical Transactions* 218 (1919): 1–118; and "Theories of Magnetic Storms," *Observatory* 42 (1919): 196–206.

21. David O. Edge and Michael Mulkay, *Astronomy Transformed: The Emergence of Radio Astronomy in Britain* (London: Wiley, 1976); Karl Hufbauer, *Exploring the Sun: Solar Science Since Galileo* (Baltimore, Maryland: Johns Hopkins University Press, 1991);

and David DeVorkin, *Science with a Vengeance: How the Military Created the U.S. Space Sciences after World War II* (New York: Springer, 1992). The ideas of inherently complex sciences in the movement of scientists from one research problem area to another are explored in Spencer R. Weart, *The Discovery of Global Warming* (Cambridge, Massachusetts: Harvard University Press, 2003, second edition, 2008) and in Gregory A. Good, "Magnetic World: The Historiography of an Inherently Complex Science, Geomagnetism, in the 20th Century," *Earth Sciences History* 26 (2007): 281–299.

22. See Gregory A. Good, "From Terrestrial Magnetism to Geomagnetism: Disciplinary Transformation in the Twentieth Century," in D.R. Oldroyd, ed., *The Earth Inside and Out: Some Major Contributions to Geology in the Twentieth Century,* (London: Geological Society Special Publications, 2002), 229–239.

23. Sydney Chapman, "Electrical Phenomena Occurring at High Levels in the Atmosphere," *Journal of the Institute of Electrical Engineers* 57 (1920): 209–222.

24. Cowling, "Sydney Chapman," 56.

25. Sydney Chapman to Adolf Schmidt, August 7, 1929, Das Archiv der Berlin–Brandenburgischen Akademie der Wissenschaften, Nachlass Adolf Schmidt. Chapman refers to Sydney Chapman and T.T. Whitehead, "The Influence of Electromagnetic Induction within the Earth Upon Terrestrial Magnetic Storms," *Proceedings of the International Mathematical Congress, Toronto* (1928): 313–317.

26. Cecil Chapman, interviewed by the author, July 28, 1999.

27. Sydney Chapman, "Tides in the Air (Presidential Address, Meteorological Association, IUGG General Assembly, Washington)," *Meteorological Magazine*, 1939, 74:273–281. This article was also published in *Procès-Verbaux des Séances de l'Association de Météorologie, II, Mémoires et Discussions* (Bergen: Imprimerie John Grieg, 1940), pp. 3–40.

28. Chapman, "Tides in the Air," 1940, pp. 8–9.

29. Chapman, "Tides in the Air," 1940, p. 36.

30. Chapman, "Tides in the Air," 1940, p. 36.

31. Sydney Chapman and V.C.A. Ferraro, "The electrical state of solar streams of corpuscles," *Monthly Notices of the Royal Astronomical Society*, 1929, 89:470–479; "A new theory of magnetic storms. Part I. The initial phase," *Terrestrial Magnetism and Atmospheric Electricity*, 1932, 37:147–156 and 421–429; and "A new theory of magnetic storms. Part II. The main phase," *Terrestrial Magnetism and Atmospheric Electricity*, 1933, 38:79–96.

32. Good, "From Terrestrial Magnetism to Geomagnetism," 2002, especially pp. 231–232.

33. Sydney Chapman, "On the Variations of Ozone in the Upper Atmosphere," *Beiträge zur Geophysik* 24 (1929): 66–68; "On the Annual Variation of Upper Atmospheric Ozone," *Philosophical Magazine* 10 (1930): 345–352; "On Ozone and Atomic Oxygen in the Upper Atmosphere," *Philosophical Magazine* 10 (1930): 369–383; "The Absorption and Dissociative or Ionizing Effect of Monochromatic Radiation in an Atmosphere on a Rotating Earth," *Proceedings of the Physical Society* 43 (1931): 26–45; and "The Absorption and Dissociative or Ionizing Effect of Monochromatic Radiation in an Atmosphere on a Rotating Earth. Part II. Grazing Incidence," *Proceedings of the Physical Society* 43 (1931): 483–501.

34. Sydney Chapman, "Some Phenomena of the Upper Atmosphere (Bakerian Lecture)," *Proceedings of the Royal Society of London* A132 (1931): 353–374. J. A. Ratcliffe gives a succinct overview of this important article in "Ionospheric Physics and Aeronomy," in Akasofu et al., *Sydney Chapman,* 27–30.

35. Chapman, "Some Phenomena of the Upper Atmosphere," 353.

36. Daniel J. Kevles, " 'Into Hostile Political Camps': The Reorganization of Science in World War I," *ISIS* 62 (1971): 47–60.

37. H.H. Turner (University Observatory, Oxford) to Col. E.H. Hills, July 10, 1917, in Papers of Sir Edward Bullard, Churchill College Archives, Cambridge, UK, Series B.1, folder "British Association Geodesy Committee/Correspondence 1916–1919

Relating to a Geodetic Institute in Great Britain." Bullard evidently collected letters and reports related to the founding of the National Committee on Geophysics.

38. Chapman appears frequently in the "Minute Books" of the British National Committee for Geodesy and Geophysics, which adhered to the IUGG, created in 1918. RSL. CMB.46, RSL.CMB.101a

39. Union Géodésique et Gèophysique International, *Deuxième Assemblée Générale, Réunie a Madrid du 1er au 8 Octobre 1924: Procès-Verbaux des Séances* (Toulouse: Imprimerie et Librairie Édouard Privat, 1925), 129–131.

40. Cowling, "Sydney Chapman," 73.

41. See Gregory A. Good, "Geophysical Travelers: The Magneticians of the Carnegie Institution of Washington," in P.N. Wyse Jackson, ed., *Four Centuries of Geological Travel: The Search for Knowledge on Foot, Bicycle, Sledge, and Camel* (London: Geological Society Special Publication 287, 2007), 395–408.

42. On some of the effects of the Great Depression on the funding of science, see Gregory A. Good, "The Rockefeller Foundation, the Leipzig Geophysical Institute, and National Socialism in the 1930s," *Historical Studies in the Physical and Biological Sciences* 21 (1991): 299–316.

43. Akasofu et al., *Sydney Chapman,* 89.

44. Cecil Chapman, interviewed by the author, July 28, 1999.

45. Akasofu et al., *Sydney Chapman,* 59–60.

46. ICSU replaced the International Research Council in 1932 as the encompassing international scientific organization. It later played a prominent role in the International Geophysical Year in the 1950s. For a detailed history of ICSU see: Frank Greenaway, *Science International: A history of the International Council of Scientific Unions* (Cambridge, UK: Cambridge University Press, 1996).

47. Akasofu, et al., *Sydney Chapman, Eighty,* 1968, p. 70. Burgers wrote a philosophical monograph, *Experience and Conceptual Activity* (Cambridge, Massachusetts: MIT Press, 1965) at the end of his career.

48. Chapman first told this story in print in "Mass Attack on the Earth's Mysteries," *Popular Mechanics* 104 (1955): 262. This story became canonical in Harold Spencer Jones, "The Inception and Development of the International Geophysical Year," in *Annals of the International Geophysical Year,* vol. 1 (London: Pergamon Press, 1959), 383. See also Fae L. Korsmo, "The Birth of the International Geophysical Year," *Physics Today* 60 (July 2007): 40.

49. See James Rodger Fleming, "Advancing Polar Research and Communicating Its Wonders: Quests, Questions, and Capabilities of Weather and Climate Studies in International Polar Years," (with Cara Seitchek), in *Smithsonian at the Poles: Contributions to International Polar Year Science,* I. Krupnik, M. A. Lang, and S. E. Miller, eds. (Washington, D.C.: Smithsonian Institution Scholarly Press, 2009), pp. 1–12. The First International Polar Year had two stations at low latitudes in the southern hemisphere, The Second International Polar year also attempted vertical observations as well and was temporally near the Byrd Antarctic expedition (although this was a separate effort).

50. Sydney Chapman, "The International Geophysical Year," *Nature* 4373 (August 22, 1953): 327–329. Chapman gives a quick overview of how he and Berkner carried this proposal through the various international scientific bodies in this article.

51. Chapman, "Mass Attack," 262.

52. Allan A. Needell, *Science, Cold War, and the American State: Lloyd V. Berkner and the Balance of Professional Ideals* (Amsterdam: Harwood Academic Publishers, 2000), especially chapters 11 and 12 "Berkner and the IGY" and "IGY Satellites and the Launch of Sputnik."

53. Needell, *Science, Cold War,* 299. Jeffrey T. Richelson examines the intelligence context of science in the United States in *The Wizards of Langley: Inside the CIA's Directorate of Science and Technology* (Boulder, Colorado: Westview Press, 2002). Chapter 1 discusses the immediate post-war period.

54. Needell, *Science, Cold War,* 299.
55. Gregory A. Good, "Ernest Harry Vestine," in *American National Biography* vol. 22 (Oxford: Oxford University Press, 1999), 343–344; and David Hounshell, "The Cold War, RAND, and the Generation of Knowledge, 1946–1962," *Historical Studies in the Physical and Biological Sciences* 27 (1997): 237–267.
56. Needell, *Science, Cold War,* 325 and following.
57. Although a wealth of correspondence survives for the mid-1950s and on, Chapman's earlier papers have disappeared. When he moved to the United States, he culled his previous records fully. Whatever he left in England is currently missing. On the V-2 Panel, later called the Upper Atmospheric Rocket Research Panel (UARRP), see DeVorkin, *Science with a Vengeance,* 167–192.
58. Quoted in Akasofu et al., *Sydney Chapman,* 68.
59. The panel's official name by this time was UARRP. DeVorkin, *Science with a Vengeance,* 287–292, 306.
60. IUGG, Association of Terrestrial Magnetism and Electricity (IATME), "Preliminary Transactions of the Brussels Meeting, August 21–September 1, 1951 (No publisher listed, no date [1951]), 5–6.
61. IUGG, IATME, "Preliminary Transactions," 6.
62. IUGG, IATME, "Preliminary Transactions," 18.
63. Spencer Jones, "Inception and Development of IGY," 383–384.
64. Spencer Jones, "Inception and Development of IGY," 383–384. See also Harold Bullis, *The Political Legacy of the International Geophysical Year, Prepared for the Subcommittee on National Security Policy and Scientific Developments of the Committee on Foreign Affairs, U.S. House of Representatives* (Washington, D.C.: Government Printing Office, 1973), 6–10.
65. Spencer Jones, "Inception and Development of IGY," 384. See also Nicolet, 1955?, Chapman Papers, Box 51, Folder 78.
66. Spencer Jones, "Inception and Development of IGY," 385.
67. Some of these letters are quoted in Spencer Jones, "Inception and Development of IGY," 386–390; Bullis, *The Political Legacy of the International Geophysical Year,* 8–10.
68. Six letters are in the Chapman Papers, Box 101, dated April 9 to June 18, 1953.
69. Berkner to Chapman, May 27, 1953, Chapman Papers, Box 101.
70. Spencer Jones, "Inception and Development of IGY," 389–391.
71. Spencer Jones, "Inception and Development of IGY," 403.
72. Chapman, "Thoughts on the Auroral Program for the International Geophysical Year," December 30, 1953, Chapman Papers, Box 81, Folder 1.
73. Chapman, "Auroras of the Past. Boulder, Jy 6/55; Nat Bur Stndrds. 3 p.m.," [typescript notes], Chapman Papers, Box 81, Folder 4.
74. Spencer Jones, "Inception and Development of IGY," 403–404.
75. Chapman, "Some Problems of International Cooperation in Science," [pencil notes], January 23, 1955, 18 pages, Chapman Papers, Box 81, Folder 4.
76. Chapman, "Some Problems of International Cooperation in Science." Page 11 of the notes indicates that Chapman reworked this material for a talk given the next day, January 24, 1955, to the local chapter of the Federation of American Scientists. For this talk he detailed more completely the structures and funding of the unions. This folder also includes a press release for the latter talk.
77. Chapman to Van Allen, February 4, 1954, Chapman Papers, Box 43, Folder 1137.
78. Chapman's speech is quoted in the report titled: "Meeting of the Special Committee of the International Geophysical Year, Rome, 30 Sept–4 Oct 1954," 29. His speech is on 17–19. This report is in the Chapman Papers, Box 53, Folder 93, and is stamped with the date 15 February 1955.
79. Chapman to Berkner, January 19, 1957, Chapman Papers, Box 52, Folder 82.
80. Chapman to Berkner, May 9, 1955, International Geophysical Year, 10.2: CSAGI General Correspondence, 1953–1960, National Academy of Sciences Archive.

81. See chapter 1 by Michael Dennis in this volume.

82. Spencer Jones, "Inception and Development of IGY," 410.

83. Father Lejay to Chapman, 8 October 1956, International Geophysical Year, 10.2: CSAGI General Correspondence, 1953–1960, National Academy of Sciences Archive. This three-page letter included a three-page draft of a reorganizational statute. Nothing came of this.

84. Spencer Jones to Berkner, October 16, 1956, International Geophysical Year, 10.2: CSAGI General Correspondence, 1953–1960, National Academy of Sciences Archive.

85. Berkner to Spencer Jones, October 24, 1956, and Kaplan to Chapman, November 30, 1956, International Geophysical Year, 10.2: CSAGI General Correspondence, 1953–1960, National Academy of Sciences Archive.

86. Berkner to Chapman, May 10, 1957, Chapman Papers, Box 52.

87. Chapman, notes, Chapman Papers, Box 52. This undated manuscript also discusses negotiations with the People's Republic of China, which places it in the spring of 1957, most likely May.

88. Chapman to Berkner, May 17, 1957, Chapman Papers, Box 52.

89. See Ronald E. Doel, "Constituting the Postwar Earth Sciences: The Military's Influence on the Environmental Sciences in the USA after 1945," *Social Studies of Science* 33 (2003): 635–666. Doel discusses the "two-Chinas" question more directly in: Ronald E. Doel, Dieter Hoffmann, and Nikolai Krementsov, "National States and International Science: A Comparative History of International Science Congresses in Hitler's Germany, Stalin's Russia, and Cold War United States," *Osiris* 20 second series (2005): 68–70.

90. Berkner to Chapman, March 8, 1957, Chapman Papers, Box 52, Folder 82.

91. Chapman to CSAGI Bureau, March 15, 1957, Chapman Papers, Box 52, Folder 82.

92. Chapman to Berkner, May 15, 1957, Chapman Papers, Box 52.

93. Chapman to Berkner and Nicolet, May 10, 1957, Chapman Papers, Box 52.

94. Odishaw to Chapman, May 13, 1957, International Geophysical Year, 10.2: CSAGI General Correspondence, 1953–1960, National Academy of Sciences Archive.

95. Chapman to Berkner and Nicolet, May 8, 1957, Chapman Papers, Box 52.

96. Berkner to Chapman, May 22, 1957, Chapman Papers, Box 52

97. Chapman to Berkner, May 17, 1957, Chapman Papers, Box 52.

98. Chapman to Berkner, Nicolet, May 17, 1957, Chapman Papers, Box 52.

# Chapter 11

# Lloyd Berkner and the International Geophysical Year Proposal in Context: With Some Comments on the Implications for the Comité Spéciale de l'Année Géophysique Internationale, CSAGI, Request for Launching Earth Orbiting Satellites

*Allan A. Needell*

## Introduction: Standard Accounts of the Origins of the IGY

In a book published in 1991, John Naugle, a former chief scientist at NASA, recited what has become the standard, and basically reliable, account of how the International Geophysical Year (IGY) was conceived. "In April 1950," Naugle wrote, "at a dinner party hosted by [James] Van Allen and his wife, Abbie, [Lloyd] Berkner suggested that the world's scientists organize a third international polar year to take place during the period of maximum solar activity expected during 1957 and 1958." He then added some details, slightly less accurate, but hardly misleading. "During the previous Polar Year (1932–1933)," he explained,

> Berkner had served as a member of Admiral Byrd's first Antarctic Expedition. In 1950, Berkner worked for the Department of Terrestrial Magnetism of the Carnegie Institution of Washington, where he conducted research in iono-spheric physics. He also had an international reputation for building scientific institutions and organizations. The dinner guests enthusiastically endorsed Berkner's idea and he and Dr. Sidney [*sic*] Chapman, another guest, promptly set out to make it happen.[1]

Actually, Berkner took part in the 1928 Byrd expedition, which took place before the 1932–33 International Polar Year; and much of the reputation Naugle attributes to Berkner in 1950 he would earn gradually during the decade that followed.

In any case, consistent with Naugle's account, by October 1952 the International Council of Scientific Unions (ICSU) established a special committee to make detailed plans for the IGY, the so-called Comité Spéciale de l'Année Géophysique Internationale (CSAGI), and elected as its president Sydney Chapman, the English physicist widely considered the world's foremost expert on the upper atmosphere (and the guest of honor at the Van Allen dinner party of 1950). Lloyd Berkner was elected CSAGI vice president.[2]

In the same book, Naugle provides a somewhat less-complete account (that has also become standard) of the process by which Earth-orbiting satellites became a major goal of the IGY:

> During 1954 the members of CSAGI recognized the possibility of using the US and USSR ballistic missiles to place satellites in orbit. On October 4, 1954, CSAGI challenged the countries participating in the IGY to place small scientific spacecraft in orbit to measure solar radiation and its effect on the upper atmosphere. The United States accepted the challenge first. On July 29, 1955, President [Dwight D.] Eisenhower announced that the United States would launch several small scientific satellites during the IGY. The Soviets waited another year before accepting the challenge.[3]

I could provide a host of other published accounts by scientists of the IGY and of satellite proposals similar in detail and implication to Naugle's.[4] Almost all emphasize unassailable scientific rationales, independence from military motivations, and commitment to the ideal of scientific internationalism.[5] Over the past 20 years it has become increasingly evident, however, at least to historians, that the origins of the IGY, and especially its role ushering in the space age, had much more complex roots, roots deeply intertwined with geopolitics, nationalism, and military and security concerns.[6] My own research has focused on the career of Lloyd Berkner, claiming that he provides a good example of the importance of the "compartmentalizing" role played by certain individuals, roles that permitted—or more accurately encouraged—the isolation of these various interests, with one consequence being contemporary scientists' belief in the apolitical and nonmilitary nature of the IGY and on that basis its usefulness as a model for future scientific endeavors.

In this essay I shall revisit the broader context of Lloyd Berkner's participation in both the dinner party at the Van Allen's and the satellite invitation, emphasizing how, with one foot planted firmly in the scientific world but the other foot just as firmly in the national security realm, Berkner, when orchestrating the IGY, worked hard to serve both.[7]

## The Context of a Dinner Party: Antarctic Experience

Naugle quite appropriately mentioned Berkner's Antarctic experience as germane to his 1950 IPY suggestion. And although, as mentioned, the first Byrd expedition preceded the second IPY by four years, it was nevertheless a formative experience for Berkner, both in terms of familiarizing him with the dearth of reliable information available about radio propagation in the polar regions and of his attitude toward managing the logistics and organization of large science-related activities. Above all else, the Byrd expedition helped prepare Berkner for the IGY because it was an undertaking in which scientific accomplishment was an important symbol, both of restless scientific striving and of national greatness.

Remarkably, before joining the 1928 Byrd expedition, Berkner had little experience as a scientist. Rather he had served as a civil service radio engineer, first at the Bureau of Lighthouses and then at the National Bureau of Standards (NBS) within the U.S. Department of Commerce. At the former he worked to perfect the use of radio technology to enhance the practice and safety of air transportation, and at the NBS he worked to improve understand of observed fading and distortions in long-distance radio communications associated with then poorly understood upper-atmospheric phenomena. Primarily because of its own interest in long-distance radio propagation, the NBS was much interested in formally participating in the privately financed Byrd expedition (which included an extensive radio-research effort and planned to make unprecedented use of aircraft), and Berkner, because of his experience as a radio engineer and as a pilot, was the obvious candidate.[8]

In an article published in 1928, just before his departure, in addition to describing the equipment he intended to deploy in Antarctica, Commander Byrd described why he was going. "The human answer is simple," Byrd explained. "Men do these things because they are men; because in the unknown lies a ceaseless challenge to man's curiosity, to his ever-expanding fund of knowledge. While anything is to be learned of this earth of ours, of its form, its history, its strange forces, men will be found who will not rest until that knowledge is complete."[9]

Until 1928 Berkner most probably viewed engineering as the career through which he might most effectively make his mark. (He dropped out of a graduate physics degree program at Minnesota to accept his first position as an engineer.) Quite possibly the Byrd expedition, its rhetoric and—to some extent—its realities, was the experience that inspired Berkner to switch emphasis from engineering to science when describing the ultimate source of human progress.

This is true, even though, as Berkner came to recognize, the scientific rhetoric of the Byrd expedition was somewhat overblown. As Byrd himself explained, a twentieth-century explorer such as himself was really in the "hero business."[10] Scientific investigations were in fact subordinate to the more glamorous aspects of the expedition: the discovery of major geographic features, the claiming of territory, and reaching the South Pole itself by air. In 1928 Berkner accepted this (as he would again in 1961 as a champion of human spaceflight). Indeed, he was eager to have direct personal involvement with the more glamorous aspects of the expedition. Still, the rhetoric of science did not ring false, at least not to a young man who simply accepted, without question, that it was science that had created and nourished his twin youthful passions: radio and flying.[11]

It is also worth noting that many of the characteristics typically associated with the IGY were operative in Antarctica in 1928: high material cost, logistical complexity, and widespread international interest.

## Berkner and International Scientific Research

In any case, following his Antarctic adventure Berkner rapidly refocused his career from engineering to scientific research. And his efforts during the next decade continued to prepare him well for his role as an IGY founding father. When, in 1932, he joined and then led the ionospheric research group at the Department of Terrestrial Magnetism (DTM) of the Carnegie Institution of Washington, Berkner's first assignment was to continue the development of ionospheric sounding equipment as a tool not only for better understanding the vagaries of long-distance radio communications but also for investigating geophysical phenomena in the upper atmosphere. The sounding equipment development work he directed had been initiated earlier at the NBS by T.R. Gilliland and his associates, but largely abandoned there because of budget cutbacks associated with the worsening depression.[12]

As a private, research-oriented organization, DTM—more directly than had the radio section of the Bureau of Standards—emphasized the ability of "pulse-echo" equipment to generate scientifically useful data, specifically information on variations over time of the heights of radio-reflecting layers in the upper atmosphere. Berkner worked hard to complete the development of a sophisticated, fully automatic, multifrequency sounder with the hope that it could facilitate such observations at widely separated locations around the globe.

DTM's organizational history is especially relevant to this goal and to the IGY connection. In the early 1920s DTM had begun to shift its original emphasis from conducting magnetic surveys (by ship and overland expeditions) to collecting a broader range of geophysical data from fixed, but geographically dispersed, observatories. Three geomagnetic observatories had already been established by the 1930s: one in Washington, D.C.; one at the geomagnetic equator (Huancayo, Peru); and one in the Southern Hemisphere (Watheroo, Australia). Gradually, magnetic measurements at these observatories were supplemented with observations of upper-atmospheric, electromagnetic phenomena.

Foreshadowing worldwide arrangements that would become part of the IGY, the Carnegie Institution of Washington's *Annual Report* for 1934 reported that, beginning that year, measurements at Huancayo and at Washington were "made to coincide in time on certain specified days each week in order that worldwide effects may be studied in some detail." In addition, "a standard method of reduction of data, applicable to ionosphere measurements obtained by different methods, was agreed upon in order that the results might be available generally."[13]

At a meeting of the International Union of Geodesy and Geophysics (IUGG) held in Edinburgh, Scotland, in September 1936, Berkner participated in discussions of the potential for a much broader coordination of ionospheric data collection. As a result of these conversations, the IUGG voted to establish a special "mixed" committee of representatives from all the existing international scientific bodies with interest in ionospheric data.

The committee was created "to encourage and to suggest methods of inter-change of data and information concerning the rapidly advancing technique of measurement, and to act in an advisory capacity in suggesting desirable experimental and theoretical problems and locations for measurement."[14] The renowned British physicist, Sir Edward Appleton, was elected to chair the special committee, and Berkner, no doubt because he controlled what was then the world's most advanced equipment for research in this area, was elected secretary. The other committee members were Sydney Chapman, S.K. Mitra of India, and David F. Martyn of Australia.[15] Over the next several years Berkner strengthened and expanded close ties with these colleagues. Years later it would be a reincarnation of this committee, with Berkner work-ing hard behind the scenes, which would provide the essential impetus for organizing the IGY.

By 1939 data that Berkner and his collaborators obtained from fixed- and multifrequency ionospheric sounding equipment provided important insight into the dynamics of solar/ionosphere interactions. In turn, that scientific insight made it possible to predict reasonably well the maximum and opti-mum frequencies for point-to-point radio communications, at least in tem-perate latitudes.

### Focus on the Polar Regions and Military Interests

As Berkner well knew, work on these problems was hardly finished. There remained vast regions of the globe, especially the polar regions, for which there were no reliable ionospheric data and within which auroral and other upper-atmospheric phenomena remained relatively mysterious. It was in just those regions that sudden solar events, such as flares or sunspots, most often led to significant disruption of radio communications. In February 1940 Berkner proposed expanding the DTM observing effort to Alaska. As jus-tification, he cited both the knowledge that could be generated and the improvements in communications that might result.[16]

Berkner had already recognized that the predictive tools of ionospheric research would be of inestimable value to the communications industry and to any government or international organization desirous of regulating radio. Shortly, he would recognize and push its value to any U.S. military organization hoping to operate on a global scale.

Not coincidentally, by the end of the 1930s a number of well-connected leaders of American science had begun to focus their attention on how the special talents and training of scientists and engineers might be applied should America become enmeshed in the war that was rapidly building over-seas. Most consequential, in 1939 Vannevar Bush relinquished his position as vice president and dean at the Massachusetts Institute of Technology (MIT) to become the third president of the Carnegie Institution of Washington (DTM's parent organization). Bush accepted the position in large part because he believed that a base in Washington, D.C., would provide him increased leverage for preparing American scientists for U.S. involvement in the European war he felt was increasingly unavoidable.[17] "Science and its applications have produced the aircraft and the bomb," he wrote in his first Carnegie president's report. "Entirely apart from all questions of national sympathies, from all opinion concerning political ideologies, we fear to witness the destruction of the treasures of civilization and the agony of peoples, by reason of this new weapon." Yet, he continued, "As science has produced a weapon, so also can it produce in time a defense against it."[18]

Such a view, both utilitarian and idealistic, of the foreseen mobilization of American science was bound to strike a powerful chord in Lloyd Berkner. Building on Richard Byrd's earlier characterization of the scientific impulse, Bush asked in that same president's report: "Can a scientist, skilled in a field such that his efforts might readily be directed to the attainment of applications which would afford protection to his fellow men against such an overwhelming peril, now justify expending his effort for any other and more remote cause?"

Bush's efforts led to his becoming head of President Franklin D. Roosevelt's National Defense Research Committee (NDRC), a civilian agency intended to supplement U.S. military activity harnessing scientific research to provide new weapons and equipment. While awaiting a decision on potential funding for his Alaska radio research proposal, Berkner took part in NDRC-sponsored work at DTM. Most notably he involved himself in the project that would soon grow into an effort, led by Merle Tuve, to design and construct radio proximity fuses to detonate antiaircraft shells near their targets. In these ventures Berkner demonstrated a remarkable ability to shift his focus, quickly and smoothly, from scientific issues and civilian applications to military requirements.[19]

Eventually, late in 1940, the Carnegie trustees agreed to grant $14,000 for Berkner's use in Alaska during 1941. Bush then obtained a commitment from the chief of the Navy Bureau of Aeronautics that Berkner—who since graduating from college had maintained his commission as an officer in the U.S. Naval Reserves—would not be reassigned to some other duty as long as he was needed for the Arctic radio project.

Late in 1940 Berkner's reserve status was changed "from general to specific service." Afterward he and a colleague packed up the automatic sounder they had operated at the DTM facility in Kensington, Maryland, and shipped it and the additional materials they needed to Alaska. Beginning that spring, measurements were taken, not only to support the forecasting effort but also in support of additional NDRC research efforts, most notably the effort to understand, and compensate for, systematic errors in the military's existing radio direction-finding equipment.

The Alaskan facility Berkner set up would provide ionospheric and radio propagation data for the entire war period. Arrangements were soon made for data to be assembled and analyzed by a special radio-condition forecasting group operating out of the Bureau of Standards. Along with the DTM observatories in Peru and Australia, the Alaska station anchored a large and growing network of stations around the globe, organized as the so-called Interservice Radio Propagation Laboratory (IRPL). The IRPL, working in conjunction with its British and Australian counterparts, provided all branches of the Allied military forces with detailed predictions of ionospheric conditions and of the most effective frequencies for communications with all theaters of the expanding global conflict.[20]

## The U.S. State Department, International Science, and Scientific Intelligence

During World War II, Berkner served as liaison between the U.S. Navy's Bureau of Aeronautics and the MIT Radiation Laboratory, which was developing radar equipment for navy fighters.[21] During the immediate postwar years, before turning his attention to resuming a civilian research career, Berkner served as executive secretary of the Joint Research and Development Board (JRDB), an organization established in 1946 to provide expert guidance on postwar development efforts for advanced weapons, and one which played a central role establishing scientific competence within the predecessor organization to what would become the Central Intelligence Agency (CIA).[22]

These activities provide important additional context for the 1950 IGY proposal. Even more directly relevant is Berkner's slightly later involvement with nonmilitary aspects of American national security policy. By 1948, with battle lines of the cold war rapidly solidifying, the now-familiar postwar national security bureaucracy of the United States began to take shape. The new National Security Council (NSC)—the civilian organization of cabinet-level officials that was established to coordinate security policy for the nation—prepared a number of important position documents. In June, NSC 10/2 established a covert operations branch within the recently established CIA. In September, the CIA produced its first coordinated intelligence estimate of Soviet capabilities and intentions. This report reinforced several earlier studies that warned the Soviet Union was bent on promoting communism throughout the world. It stated forthrightly that, in lieu of military force, the Soviet Union was likely to make increasing use of propaganda and covert operations. The aim of the Communists would be to entice other nations to their side of the struggle and disrupt the functioning, and therefore the appeal, of Western-style democracy.[23]

With these events as background, and based on high-level contacts within the U.S. government, Berkner sought avenues to enhance his connections with various government and quasi-governmental organizations. At the same time, in 1948, it was with great satisfaction and professional pride that Berkner accepted formal nomination and election to the U.S. National Academy of Sciences.[24] He especially welcomed the prospect that, in addition to imparting status beyond that attached to his own prior achievements, the National Academy could provide him additional connections and further opportunities to organize scientists to address national problems. Indeed, the National Academy—which had been established during the Civil War—and its research arm, the National Research Council (NRC)—created during

World War I expressly for that purpose—remained the most direct organizational conduits among the nation's scientists and the civilian side of the federal government.[25]

Berkner's first academy task had special relevance to the upcoming IGY proposal. In July 1948 Isaiah Bowman, president of the Johns Hopkins University, agreed to a State Department request to chair a special academy conference on research in Antarctica. Aware of Berkner's experience with the Byrd expedition and of his close association with military research, Bowman asked him to contribute a report on ionospheric research in that region. Berkner agreed and proposed a relatively ambitious Antarctic research program to study atmospheric electricity and magnetism. In his report Berkner emphasized that such research could best be accomplished by the combined efforts of scientists from many nations.[26]

Berkner's recommendations, especially for international involvement, made a great deal of scientific sense. But they also provided the State Department with just the sort of diplomatic ammunition it was looking for. Contemporary documents from the military's Research and Development Board reveal that it was, in fact, the military that had sparked this sudden State Department concern with Antarctica. Navy analysts feared the Soviet Union might try to use bases established there to gain control over important shipping lanes. Unwilling to allow the Soviets to pursue sovereignty claims on the Antarctic continent, military planners informed the State Department of their desire that diplomatic efforts to forestall such action be increased. It is not clear from the records how widely known these military concerns were among scientists. But, as Bowman put it the following February in a letter to Merle Tuve, he (and certainly Berkner) knew quite well that the State Department, at the time of the academy conference, believed "that more emphasis upon cooperative international research in Antarctica and less on national ownership of particular sectors would be an advantage all around."[27]

Even without these motivations, Berkner would undoubtedly have championed international cooperation in Antarctica, having worked throughout the 1930s cultivating just such cooperation in ionospheric research. What is remarkable is how adept he was at taking advantage of the revival of such efforts, making them mesh with the largely secret national security agendas then emerging. State Department officials were doubtless pleased with whatever boost to their efforts the academy conference provided.

Berkner found it easy to associate interests of gathering knowledge and of promoting intellectual freedom around the world with his Antarctic assignment. But the matter was to become more complicated. During the next several years Berkner would involve himself deeply in several other, much more problematic, civilian national-security initiatives. One was a conscious American effort to spur the rehabilitation of science in Europe as a component of an overall effort to create a strong anti-Soviet alliance.[28] Another, with direct relevance to the IGY proposal, was a highly secret examination of the potential of U.S. intelligence organizations to exploit international scientific programs and organizations.

Berkner was experienced enough to realize the involvement of scientists in efforts like these involved potentially serious conflicts with the norms, values, and expectations of much of the international scientific community. He realized also that the challenge of containing or hiding such conflicts would be greater in the civilian realm than it would be within military organizations. Military research assignments accepted by academic scientists tended (in perception if not entirely in reality) to be part-time, focused on specific weapons systems or problems, and carried out in isolation from other professional activities.[29] Achieving goals, such as strengthening European science and enlisting it as a bulwark against communism, in contrast, involved conscious exploitation of important aspects of ordinary scientific work. Successful outcomes of such efforts (whether success was defined in terms of national geopolitical advantage or of the ability to protect and maintain the productivity and integrity of American science) would, therefore, require a subtle form of leadership and entail much greater professional risk. Lloyd Berkner, it would turn out, was one of the few men with both the will and connections to accept the risk and with the skills to be effective in the sensitive role of liaison between the world of science and the emerging national security state.[30]

## "Science and Foreign Relations": The Berkner Report of 1950[31]

Most diplomats and foreign relations experts employed by the State Department in 1950 had little experience with science. Dean Acheson, who in 1949 replaced General George Marshall as President Harry Truman's secretary of state, was no exception. On his appointment Acheson readily agreed with the president's suggestion that he accept as his undersecretary James E. Webb (the future leader of the Apollo-era American civilian space program). A friend of Lloyd Berkner's and a fellow Naval Reserve officer, Webb was serving as director of the Bureau of the Budget. He was one of the Truman administration's leading advocates of public policy and management reform, and had taken a special interest—working with Berkner, Vannevar Bush, and the JRDB—in the problems of enhancing military/science relations and providing the government with use of the nation's scientific and technical capabilities. Acheson looked to Webb to manage the growing set of science-related responsibilities facing his new department.

In many ways Webb chose to follow the model set (largely by Bush) within the postwar military, including recruitment of Lloyd Berkner as an organizer and project leader. Intimately familiar with the nation's industrial and academic brain power as it related to the military, Berkner was a natural choice to serve as consultant to the State Department as it tried to streamline its own science-related advisory, oversight, and operational activities.

During the early part of 1949, with Acheson's concurrence, James Webb established an internal review of "State Department Responsibilities in the Field of Science." Chief among the science-related concerns were: gathering scientific intelligence, carrying out the president's so-called "Point IV initiative" to provide technical assistance to emerging postcolonial nations, and managing the exchange of scientific information and personnel abroad.

Rather than tackle all the State Department's science-related problems at once, Berkner chose to divide the task. Most visibly—and most revealing of State Department priorities—Berkner arranged to survey and make recommendations on the "International Flow of Scientific Information." Writing (over Webb's signature) to A.N. Richards, president of the National Academy of Sciences, Berkner invited the academy to assist "in whatever manner it may consider appropriate," pointedly suggesting that it appoint an advisory committee within the NRC to review the State Department's analysis and recommendations once they were prepared by the internal government working groups. In December Berkner wrote to Detlev Bronk, the NRC chair, suggesting the NRC prepare one of the position papers upon which he and his survey group at the State Department would draw in drafting their recommendations. The NRC paper was to reflect "the views of universities, research organizations and other nongovernmental entities concerned with the problem."[32]

By employing the National Academy and the NRC to prepare his State Department report, Berkner added prestige and influence to its conclusions and helped ensure that American scientists would feel they had had at least some say in recommendations that would be made. To bolster that impression, in April Berkner attended the business session of the National Academy's annual meeting to report formally on the results of the State Department study and the contributions made by the academy.

"It is hardly necessary," he began, "to say that there has been a steadily increasing relationship between science and our international relations and foreign policy." According to Berkner, the "Survey Group" he had directed agreed at the outset that its objective was "to develop detailed recommendations on the most effective means of utilizing the functions and facilities of the [State] Department for meeting the needs of United States science and for strengthening national security." To make such recommendations the group had first been asked to set forth a set of objectives for a U.S. "international science policy." Among those objectives were "the furtherance of understanding and cooperation among the nations of the world" and the "promotion of scientific progress and the benefits to be derived therefrom." But Berkner also emphasized the stake he believed his academy audience had in the more controversial aspects of the forthcoming report. International science policy, he stated, should also be devoted to "the maintenance of that measure of security of the free peoples of the world required for the continuance of their intellectual, material and political freedom."[33]

To meet all these objectives, continued Berkner, the Survey Group had made 19 specific recommendations on how to change the organization and procedures of the State Department to encourage the regular injection of scientific competence into the conduct of foreign affairs. Among the recommendations of most interest to his present audience were calls to establish a science office in the State Department "at the policy level," to place scientific attachés at various embassies abroad, and to create a National Academy committee of eminent scientists to serve as a State Department advisory board.

As the final public report was being prepared, and Berkner was communicating its substance to scientists at the academy and elsewhere, working groups composed of State Department and intelligence agency personnel were addressing the more sensitive aspects of the topic of science and foreign relations. In May 1950 (just three weeks after the Van Allen dinner and IGY proposal) the entire study was the subject of one of the regular meetings the undersecretary held with various State Department assistant secretaries. According to the minutes of that meeting, Berkner was present and informed the attendees that "while the unclassified portion [of the report] has been designed to stand alone, it should be considered as a cover for the classified section."[34]

It was not until July 1998 that the State Department declassified the appendix referred to by Berkner. This material reviewed the overall status of scientific intelligence-gathering, both inside and outside the State Department, and emphasized its growing importance. "The determining factor," the secret section declared, "in a decision by the U.S.S.R. either to make war or to resort to international blackmail may well be the state of its scientific and technological development in weapons of mass destruction. It is therefore imperative that, in the Department, the scientific potential and technical achievements of the Soviet Union and their implications be integrated with the other elements of a balanced intelligence estimate for foreign policy determination."[35]

To the assembled State Department managers, and in writing, Berkner characterized the "present over-all collection of scientific raw intelligence" as "woefully and dangerously inadequate." The only success he noted with respect to the Soviet Union was the limited accomplishments of the State Department's "publication procurement effort." Other than that, the appendix declared, the "collection of information on U.S.S.R science by conventional intelligence methods has failed."

In the May undersecretary's meeting Berkner insisted the Soviet Union was not the only target, noting that such information from Austria and Germany was now "non-existent," and "the UK [United Kingdom] flow is about five years behind." In the classified appendix he had gone further and set forth why such information was also crucial. "First," the report asserted, it was "because research and development results in those countries may contribute to our own scientific and technical advancement, and second, because such discoveries may become known to the Soviet Union and so be of potential use against this country." The classified appendix applauded recent accomplishments of some unnamed "new techniques for acquiring scientific information" but added that "other new and non-conventional methods must be developed."

During the undersecretary's meeting, Berkner insisted to the State Department managers that "the Department should not interfere with, but

rather encourage, the private international flow of scientific information."
Ways of stimulating that flow might include encouragement of travel abroad
by American scientists. He also cited international scientific congresses and
conferences as "a good medium." Berkner commented that "this problem
involves overt operation in terms of science; but in terms of intelligence, it is
an unconventional operation." By that, he meant, for example, that traveling scientists could be debriefed on their return to the United States. But,
he continued, "the debriefing should be handled carefully by scientists in
such a way as not to suggest that the information is to be used merely in the
nationalistic sense."

The classified appendix elaborated on this point. "In regard to the
U.S.S.R," it claimed,

> unpublished data on research in progress, access to prepublication reports,
> information on the trends of thinking of Russian scientists, etc. can only be
> secured through personal contact. With such contacts impossible within the
> U.S.S.R. itself, there remain a number of perfectly open and well accepted
> methods of making such contact. Organizations such as UNESCO, the inter-
> national scientific unions, and international scientific congresses and conven-
> tions provide frequent opportunity for American science to make effective
> contact with the Soviet counterpart.

The appendix also explicitly addressed the sensitivity of using scientists
as intelligence-gatherers: "The opportunities for this contact," it argued,
"should be encouraged openly through full support by the Department of
State to American participation, and by application of whatever measures
are necessary to ensure a fully competent American delegation." But, the
appendix continued, "advance briefing of delegations on intelligence matters
is dangerous and should not be done, except under the most unusual situations. Such a procedure would defeat its purpose in most cases."

In fact, in his work for the State Department—as in his contemporary
proposals to colleagues for international scientific programs—Berkner was
walking a delicate line, attempting to serve as a broker between the national
security bureaucracy and the scientific community. Although to the National
Academy audience he had emphasized the service to science the State
Department could provide, in the confines of the undersecretary of state's
meeting room, as in the secret report, he emphasized the government's hidden agendas and the covert uses to which scientists were to be put.

The revelation of Berkner's secret agenda should not, I think, be taken
to indicate a lack of sympathy for the goals, needs, and ideals of his fellow scientists. Subsequent actions suggest he remained sincerely interested in
furthering those goals and in pursuing the proposals made in the unclassified report, not only for "nationalistic" reasons but also because of an abiding personal interest in international science. Reinforcing that interest, of
course, was Berkner's recognition that his own effectiveness as a liaison, and
therefore as an agent of the U.S. bureaucracy, was in large part dependent on
the perception among scientists of his commitment to scientific idealism.

This, then, provides a broader context for the IGY proposal of April 1950 than has generally been given, especially by scientists. At this very juncture, in part to follow up on his dinner proposal and presumably his government work, Berkner was making plans for a trip to Europe to consult with fellow ionospheric physicists and to attend (as president of the American section) a meeting of the International Scientific Radio Union in Zurich as well as a meeting in Brussels of the so-called Mixed Commission on the Ionosphere. His trip would include stops in Liverpool, Cambridge, London, Paris, Eindhoven, and Rome. What is most remarkable (and characteristic of the immediate postwar era) is how successful Berkner was in keeping the various compartments of his professional relations separate, yet mutually reinforcing.

### Compartmentalization and the 1954 Satellite Resolution[36]

Many books have been written on the origins of the U.S. civilian space program. Lloyd Berkner's name invariably is mentioned (if not highlighted) as one of its leading scientist advocates. In fact, organizing to obtain access to outer space and to exploit that access for scientific purposes was a natural for Berkner. In 1954 it was Berkner—as a leading official within the international organizing committee of the IGY—who acted forcefully, behind the scenes, to have the IGY include among its major goals the launching of scientific satellites. In doing so Berkner drew on his relationships with interested scientists and on his military and broader "national security" experiences. And historically just as in the case of the 1950 dinner proposal, ranking the two motivations—scientific and national security—in terms of influence, is nearly impossible.[37]

Berkner's interest in space grew both out of his interest in geophysics and his conviction that international science offered a potentially powerful (yet seriously neglected) vehicle for promoting American interests in the postwar world. By the mid-1950s the capabilities of Soviet science and technology had become a deep and disturbing mystery to American military planners. Enticing the Soviet Union to make public statements and analyzable demonstrations of just what it had been able to accomplish, especially in rocketry, became a high-priority goal of American intelligence organizations. Additionally, orbiting satellites promised to provide a reconnaissance platform unmatched in its potential to reveal otherwise-hidden details of Soviet military activity, while being far less obtrusive than high-flying aircraft and less subject to objections or even countermeasures.

Lloyd Berkner was in an excellent position to know about all these security requirements, the potential benefits to science, and how both could be advanced if spaceflight were introduced as a peaceful, internationally sanctioned activity. Not surprisingly, Berkner therefore became a central figure the 1954 decision-making process through which the international scientific unions requested that capable countries attempt to launch scientific satellites during the IGY.

A "minimal Earth satellite" had long been the dream of the young geophysicist S. Fred Singer, who had worked at the Johns Hopkins Applied Physics Laboratory with James Van Allen, using Aerobee rockets to measure magnetic fields in the upper atmosphere. Singer had been one of the invited guests at the 1950 Van Allen dinner party. Berkner followed the Aerobee rocket experiments closely and had maintained contact with Singer even after he moved to London in late 1950 to assume the responsibilities of scientific liaison officer for the Office of Naval Research.[38]

Once in Europe, Singer became isolated from the rocket research coordinated by Van Allen's unchartered but influential panel of experts.[39] Unable to conduct actual experiments, Singer indulged in a great deal of speculation, much of which focused on how an instrumented Earth-orbiting satellite might be constructed and launched. In November 1951 Singer lectured on rocket research to the British Interplanetary Society, emphasizing the advantages that could be obtained by placing instruments on a rocket-launched "earth satellite vehicle."[40] He delivered several similar lectures during the next two years to organizations such as the British Association for the Advancement of Science, the International Astronautics Federation, and the American Rocket Society. In the course of making these presentations Singer became acquainted with the active group of space advocates surrounding Wernher von Braun in Huntsville, Alabama.

Owing to his outspokenness in an area that other American scientists felt reluctant to speak about, however, Singer came to be viewed by some of his colleagues as something of a self-promoter and loose cannon.[41] Nevertheless, he was appointed, at Berkner's behest, to the U.S. National Committee Panel on Rocketry created to oversee the development of the IGY rocket-observation program. Afterward, Singer arranged to deliver a paper, "Geophysical Applications of Earth Satellites," during the eleventh URSI assembly held at The Hague, from August 22 to September 2, 1954, a step that, if anything, tended to reinforce the negative image he had among active but less flamboyant rocket experimenters.

Berkner, of course, was familiar with satellite proposals dating back to his wartime service in the Navy Bureau of Aeronautics and to his stint in 1946 as executive secretary of the military's Joint Research and Development Board. There is no evidence, however, as far as I know, that an IGY satellite was discussed at the meeting of the Mixed Commission on the Ionosphere that Berkner attended in Brussels, beginning on August 15, 1954. A week later, however, following Singer's URSI presentation at The Hague, Berkner endorsed a draft resolution on the desirability of launching a satellite some time during the IGY. In an unpublished account of his own role in establishing the satellite program, Singer described Berkner's encouragement and claimed his satellite resolution was discussed and passed at the last URSI "Commission 3" (ionospheric radio) session. Singer also claimed Sir Edward Appleton, the commission chair, endorsed and seconded the resolution.[42]

Commission 3 passed the resolution on to the URSI general assembly. Although he knew Singer had little support among well-connected rocket researchers, Berkner continued to express interest in, and support for, his proposal. And in the end, URSI adopted the following formal statement:

> U.R.S.I. recognizes the extreme importance of continuous observations, from above the E region[,] of extra terrestrial radiations, especially during the forthcoming [IGY]. U.R.S.I. therefore draws attention to the fact that an extension of present isolated rocket observations by means of instrument[ed] earth satellite vehicles would allow the continuous monitoring of solar ultra-violet and X radiation intensity and its effects on the ionosphere, particularly during solar flares, thereby greatly enhancing our scientific knowledge of the outer atmosphere.[43]

Following the URSI assembly, Berkner took part in a conference on the ionosphere held at the Cavendish Laboratory in Cambridge, England. He then attended the tenth general assembly of the International Union of Geodesy and Geophysics in Rome. Singer also took part in the IUGG assembly and brought with him the URSI satellite resolution. American experimenters immediately raised objections, for many believed a U.S. satellite initiative was bound to take resources away from the already-planned program of high-altitude observations using sounding rockets.[44] Negotiations took place in a panel of rocket experimenters convened explicitly to consider the form of a possible IUGG-satellite proposal to go with the one passed by URSI. Chaired by Homer E. Newell, an active rocket experimenter at the Naval Research Laboratory and a long-term member of Van Allen's influential American Rocket Panel, the group considered the implications of the proposal for existing programs and the likelihood there would be a positive response from American organizations with the ability to carry it out. By the end of the assembly, on September 25, 1954, an IUGG resolution calling for an IGY satellite program was also adopted.

One major step remained. The next plenary session of the CSAGI was scheduled to begin in Rome immediately after the IUGG adjournment. During the CSAGI meeting, which was attended by virtually all of the IGY international and American leaders, Berkner took it on himself to engineer a consensus on the satellite question. Late in the CSAGI session, he called an impromptu evening meeting of concerned Americans in his hotel room. Gathered there, in addition to Singer and Newell, was virtually the entire U.S. National Committee (USNC) for the IGY, in addition to Hugh Odishaw, J.W. Joyce (who had joined the staff at the National Science Foundation to serve as liaison to the USNC), Wallace Atwood of the National Research Council, and even Earl Droessler, representing the assistant secretary for research and development of the U.S. Defense Department.

After a technical presentation, given for those who had not attended earlier discussions, there followed "a discussion of the political aspects of satellites lead [*sic*] mainly by Berkner and [Athelstan] Spilhaus." According to the notes Singer dictated a year later, Berkner and Spilhaus emphasized there was an element of competition with the Soviet Union (which was considering joining the IGY program). The fears that the satellite proposal would affect the rocket effort were apparently laid to rest, probably by carefully worded assurances from the Defense Department representative. Thus, the way was cleared for the powerful American contingent to push for passage of a formal CSAGI resolution asking that "thought be given to the launching of small satellite vehicles, to their scientific instrumentation, and to the new problems associated with satellite experiments, such as power supply, telemetering, and orientation of the vehicle."

That basically is all the existing documents reveal about the maneuverings that led to the IGY satellite proposal. The timing, however, suggests much more going on behind the scenes. For example, it is possible that by the time Berkner traveled to Rome he was in a position to know about the recent RAND Project Feedback report, which made detailed proposals in spring 1954 about a possible reconnaissance satellite. It is also possible, yet undocumented, that he was familiar with a recently declassified CIA proposal from the fall of 1954 recommending the United States undertake a small scientific satellite project to help establish the principle of "freedom of space," thus opening the way for reconnaissance satellites.[45] And it is likely that he was familiar with the deliberations that summer of the so-called Technological Capabilities Panel (TCP) that had been appointed in April by the Science Advisory Committee of the Office of Defense Mobilization. Significantly, that panel, chaired by close Berkner associate James Killian, considered, among other matters, the utility and feasibility of reconnaissance satellites and would soon, like the CIA, recommend that, to ease the way for their eventual use, the United States should undertake a small scientific satellite project.

Although no documentary evidence of Berkner's direct involvement in the RAND study, the CIA proposal, or the TCP report has so far turned up in the unclassified or declassified literature, we do know that Berkner was formally briefed before his departure for Europe.[46] In Rome, therefore, Berkner may have based his actions primarily on his ambitions for science; he may have been acting as a concerned and well-informed American scientist with a complex understanding of science and its relations to government plans and policies; or he may have been acting as a covert agent of one or more national security agencies of the U.S. federal government. But given how easily Berkner moved between the worlds of science, government, and national security, and given the vision and experiences he developed during 25 years of organizing scientific activity for multiple, reinforcing purposes, there is little reason to expect that he would have acted differently in any of those capacities.

Berkner had worked energetically for more than a decade, motivated largely by the presumption that the goals of the American scientific, military, and national security establishments closely overlapped with the interests of scientists and peoples throughout the globe. And that, I suggest, is the larger context within which we can best understand the origins and conduct of the 1957–58 IGY.

# Notes

1.  John E. Naugle, *First among Equals: The Selection of NASA Space Science Experiments*, SP 4215 (Washington, D.C.: National Aeronautics and Space Administration, 1991), online version http://www.hq.nasa.gov/office/pao/History/SP-4215/ch1–1.html#1.1.3 (accessed August 7, 2008).

2.  A recent, concise summary of the details of the IGY organization can be found in Rip Bulkeley, "Aspects of the Soviet IGY," *Russian Journal of Earth Sciences* 10, ES1003, doi:10.2205/2007ES000249 (February 12, 2008): 1.

3.  Naugle, *First among Equals*.

4.  One example is Herbert Friedman, *Proceedings of the American Philosophical Society* 140, no. 1 (March 1996): 12. "In April 1950, at a small dinner party in the home of James A. Van Allen, Lloyd Berkner and Sidney [*sic*] Chapman conceived of the International Geophysical Year (IGY). Years of international, planning followed, with commitments to launch earth-orbiting satellites. In the U.S., the IGY was planned as a civilian program conducted independently of military goals."

5.  On the ideal of scientific internationalism, see Paul Forman, "Scientific Internationalism and the Weimar Physicists: The Ideology and Its Manipulation in Germany after World War I," *Isis* 64, no. 2 (June 1973): 150–180; Joseph Manzione, "'Amusing and Amazing and Practical and Military': The Legacy of Scientific Internationalism in American Foreign Policy," *Diplomatic History* 24, no. 1 (Winter 2000): 21–55; Ronald E. Doel, Dieter Hoffmann, and Nikolai Krementsov, "National States and International Science: A Comparative History of International Science Congresses in Hitler's Germany, Stalin's Russia, and Cold War United States," *Osiris* 20 (2005): 49–76.

6.  For example, see Rip Bulkeley, *The Sputniks Crisis and Early United States Space Policy: A Critique of the Historiography of Space* (Bloomington: Indiana University Press, 1991); and Walter A. McDougall, *The Heavens and the Earth: A Political History of the Space Age* (New York: Basic Books, 1985).

7.  Much of what follows is excerpted or based on the account provided in my book on Lloyd Berkner published in 2000. Allan A. Needell, *Science, Cold War, and the American State: Lloyd V. Berkner and the Balance of Professional Ideals* (Amsterdam: Harwood Academic Publishers, 2000). Detailed references and a bibliography can be found there.

8.  Needell, *Science, Cold War,* chapter 1.

9.  Richard E. Byrd, "Into the Home of the Blizzard: On the Eve of His Departure for the Antarctic, Commander Byrd Explains Why He Attempts the Exploration of the Frozen Continent by Air and Discusses Problems He Must Surmount," *New York Times* (1857–current file), (September 23, 1928), online at http://www.proquest.com (accessed August 7, 2008).

10.  Eugene Rodgers, *Beyond the Barrier: The Story of Byrd's First Expedition to Antarctica* (Annapolis, Maryland: Naval Institute Press, 1990), 24.

11.  Berkner was hardly unique in accepting the primary role of science. See Paul Forman, "The Primacy of Science in Modernity, of Technology in Postmodernity, and of Ideology in the History of Technology," *History and Technology* 23, nos.1/2 (March/June 2007): 1–152.

12.  Needell, *Science, Cold War,* chapter 2.

13.  Carnegie Institution of Washington, *Year Book No. 33* (1934), 215–216.

14.  W.F. Evans, "History of the Radio Research Board," typescript (Melbourne, Australia: Commonwealth Scientific and Industrial Research Organization, 1973), 132. Library of Congress: LCCN 76–373698 r84).

15.  Needell, *Science, Cold War,* chapter 2.

16.  Needell, *Science, Cold War,* chapter 3.

17.  A. Hunter Dupree, "The Great Instauration of 1940: The Organization of Scientific Research for War," in Gerald Holton, ed., *The Twentieth-Century Sciences: Studies in the Biography of Ideas* (New York, 1972), 448–450.

18. Vannevar Bush, "Report of the President," Carnegie Institution of Washington, *Year Book No. 38* (1939), 6.

19. Needell, *Science, Cold War,* chapter 3.

20. Wilbert F. Snyder and C.L. Bragaw, *Achievement in Radio: Seventy Years of Radio Science, Technology, Standards, and Measurement at the National Bureau of Standards,* Special Publication 555 (Washington, D.C.: National Bureau of Standards, 1986), 317, 404–420.

21. Needell, *Science, Cold War,* chapter 3.

22. Needell, *Science, Cold War,* chapter 4.

23. Needell, *Science, Cold War,* chapter 5. On the establishment of the CIA, see Rhodri Jeffreys-Jones, *The CIA and American Democracy* (New Haven, Connecticut: Yale University Press, 1989).

24. Frederick Seitz, "Berkner and the National Academy of Sciences," *Proceedings of the Dedication Ceremony, Lloyd V. Berkner Hall, The University of Texas at Dallas* (opening session of the First Berkner Memorial Symposium of the American Geophysical Union), 42.

25. A. Hunter Dupree, *Science in the Federal Government: A History of Policies and Activities,* 2nd edition (Baltimore, Maryland: Johns Hopkins University Press, 1986), 135–141, 305–312.

26. Lloyd V. Berkner, "Some Significant Polar Observations and Experiments in Electricity and Magnetism," dated September 27, 1948, File Series "Org, NAS, Committee on Antarctica, 1948," NAS/NRC Central File, National Academy of Sciences Archives, Washington, D.C.

27. References are in Needell, *Science, Cold War,* chapter 5.

28. Allan A. Needell, "Rabi, Berkner, and the Rehabilitation of Science in Europe: The Cold War Context of American Support for International Science 1945–1958," in Francis H. Heller and John Gillingham, eds., *The United States and the Integration of Europe: Legacies of the Post War Era* (New York: St Martin's Press, 1995), 289–306. On the rehabilitation of science in Europe, more generally, see John Krige, *American Hegemony and the Postwar Reconstruction of Science in Europe* (Cambridge, Massachusetts: The MIT Press, 2006).

29. Needell, *Science, Cold War,* 155.

30. Needell, *Science, Cold War,* chapter 6.

31. Lloyd V. Berkner, "Science and Foreign Relations: International Flow of Scientific and Technological Information," (Washington, D.C.: Department of State, 1950). What follows draws directly from the account in Needell, *Science, Cold War,* chapter 5. For detailed references, see that chapter.

32. Douglas Whitaker to Warren Weaver, December 29, 1949, National Academy of Sciences/National Research Council Central Files, National Academy of Sciences Archives, Washington, D.C. Douglas Whitaker was the scientist Bronk appointed to head the NRC survey for Berkner.

33. Transcript, Annual Meeting NAS Business Session, File "International Relations, International Science Policy Study of State Dept., NRC Portion General, 1949," NAS/NRC Central File, National Academy of Sciences Archives, Washington, D.C.

34. Meeting Minutes, May 8, 1950; Minutes of the Undersecretary's Meetings, General Records of Office of the Executive Secretariat, Entry 396.3, General Records of the Department of State, Record Group 59, National Archives at College Park, College Park, Maryland.

35. "Scientific Intelligence," attachment to memorandum, Lloyd V. Berkner to R. Gordon Arneson, "Reporting on Intelligence," Box 64, Lot File 57 D 688, General Records of the Department of State, Record Group 59, National Archives at College Park, College Park, Maryland.

36. What follows draws directly from the account in Needell, *Science, Cold War,* chapter 12. For detailed references, see that chapter.

37. There are several works on the history of the IGY and Sputnik that offer these and other observations about the satellite. See Robert A. Divine, *The Sputnik Challenge: Eisenhower's Response to the Soviet Satellite* (New York: Oxford University Press, 1993), Walter A. McDougall, *The Heavens and the Earth: A Political History of the Space Age* (New York: Basic Books, 1985, reprint edition, Baltimore, Maryland: Johns Hopkins University Press, 1997); Rip Bulkeley, *The Sputniks Crisis and Early United States Space Policy: A Critique of the Historiography of Space* (Bloomington: Indiana University Press, 1991); Roger D. Launius, John M. Logsdon, and Robert W. Smith, eds., *Reconsidering Sputnik: Forty Years Since the Soviet Satellite* (Amsterdam: Harwood Academic Publishers, 2000); Matt Bille and Erika Lishock, *The First Space Race: Launching the World's First Satellites* (College Station: Texas A&M University Press, 2004); Paul L. Dickson, *Sputnik: The Shock of the Century* (New York: Walker and Co., 2001); Matthew Brzezinski, *Red Moon Rising: Sputnik and the Hidden Rivalries that Ignited the Space Age* (New York: Times Books, 2007). Other aspects are recounted in Martin Collins, ed., *After Sputnik: 50 Years of the Space Age* (New York: Collins, 2007); Asif A. Siddiqi, *Sputnik and the Soviet Space Challenge* (Gainesville: University Press of Florida, 2003); James J. Harford, *Korolev: How One Man Masterminded the Soviet Drive to Beat America to the Moon* (New York: John Wiley & Sons, 1997); Philip Nash, *The Other Missiles of October: Eisenhower, Kennedy, and the Jupiters, 1957–1963* (Chapel Hill: University of North Carolina Press, 1997); Peter J. Roman, *Eisenhower and the Missile Gap* (Ithaca, New York: Cornell University Press, 1995); and Kenneth Osgood, *Total Cold War: Eisenhower's Secret Propaganda Battle at Home and Abroad* (Lawrence: University Press of Kansas, 2006).

38. Homer E. Newell, *Beyond the Atmosphere: Early Years of Space Science* (Washington, D.C.: National Aeronautics and Space Administration, 1980), 52–53. Berkner's interest is documented in Singer to Berkner, February 28, 1951, Folder "ONR London Sci. Correspondence 1950–53," Box 1108, S. Fred Singer Papers, National Air and Space Museum Archives, Washington, D.C. [henceforth Singer Papers].

39. See David H. DeVorkin, *Science with a Vengeance: How the Military Created the U.S. Space Sciences after World War II* (New York: Springer-Verlag, 1992).

40. "Research in the Upper Atmosphere with Sounding Rockets and Earth Satellite Vehicles," *Journal of the British Interplanetary Society* 11, no. 2 (March 1952): 61.

41. Homer E. Newell, *Beyond the Atmosphere: Early Years of Space Science* (Washington, D.C.: National Aeronautics and Space Administration, 1980), 52–53.

42. Memorandum by Fred Singer dated November 1955, Box 1107, Folder "Mouse," Singer Papers, NASM.

43. *Proceedings of the 11th General Assembly* 10, part 3, Commission III on Ionospheric Radio (Brussels: URSI, 1954), 194.

44. These details and the remaining discussion of the 1954 satellite proposal are based in large part on the above-cited memorandum by Fred Singer dated November 1955, Box 1107, Folder "Mouse," Singer Papers, NASM.

45. Dwayne Day has presented and written about recently declassified CIA documents relevant to the origins of the Freedom of Space concept. An informal presentation of his preliminary analysis (which includes no reference at all to Berkner's activities in Europe) has appeared in the online journal: *The Space Review* as "Tinker, Tailor, Satellite, Spy," October 29, 2007. See http://www.thespacereview.com/ article/989/1 (accessed April 10, 2008).

46. In a March 1955 letter to Walter Rudolf of the State Department Science Advisors' office about the desirability of Sydney Chapman, president of CSAGI, traveling to Moscow to discuss the details of Soviet participation in the IGY. Berkner prefaced his remarks: "As you know, after personal consultation between myself and representatives of the Department of State, the Comité Spéciale de l'Année Géophysique Internationale 1957–58 established the policy of inviting all nations of the world including those behind the iron curtain to participate in the Année Géophysique

Internationale." Berkner to Rudolf, March 16, 1955; copy in Box 52, Sydney Chapman Papers, University of Alaska Archives, Fairbanks, Alaska. It was at the Rome CSAGI meeting that the Soviet Union formally agreed to participate in the IGY. Similar "consultations" with other agencies (such as the CIA and Department of Defense) can only be surmised.

# Chapter 12

# Polar and Global Meteorology in the Career of Harry Wexler, 1933–62

*James Rodger Fleming*

On April 27, 1955, Harry Wexler was summoned to a meeting at the International Geophysical Year (IGY) headquarters by Joseph Kaplan, chair of the U.S. National Committee, for an "important" matter. "Upon arriving," Wexler noted in his diary, "I found Larry Gould, Lloyd Berkner, Joe Kaplan, Lincoln Washburn, Wally Joyce, and Hugh Odishaw. They asked me if I would accept the post of 'Chief Scientist' of the US/IGY Antarctic Expedition—duties to begin immediately."[1] Wexler, chief of Scientific Services at the U.S. Weather Bureau (USWB), was intrigued, yet concerned about the effect that this assignment might have on his own research and responsibilities.

In the gray dawn of the following day, after tossing and turning most of the night, Wexler asked himself, "What am I getting into? Am I losing five years of my scientific life with global meteorological problems to concentrate on a small area to act as father confessor to frustrated Antarctic scientists—to battle with the Navy for scientific program priorities?" Wexler did not agonize for long. That day he consulted with polar meteorologist Paul Humphrey, who gave him some "useful background about Antarctic cultists" and then spoke to Princeton mathematician John von Neumann about his new program on computer models of the general circulation of the atmosphere. After calling his boss, F.W. Reichelderfer, Wexler accepted the IGY position on the condition that he find fieldwork as enjoyable and productive as his desk-and-lecture duties. He had convinced himself that studying Antarctica's weather and climate and the Southern Hemisphere's oceanic and atmospheric circulation held the keys to a better understanding of global meteorology.

For almost two years, until January 19, 1957, Wexler did not reach the Antarctic circle, see an iceberg, or even try on his parka, but getting there was in a way anticlimactic, because he had long been thinking about the need to pay greater attention to geophysical interactions between the hemispheres.

Throughout his career Wexler was involved in, and subsequently led, a technological revolution that established the modern atmospheric sciences. He supported the continued development of high-altitude sounding rockets, studied global fallout patterns from atmospheric atomic tests, and promoted the use of digital computers in operational weather forecasting and general circulation studies. He established radiation, ozone, and, notably, $CO_2$ measurements at the Mauna Loa Observatory and later was in charge of the Tiros program and supported the first global heat budget experiment on *Explorer 7*.[2] For Harry Wexler, nuclear testing, digital computing, observations from space, from the top of a Hawaiian volcano, or from the Antarctic ice cap were all part of "making meteorology global."

## Early Career

Harry Wexler was born in 1911 in Fall River, Massachusetts, the third son of Russian emigrants Samuel and Mamie (Hornstein) Wexler.[3] As a child he was interested in science and enjoyed both "mathematical recreations" and baseball—his high school physics teacher was his baseball coach. Wexler said his fascination with meteorology developed while delivering newspapers through fair weather and foul. It was an interest he shared with his childhood friend, future brother-in-law, and noted meteorologist Jerome Namias. Wexler majored in mathematics at Harvard University, graduating magna cum laude in 1932. He then attended the Massachusetts Institute of Technology (MIT), studying in the department of aeronautical engineering under the mentorship of Carl-Gustaf Rossby, arguably one the most influential meteorologists of the twentieth century and "instrumental in bringing American meteorology to a position of world leadership."[4] Experimental aircraft observations and the new technology of balloon-borne radiosondes to measure vertical profiles of atmospheric temperature, pressure, and humidity were under development at MIT and undoubtedly made an impression on Wexler. Under Rossby's mentorship, Wexler soon became proficient in Norwegian methods of air-mass, frontal, and isentropic analysis. He published his first article on the turbidity of air masses in 1933 during the second International Polar Year (IPY).[5]

After receiving a master's degree in meteorology in 1934, Wexler began his lifelong affiliation with the USWB as an assistant meteorologist in Chicago and Washington. He married Hannah Paipert in Chicago in December 1934, and they had two daughters. In 1935 Dr. Willis L. Gregg, new chief of the USWB, established an Air Mass Analysis Section under the leadership of another Rossby student, Dr. Horace Byers, and assigned Wexler to help with the task of developing operational techniques of air mass and frontal analysis. His duties involved teaching the methods to others, including older forecasters who were already set in their ways. Wexler also conducted original research and published papers on weather fronts, lower atmospheric cooling, and the formation and structure of cold air masses. His 1936 article, "Cooling in the Lower Atmosphere and the Structure of Polar Continental

**Figure 12.1**    Harry Wexler, 1911–1962. (Courtesy of Susan Wexler Schneider)

Air" in the *Monthly Weather Review*, explains how, apart from conduction and mechanical turbulence, cooling from the surface and lower levels of the atmosphere extends to higher levels, generating a polar air mass. The article demonstrated Wexler's complete command of the theory of atmospheric infrared radiative transfer by carbon dioxide and water vapor and his understanding of the evolution of air masses and their theoretical structure and minimum temperatures. Wexler followed this study with articles on the formation of polar anticyclones and an analysis of aerological observations and measurements of outgoing radiation at Fairbanks, Alaska, and Fargo, North Dakota, during the winters of 1936–37 and 1937–38, as part of an investigation of the formation and structure of polar continental air.[6]

By this time it was quite clear that, compared to the North Atlantic region where the Bergen School methods were first developed, air mass and frontal analysis were much more complex over the North American continent. Yet, beyond the task of standardizing and objectifying the techniques of weather-map analysis and forecasting lay a greater challenge in understanding the basic dynamics of atmospheric flows and their interaction on continental, hemispheric, and global scales. For example, Rossby had long advocated measurements in the southern ocean, because the Antarctic current was unobstructed by continents and might therefore be expected to be similar to the atmospheric Westerlies, possibly exhibiting jet-stream structure, meandering waves, convergence and divergence patterns, and eddy formation. Such studies were pursued by Rossby and his students, notably by Harry Wexler.

Before the first IPY in 1882–83, Austrian naval officer and polar explorer Carl Weyprecht had pointed out the need for comparable, simultaneous measurements in polar regions: "Decisive scientific results can only be attained through a series of synchronous expeditions, whose task it would be to distribute themselves over the Arctic regions and to obtain one year's series of observations made according to the same method." More than 50 years later the noted American geographer Isaiah Bowman observed that the plan of the second IPY (1932–33) was "inspired by a profound curiosity as to the suspected influence of weather conditions in high latitudes upon (or interaction with) those of the temperate regions as well as the tropics."[7] These two themes—comparable, simultaneous measurements and interactions between high and low latitudes—shaped the research agenda for global meteorology in the twentieth century and are prominent in the career of Harry Wexler. Early evidence of this appears in his 1935 article, "Deflections Produced in a Tropical Current by Its Flow over a Polar Wedge," in the *Bulletin of the American Meteorological Society*. This was but the first in a long series of studies of the general circulation of the atmosphere.

In 1937 the U.S. Department of Agriculture's Bankhead-Jones Special Research Fund provided funding for Wexler to return to MIT for further study. There he conducted additional research on the formation of polar air masses, supported Rossby's program on the general circulation, and developed techniques useful in extended, or long-range, forecasting, a shared topic of interest with his childhood friend and MIT colleague Jerome Namias. In 1938 he was promoted to associate meteorologist in the USWB, and in 1939 received a doctor of science degree from MIT; his dissertation, supervised by Rossby, combined observations of transverse circulations in the atmosphere with theoretical insights into their climatological implications. The USWB assigned Wexler to supervise a forecasting team at LaGuardia Field, New York. From June to September 1940 he tested his theories against practice.

Following the 1939 outbreak of war in Europe, the U.S. military began a program to train weather forecasters. Wexler took a leave of absence from the USWB in 1940 to teach in this program at the University of Chicago as an assistant professor of meteorology. In 1941 he returned to the USWB in Washington as senior meteorologist in charge of training and research, working to assist in defense preparations. He accepted a captain's commission in the U.S. Army in 1942 and served as the senior instructor of meteorology to the Army Air Force (AAF) Aviation Cadet School at Grand Rapids, Michigan. While in this position, he joined the University Meteorological Committee, established to assist the military services in matters related to

**Figure 12.2**    Visitors and some participants in the 1950 ENIAC computations. (L–r): Harry Wexler, John von Neumann, M. H. Frankel, Jerome Namias, John Freeman, Ragnar Fjörtoft, Francis Reichelderfer, and Jule Charney. (Courtesy of Defense Advanced Research Projects Agency, Washington, D.C.)

meteorological services. Following his promotion to major in 1943, Wexler worked in the Pentagon Weather Division, AAF headquarters, in charge of research and development. New technologies for weather research were being opened up by improvements in aviation and its extension to new altitudes and into new areas of the world. The task at hand was to utilize meteorology more effectively in aerial navigation, bombing ballistics, and weather forecasts for military operations.

On September 14, 1944, Major Harry Wexler participated in what may be called the first research reconnaissance flight into a hurricane, which occurred off Cape Henry, Virginia, in a Douglas A-20 Havoc flown by Colonel Floyd Wood. In his published account of the mission, Wexler described the data collected, concluding, "that the major portion of this hurricane cloud was caused by a strong but narrow area of ascending air near the center of the storm and that outside this area, descending air was found."[8]

## Head of Research, USWB

Following his honorable discharge with the rank of lieutenant colonel in January 1946, Wexler returned to the USWB as chief of its special Scientific

Services division. As chief of research at the USWB, Wexler encouraged the development of new technologies, including airborne observations of hurricanes, radioactive tracers, the use of electronic computers for numerical weather prediction, and satellite meteorology. He also supported the study of Earth's upper atmosphere and the atmospheres of other planets. He served on numerous federal and international panels and committees, including the military's Research and Development Board and the Subcommittee on Meteorological Problems of the National Advisory Committee for Aeronautics.

Wexler joined the nuclear age on July 16, 1945, when he analyzed the pressure waves generated by the Trinity atomic bomb test. He studied the weather's effects on reactor safety and developed techniques for following radioactive tracers downwind and around the globe after atmospheric nuclear tests. He was a member of the Advisory Committee on Reactor Safeguards for the U.S. Atomic Energy Committee, chaired the National Academy study group on Meteorological Aspects of the Effects of Atomic Radiation, and served as a U.S. delegate to the 1955 Atoms-for-Peace Conference in Geneva, Switzerland. The USWB publication *Meteorology and Atomic Energy* was prepared under his supervision. Wexler often responded to public concerns linking nuclear tests with adverse weather events, pointing out that the energy released by atmospheric events was far greater than that of even the largest bomb and that, while the immense heat and towering mushroom cloud produced by a nuclear blast had dramatic short-term and local effects, he had found no evidence of any long-term effects on the weather and climate. He supported, however, continued studies of radioactive fallout as a potential health concern and as a global tracer, even in the snowfields of the South Pole.[9]

Radar meteorology was an important byproduct of wartime electronics research. Wexler's brother Ray had worked on weather radar with the U.S. Army Signal Corps Engineering Laboratories in New Jersey during the war, and in 1946 had detected and analyzed a frontal storm in the Atlantic, publishing useful photographs of the event. A year later, Harry Wexler included these techniques in an early article on the structure of hurricanes as determined by radar. By 1960 radar images were being combined with photographs from the Tiros satellites to reveal both the large-scale and detailed structure of weather systems on a variety of spatial scales, including mesoscale convection on the order of ten miles in diameter, spiral bands in hurricanes spanning 250 miles, and similar cloud structures in mid-latitude cyclones up to a thousand miles across.[10]

Air pollution, broadly defined, was also one of Wexler's interests. Early in his career he had studied the turbidity of the air and the trajectories of the dust clouds of the 1930s as they made their way from the American heartland to the cities of the East Coast. He also prepared a meteorological analysis of the air pollution episode of October 1948 in Donora, Pennsylvania, and examined the effects of Canadian forest fires (September 1950) that generated the "great smoke pall" over the Northeast. The latter study is still cited in the literature on nuclear winter, because it generated a widespread regional cooling effect.[11] Since volcanic eruptions also could cool the entire planet by blanketing it with dust and sulfate aerosols, Wexler took up the suggestion of William Jackson Humphreys that a series of such events could have caused the Quaternary ice ages. Assuming a diminution of available incoming radiation of up to 20 percent, caused by volcanism, Wexler derived a possible climatic scenario with increased meridional circulation in which an upper cold trough dominated the weather over central North America. While this ice-age pattern could not be sustained by the typical volcanic eruption frequency of four per century, researchers could not rule out a ten-fold increase at least once every 10,000 years.[12] Such evidence might be revealed in deep ice cores.

In the immediate postwar period digital computing constituted the cutting edge of meteorology. At the birth of modern computing, in 1945, Dr. V.K. Zworykin, research director of the RCA (Radio Corporation of America) Laboratories in Princeton, imagined a perfectly accurate machine that would predict the immediate future state of the atmosphere and identify the precise time and location of leverage points or sensitive conditions, ultimately "for determining what control measures should be taken, and at what point in space and time this control should be applied"—literally to pour oil on troubled ocean waters or even set fires or detonate bombs to disrupt storms before they formed, deflect them from populated areas, and otherwise control the weather. According to Zworykin: "The eventual goal to be attained is the international organization of means to study weather phenomena as global phenomena and to channel the world's weather, as far as possible, in such a way as to minimize the damage from catastrophic disturbances, and otherwise to benefit the world to the greatest extent by improved climatic conditions where possible."

Wexler's close colleague, the multitalented mathematician extraordinaire John von Neumann, who had initiated a meteorology project at the Institute for Advanced Study, agreed: "All stable processes we shall predict. All unstable processes we shall control."[13]

The Princeton group was informed by theoretical work done at the turn of the century by Felix Exner and Vilhelm Bjerknes, who had identified the basic equations of atmospheric dynamics. In 1922 L.F. Richardson had actually tried to solve the equations numerically, but with rather poor data and without the use of a computer. Inspired by the subsequent work of Rossby and his students, and led by meteorologist Jule Charney, the institute meteorology project laid the foundation for numerical weather prediction and, eventually, for general circulation modeling.

Wexler was the USWB official liaison, commuting from Washington to Princeton for monthly meetings to get the Institute of Atmospheric Sciences Meteorology Project off the ground, seeing it institutionalized in 1954 as the U.S. Joint Numerical Weather Prediction Unit in Suitland, Maryland. This was a partnership of the USWB with the U.S. Air Force and U.S. Navy "to produce prognostic weather charts on an operational basis using numerical techniques."[14] A year later, based on a successful numerical experiment by Norman Phillips, in which he was able to simulate realistic features of the general circulation of the atmosphere, von Neumann and Wexler argued for the creation of a General Circulation Research Section (later Laboratory) to be located nearby in Suitland. This was really the beginning of computer climate-modeling.[15]

Rockets and satellites also came under Wexler's purview as scientific probes and observational platforms to investigate the upper atmosphere. He served as chair of several influential committees on this subject, including the Upper-Atmosphere Committee of the American Geophysical Union, the National Advisory Committee for Aeronautics (NACA) Special Committee for the Upper Atmosphere, and—after the launch of *Sputnik*—the National Research Council's Space Science Board.

In his 1954 lecture presented at a visionary symposium on space travel, held at the Hayden Planetarium in New York, Wexler explored the possibilities of Earth observations from space. He pointed out that a V-2 flight in 1947 had photographed clouds from an altitude of 100 miles, and an Aerobee rocket launched in 1954 had identified a previously unknown tropical storm in the Gulf of Mexico. Wexler displayed an artist's impression of Earth from space, in full color, showing clouds, land, and ocean, and depicting weather features, such as a family of three cyclonic storms along the polar front, a small hurricane embedded in the trade winds, evidence of jet-stream winds, and fog off the coasts.[16] Encouraged by novelist and futurist Arthur C. Clarke, Wexler published versions of his remarks in the *Journal of the British Interplanetary Society* and in the *Journal of Astronautics,* where he made strong claims for the utility of the meteorological satellite, not only as a "storm patrol," but also as a potentially revolutionary new tool with global capabilities.

> Since the satellite will be the first vehicle contrived by man which will be entirely out of the influence of weather it may at first glance appear rather startling that this same vehicle will introduce a revolutionary chapter in meteorological science—not only by improving global weather observing and forecasting, but by providing a better understanding of the atmosphere and its ways. There are many things that meteorologists do not know about the atmosphere, but one thing they are sure of is this—that the atmosphere is indivisible—that meteorological events occurring far away will ultimately affect local weather. This global aspect of meteorology lends itself admirably to an observation platform of truly global capability—the Earth satellite.[17]

This was by no means idle speculation, because his ideas about what was possible were derived from careful research into the technical capabilities of a future satellite.

The invitation Wexler received in 1955 to serve as chief scientist for the U.S. expedition to the Antarctic for the IGY was based on his lifelong engagement with the atmospheric sciences: solid training and mentorship by Rossby, early theoretical insights on Arctic air mass formation and storm dynamics, experience in World War II and postwar military-advisory circles, and, as chief of Scientific Services at the USWB, a plethora of new techniques and technologies to collect data from equator to pole and from surface to outer space. By accepting the challenge of the IGY, Wexler knew he could shape the programs intended to collect critical new information about both the South Pole and the Southern Hemisphere and integrate it into a global picture of circulation and dynamics of the entire atmosphere.

## The IGY

In accepting the duties of chief scientist of the U.S. IGY Antarctic program and consultant to the Antarctic committee, Wexler was merely adding to his considerable workload. In addition to continuing duties with the USWB, he served in a number of positions for the U.S. National Committee for the IGY, including as a member of its Executive Committee, its Technical Panel on Meteorology, its ad hoc Arctic and Equatorial Committee, and its Panel for Radioactivity of the Air. He was also centrally involved in planning the launches of the first U.S. Earth satellites.

Wexler was a good communicator and was often asked to write for the general public. In 1957 Wexler published a classic article on "Meteorology in the International Geophysical Year" in the *Scientific Monthly* that highlighted some of the fundamental issues in the understanding of the atmosphere, meteorology's relationship to other geophysical sciences, and the importance of Antarctic science, climate science, and weather satellites. He concentrated on how, with new tools and technologies such as weather satellites, scientists might better be able to understand the general circulation of the atmosphere. The essay began:

> Here is a rotating sphere named the "Earth" moving through space and separated from it by a thin mantle of air that is proportionately as thin as the skin of an orange. Boring down into the mantle are cosmic rays, meteors, radiations from the stars, and, most important, radiations and energetic particles from the sun. The bottom of the mantle of air, which is deformed by mountains and plateaus, absorbs heat and water vapor, and its movement is opposed by friction as the air moves over land and water. The atmosphere repays its debt to the sun and space by passing energy upward in the form of infrared radiation.[18]

In detailing the specific benefits meteorologists were expecting to gain from the IGY, Wexler first listed the extension—laterally and vertically—of existing meteorological observing networks into the unknown. "Several generations of meteorologists have exploited this relationship by supporting wholeheartedly the First and Second Polar Years in 1882–83 and 1932–33 and now the International Geophysical Year of 1957–58." Wexler's colleague, noted geophysicist Sydney Chapman, served as president of Comité Spéciale de l'Année Géophysique Internationale (CSAGI), which coordinated the IGY effort internationally. In his presidential remarks, Chapman emphasized Earth's fluid envelope and the continuing need for widespread simultaneous observations: "The IGY's main aim is to learn more about the fluid envelope of our planet—the atmosphere and oceans—over all the earth and at all depths.... These researches demand widespread simultaneous observation."[19]

Expanded measurements during the IGY were critical to Wexler's research program. They would permit more accurate estimates to be made of trace gases, global wind systems, and the poleward transports of atmospheric energy, momentum, and water vapor. This study was particularly needed in the mostly marine Southern Hemisphere. Wexler anticipated that, when combined with older and ongoing studies of the Northern Hemisphere (which was dominated by land masses), such research would greatly facilitate understanding of complicated worldwide "feedback" mechanisms. The continent of Antarctica was the key to understanding the Southern Hemisphere's weather systems and ocean dynamics and, as Rossby had predicted, that hemisphere was in turn critical for understanding global climate and circulation systems.

This was a timely article, for, as Wexler states, "the answers to many of these problems will have to wait for the digestion of data obtained during the 1957–58 International Geophysical Year and probably also for the obtaining of additional data from future International Geophysical Years. The world's geophysicists bear a heavy responsibility to future generations in insuring [*sic*] that basic environmental measurements are carefully recorded and that new ones are initiated."[20]

On June 27, 1957, at the opening ceremony of the U.S. IGY at the National Academy of Sciences Building in Washington, D.C., Harry Wexler joked with characteristic style and good humor about the global role of meteorology: "My friends, the oceanographers like to put us landlubbers in our proper places by pointing out that the oceans cover three-fourths of the earth's surface. In meteorology we have no need for such defense mechanisms— everyone knows that the atmosphere covers *all* of the earth's surface."[21]

The serious point Wexler was making is that the atmosphere plays a role as the largest common denominator of all the geophysical sciences—from those dealing with the oceans to those concerned with near-space. The exchange of heat, moisture, momentum, gases (such as carbon dioxide and oxygen), elements (such as sodium, chlorine, potassium, calcium, and magnesium), and nutrients across the air-sea interface has profound influences on the subsequent behavior of both atmosphere and ocean. At the other extreme, the physics of the upper atmosphere involves studies of the ionosphere, aurora and airglow, geomagnetism, and cosmic rays with effects that might propagate to lower atmospheric levels, and vice versa.

One of the main aspects of the IGY that distinguished it from its predecessors was the attention focused on Antarctica—the bull's-eye of the Southern Hemisphere—and studies of its geophysics in general and its meteorology in particular. As chief scientist of Antarctic programs for the IGY, Wexler supervised the first census of Antarctic weather and formulated a complete and comprehensive model of the circulation of air above the high ice dome of that continent. He complemented this study with an analysis of the winds and temperatures of the Arctic stratosphere.[22]

Wexler used the IGY investigations as a benchmark and a springboard for further research in both polar and global meteorology. One of his first efforts following the IGY was his article on the Antarctic circumpolar ocean current, published as a tribute to his mentor, Rossby, who had died suddenly in the summer of 1957.[23] Although the IGY could not muster sufficient resources to complete the study of Southern Hemisphere ocean currents, Wexler encouraged future international cooperation on this fruitful ocean-atmosphere research problem. Further analysis of IGY and earlier Antarctic temperature records allowed Wexler to detect a significant 2°C warming trend at the Little America station on the edge of the Ross Ice Shelf, from 1912 to 1958, in line with the first wave of climate warming in the first half of the twentieth century.[24] He also reported the early results of ozone measurements from Little America.[25]

## Beyond the IGY

Three themes demanded Wexler's attention in the early 1960s: satellite meteorology, peaceful international cooperation, and the possibilities of global climate control. As mentioned earlier, Wexler was a big promoter of the use of satellites in meteorology. At the First National Conference on Peaceful Uses of Space, he commented on the global-observing system possible with satellite coverage: "A system of satellites of two types would be ideal for charting the world's weather. One system would circle the earth over the poles; the other would circle around the equator. Both types of satellites could send their observations into a central office. They could also pick up and transmit information from automatic weather stations located in uninhabited areas."[26] Wexler believed that information gathered from satellites on an international scale would be of great value to everyone in the world for warning about severe weather and other weather changes.

In his first State of the Union address, President John F. Kennedy promised to explore promptly all possible areas of international cooperation "to invoke the wonders of science instead of its terrors." Specifically, he invited all nations—including the Soviet Union—to join in developing a weather-prediction program, a new communications-satellite program, and interplanetary scientific probes. A month later, when asked at a press conference about special areas of cooperation with the Soviet Union that would not harm national security, Kennedy singled out meteorology, specifically weather control, as a possible focus of common effort.[27] By May he had promised $53 million to the USWB for "a satellite system for worldwide weather observation." By September 1961, in the wake of successful Soviet cosmonaut flights, Kennedy proposed peaceful and humanitarian uses for outer space, including cooperative international efforts in weather prediction and eventually in weather control.

In 1961 the Kennedy administration appointed Wexler as lead negotiator for the United States in talks with the Soviet Union concerning the joint use of meteorological satellites. The negotiations resulted in a multinational effort to institute a World Weather Watch with Wexler and Soviet academician V.A. Bugaev as lead architects for a new program to be administered by the World Meteorological Organization in Geneva. Formally established in 1963, and still in existence, the World Weather Watch coordinates the efforts of member nations by combining observing systems, telecommunications facilities, and data-processing and forecasting centers to make available meteorological and related environmental information needed to provide efficient weather services in all countries.

The potential manipulation of Earth's heat budget and stratospheric ozone shield was a final concern of Wexler's. In 1962 he lectured to technical audiences on inadvertent and purposeful damage to the ozone layer involving catalytic reactions of chorine and bromine; he also lectured on climate-engineering through purposeful manipulation of Earth's heat budget.[28] Wexler discussed his concerns about increasing industrial pollution in the lower atmosphere, and from the increased use of sounding rockets in the thin, upper atmosphere. As an example he mentioned rising carbon dioxide emissions and cited a 1961 study by the Geophysics Corporation of America on modification of Earth's upper atmosphere by rockets. He also reviewed recent developments in computing and satellites that led him to believe that manipulating and controlling large-scale phenomena in the atmosphere were distinct possibilities.

In his lecture Wexler focused on both inadvertent and planned manipulations of "*Earth's radiative balance on a rather large scale* [original emphasis]." After reviewing the dangers—to human health and to the climate system—of releasing huge quantities of carbon dioxide and other gases and particles to the lower atmosphere, Wexler turned to purposeful techniques of climate control.[29] He presented some 20 technical slides of the atmosphere's radiative heat budget and discussed means of manipulating it, concluding with a grand cautionary summary of various techniques to:

a. *increase* global temperature by 1.7 C by injecting a cloud of ice crystals into the polar atmosphere by detonating 10 H-bombs in the Arctic Ocean;
b. *lower* global temperature by 1.2 C by launching a ring of dust particles into equatorial orbit, a modification of an earlier Russian proposal to warm the Arctic; and
c. *destroy all stratospheric ozone,* raising the tropopause, and cooling the stratosphere by up to 80°C by an injection of a catalytic agent such as chlorine or bromine.[30]

He estimated that a mere 0.1 megaton bromine "bomb" would destroy all ozone in polar regions and 0.4 megaton would be needed near the equator.

Wexler was concerned that inadvertent damage to the ozone layer might occur if increased rocket exhaust polluted the stratosphere. He was also concerned that future near-space experiments could go awry, citing: Operation Argus (nuclear blasts in near space, 1958); Project West Ford (a ring of small copper dipole antennas in orbit; launch failed 1961, launched 1963); and Project High Water (ice crystals injected into the ionosphere, 1962) as significant recent interventions with unknown risks. Purposeful damage was also not out of the question. In 1934 Sydney Chapman had proposed making a temporary "hole in the ozone layer" using a yet-to-be-identified catalytic "deozonizer." According to Chapman, a small hole cut at a remote location might enable astronomers to make observations at ultraviolet wavelengths where radiation was otherwise absorbed by ozone. Much more sinister and relevant to the cold war was possible military interest in waging geophysical warfare by attacking the ozone layer above a rival nation.

Seeking advice on how to cut a "hole" in the ozone layer, Wexler turned to chemist Oliver Wulf at Caltech, who suggested that chlorine or bromine

**Figure 12.3**   Harry Wexler in Antartica, 1957. (Courtesy of Susan Wexler Schneider)

atoms might act in a catalytic cycle with atomic oxygen to destroy thousands of ozone molecules. In a handwritten note, composed in January 1962, Wexler scrawled the following: "UV decomposes $O_3 \rightarrow O$ in presence of halogen like Br. $O \rightarrow O_2$ recombines and so prevents more $O_3$ from forming." On another slip of paper: "$Br_2 \rightarrow 2$ Br in sunlight destroys $O_3 \rightarrow O_2 + BrO$." These are essentially the basis of modern ozone-depleting chemical reactions. Using Manabe and Möller's model, Wexler was able to calculate an 80 C stratospheric cooling with no ozone layer.[31]

In the summer of 1962 Wexler accepted an invitation from the University of Maryland Space Research and Technology Institute to lecture on "The Climate of Earth and Its Modifications." Under normal circumstances, he would have prepared his ideas for publication, perhaps as he had done in his 1958 article in *Science*, "On Modifying the Weather on a Large Scale." On August 11, 1962, however, during a working vacation at Woods Hole, Massachusetts, Wexler was cut down at the age of 51 by a sudden heart attack. A pile of papers on the upcoming World Weather Watch cluttered his desktop. The documents relating to his career, from his early work at MIT to his final speeches on ozone depletion and climate control, headed into the archives, probably not to be seen and certainly not to be reevaluated until today. The well-known and well-documented Supersonic Transport (SST) and ozone depletion issues developed about a decade later. The idea that bromine and other halogens could destroy stratospheric ozone was published in 1974, while CFC production expanded rapidly and dramatically between 1962 and its peak in 1974. Had Wexler lived to publish his ideas, they would certainly have been noticed and could have led to a different outcome and perhaps an earlier coordinated response to the issue of stratospheric ozone depletion.

## Conclusion

Polar and global issues dominated the career of Harry Wexler. His first investigations into the formation of Arctic air masses, and his final speculations on the possibilities of ozone holes and modification of polar climates, were grounded in his life-long career and central involvement in the atmospheric sciences. He was trained under C.G. Rossby at MIT, spent his career as a technical advisor to government, played a central scientific role in the IGY, and used all his influence to promote the development of new tools, such as radar, satellites, and digital computers. Wexler served on numerous governmental scientific panels and advisory boards, had access to, and helped collect, global climate data, and understood the theoretical issues and their complexity. His sources of information included scientists on the leading edge of meteorology, climatology, digital computing, and space experimentation.

I am not arguing here that either Wexler or the IGY "made meteorology global." That is a much longer and more nuanced story. I am claiming that for several key decades in the middle of the twentieth century, Harry Wexler was at the center of action in the atmospheric sciences and was able to combine interests in polar and global meteorology in a productive manner.

A summary of Wexler's interests covered in this article and the approximate dates of first involvement helps make this point.

Arctic air mass analysis (1936)
World War II weather officer training (1940)
Nuclear tracers (1946)
Air pollution and biomass burning (1950)
Rocket and satellite meteorology (1954)
Computer models, NWP, GCMs (1950s)
Mauna Loa $CO_2$ monitoring (1957)
Antarctic warming (1958)
Climate-engineering (1958)
Satellite meteorology (1960)
World Weather Watch (1962)
Stratospheric ozone depletion (1962)[32]

As we celebrate the anniversaries of the IPYs and the IGY, and the fiftieth anniversary of the space age, we are at the same time acutely aware that current changes in polar environments are the leading indicators of global climate change. As international teams of scientists seek evidence of past and ongoing global changes, as satellite sensors and other technologies continually monitor the planet, and as global climate modelers attempt to understand it all, it is historically rewarding to follow the career of Harry Wexler as a central figure and exemplar of the processes that made meteorology the kind of global science it is today.

## Notes

1. Antarctic Diary, Harry Wexler Papers, Box 27, U.S. Library of Congress. Joseph Kaplan (1902–91), chair of the U.S. National Committee (USNC) for the IGY; Laurence McKinley "Larry" Gould (1896–1995), director of the IGY Antarctic program; Lloyd V. Berkner (1905–67), one of the architects of the IGY; A. Lincoln Washburn (1911–2007), USNC member; J. Wally Joyce (dates unknown), head of the IGY office at NSF; Hugh Odishaw (1916–84), executive director of the USNC.
2. Chicago Conference 1946, Notes on Problems of Meteorological Research, 9–13 December, 1946, Wexler Papers, Box 3; Harry Wexler, "TIROS I." *Monthly Weather Review* 88 (1960): 79–87.
3. Biographical information from the Harry Wexler Papers, Library of Congress and from the *New Dictionary of Scientific Biography* (New York: Scribner's, 2007).
4. Gisela Kutzbach, "Rossby, Carl-Gustaf Arvid," *Complete Dictionary of Scientific Biography* 11 (Detroit: Charles Scribner's Sons, 2008), 557–559.
5. Harry Wexler, "A Comparison of the Linke and Ångstrom Measures of Atmospheric Turbidity and Their Application to North American Air Masses," *Transactions, American Geophysical Union* 14 (April 1933): 92–99.
6. Harry Wexler, "Cooling in the Lower Atmosphere and the Structure of Polar Continental Air," *Monthly Weather Review* 64 (April 1936): 122–136; "Formation of Polar Anticyclones," *Monthly Weather Review* 65 (June 1937): 229–236; "Observations of Nocturnal Radiation at Fairbanks, Alaska, and. Fargo, N. Dak.," *Monthly Weather Review Supplement* 46 (1941).

7. K. Weyprecht, "Scientific Work on the Second Austro-Hungarian Polar Expedition, 1872–4," *Royal Geographical Society Journal* 45 (1875):19–33; I. Bowman, "Polar Exploration" *Science* n.s. 72 (1930): 439–449. For further details, see James Rodger Fleming and Cara Seitchek, "Advancing Polar Research and Communicating Its Wonders: Quests, Questions, and Capabilities of Weather and Climate Studies in International Polar Years," in *Smithsonian at the Poles: Contributions to International Polar Year Science*, I. Krupnik, M.A. Lang, and S.E. Miller, eds. (Washington, D.C.: Smithsonian Institution Scholarly Press, 2009), 1–12.

8. Harry Wexler, "The Structure of the September 1944 Hurricane When Off Cape Henry, Virginia," *Bulletin* of *the American Meteorological Society* 26 (May 1945): 156–159.

9. Harry Wexler, "Meteorological Aspects of Atomic Radiation," *Science* 124 (July 20, 1956): 105–112.

10. Ray Wexler, "Radar Detection of a Frontal Storm 18 June 1946," *Journal of Meteorology* 4 (1947): 38–44; Harry Wexler, "Structure of Hurricanes as Determined by Radar," *Annals of the New York Academy of Sciences* 48 (September 15, 1947): 821–844.

11. Harry Wexler, "Meteorology and Air Pollution," *American Journal of Public Health,* Part II, *Yearbook* 41 (May 1951); "Turbidities of American Air Masses and Conclusions Regarding the Seasonal Variation in Atmospheric Dust Content," *Monthly Weather Review* 62 (November 1934): 397–402; (with H.H. Schrenk, H. Heimann, G.D. Clayton, and W.M. Gafafer), "Air Pollution in Donora, Pa.: Epidemiology of the Unusual Smog Episode of October 1948," *Public Health Bulletin* No. 306 (Washington, D.C.: U.S. Public Health Service, 1949). "The Great Smoke Pall— September 24–30, 1950," *Weatherwise* 3 (December 1950): 129–134, 142.

12. Harry Wexler, "Variations in Insolation, General Circulation, and Climate," *Tellus* 8 (November 1956): 480–494.

13. Vladimir Zworykin, "Outline of Weather Proposal," 1945, copy in the Wexler Papers.

14. Facts sheet, U.S. Joint Numerical Weather Prediction Unit, 1955, rare book, NOAA Central Library. Wexler's monthly reports are in the Wexler Papers, Library of Congress.

15. Warren M. Washington, "Computer Modeling the Twentieth- and Twenty-First-Century Climate," *Proceedings of the American Philosophical Society* 150 (September 2006): 414–427.

16. James R. Fleming, "A 1954 Color Painting of Weather Systems as Viewed from a Future Satellite," *Bulletin of the American Meteorological Society* 88 (October 2007): 25–27.

17. Harry Wexler, "Observing the Weather from a Satellite Vehicle," *Journal of the British Interplanetary Society* 13 (1954): 269–276; "The Satellite and Meteorology," *Journal of Astronautics* 4 (Spring 1957): 1–6; Wexler had made considerable progress since 1954 in his thinking about the role of meteorological satellites. See "Meteorological Satellites," *Exploring the Unknown: Selected Documents in the History of the U.S. Civil Space Program,* vol. III: *Using Space,* John M. Logsdon, ed., (Washington, D.C.: NASA History Office, 1998), 156, and http://history.nasa.gov/SP-4407/vol3/cover. pdf.

18. Harry Wexler, "Meteorology in the International Geophysical Year," *Scientific Monthly* 84 (March 1957): 141–145.

19. Sydney Chapman, "Presidential Address 28 Jan. 1957," *Annals of the International Geophysical Year* 1 (1959): 3.

20. Wexler, "Meteorology in the International Geophysical Year."

21. Remarks at IGY opening ceremony, typescript, Wexler Papers.

22. IGY-era publications on polar topics by Harry Wexler include the following: "Antarctic Research during the International Geophysical Year," American Geophysical Union *Geophysical Monograph* 1 (1956): 7–12; "Antarctic Climatology and Meteorology,"

American Geophysical Union *Geophysical Monograph* 1 (1956): 36–43; "Antarctic Research—International Geophysical Year," *Geophysica* 21 (July 1956): 681–690; (with W.B. Moreland), "Winds and Temperatures in the Arctic Stratosphere," *Proceedings of the Polar Atmosphere Symposium*, Oslo, July 2–8, 1956, Part 1, Meteorology Section (New York: Pergamon Press, 1958), 71–84; "Some Aspects of Antarctic Geophysics," *Tellus* 10 (February 1958): 76–82; "IGY Meteorology in Antarctica," *Weatherwise* 11 (August 1958): 114–115; "The 'Kernlose' Winter in Antarctica," *Geophysica* 6 (1958): 577–595.

23. "The Antarctic Convergence—or Divergence?" *The Atmosphere and the Sea in Motion, Scientific Contributions to the Rossby Memorial Volume*, Bert Bolin, ed. (New York: Rockefeller Institute Press with the Oxford University Press, 1959), 107–120.

24. Harry Wexler, "A Warming Trend at Little America, Antarctica," *Weather* 14 (June 1959): 191–197; "Additional Comments on the Warming Trend at Little America, Antarctica," *Weather* 16 (February 1961): 56–58. For a summary of the issues, see Harry Wexler (with M.J. Rubin and J.E. Caskey Jr., eds.), "Antarctic Research: the Matthew Fontaine Maury Memorial Symposium," American Geophysical Union *Geophysical Monograph* 7 (1962).

25. Harry Wexler (with W.B. Moreland and W.S. Weyant), "A Preliminary Report on Ozone Observations at Little America, Antarctica," *Monthly Weather Review* 88 (February 1960): 43–54.

26. Wexler, "The Satellite and Meteorology."

27. John F. Kennedy, State of the Union address, January 30, 1961; JFK, press conference, March 1, 1961; JFK, special message to the Congress on urgent national needs, May 25, 1961; JFK, address in New York city before the General Assembly of the United Nations, September 25, 1961.

28. Harry Wexler, "On the Possibilities of Climate Control," Boston chapter of the American Meteorological Society, January 9, 1962; Traveler's Research Corporation, Hartford, Connecticut, January 11, 1962; and UCLA Department of Meteorology (as Regent's Lecturer in meteorology), February 28, 1962. Notes in Manuscript Division, Library of Congress, Wexler Papers, Box 18, Speeches and Lectures, 1962.

29. Wexler Papers, Box 18, "Further Justification for the General Circulation Research Request for FY 63," draft, February 9, 1962.

30. H. Wexler, "Modifying Weather on a Large Scale," *Science* n.s. 128 (October 31, 1958): 1059–1063.

31. S. Manabe and F. Möller, "On the Radiative Equilibrium and Balance of the Atmosphere," *Monthly Weather Review* 89 (1961): 503–532.

32. See "Bibliography of the Publications of Harry Wexler," Malcom Rigby and Pauline A. Keehn, compilers, *Monthly Weather Review* 91 (1963): 477–481.

**Part Four**

# National Roles for International Science: Quests and Questions in the International Geophysical Year

# Chapter 13

# Science, Environment, and Sovereignty: The International Geophysical Year in the Antarctic Peninsula Region

## Adrian Howkins

The continent of Antarctica was a major focus of global research during the International Geophysical Year (IGY) of 1957–58. Even as late at the 1950s, remarkably little was known about this vast frozen region. The heroic era of the early twentieth century—which culminated in the famous race to the South Pole between Roald Amundsen and Robert Scott—had barely scratched the surface of scientific knowledge about the Antarctic environment. Into the 1950s, it remained unknown for certain whether Antarctica was one continent or two.[1] Nobody really knew the depth of the ice, and meteorologists pondered the apparent paradox of a climate that seemed too cold for snow.[2] On the eve of the IGY, Antarctica remained a virtual terra incognita, with speculation and fantasy often substituting for any real knowledge about the region. During the 18-month period between 1957 and 1958, scientists instituted "the most comprehensive scientific program ever undertaken by man," and the scientific understanding of Antarctica increased dramatically.[3]

The IGY came at a particularly tense time in Antarctica's political history.[4] By the 1950s, seven countries—Great Britain, New Zealand, France, Australia, Norway, Chile, and Argentina—had advanced sovereignty claims to parts of the Antarctic continent (figure 13.1). These claims had been motivated, at least in part, by the belief that Antarctica might prove to be a frozen El Dorado, brimming with vast mineral wealth.[5] The United States and the Soviet Union refused to recognize any claims to sovereignty in Antarctica, while reserving their rights to possession of any part of the continent. In the Antarctic Peninsula region, directly to the south of South America, the claims of Great Britain, Argentina, and Chile substantially overlapped (figure 13.2). During the 1940s and 1950s, these overlapping claims led to a 20-year period of intense dispute, which on more than one occasion threatened to turn violent.[6]

During the IGY, 12 nations participated directly in Antarctic research: the seven claimant countries, the two superpowers, South Africa, Japan, and Belgium. Within a year of the conclusion of the IGY, these same 12 nations met in Washington, D.C., to sign the Antarctic Treaty, which effectively "internationalized" Antarctica as a "continent dedicated to peace and science."[7] Article IV of the Antarctic Treaty suspended (or "froze" in the official pun of the conference) all sovereignty claims to the continent for the treaty's duration, bringing to an end the active phase of the dispute among Great Britain, Argentina, and Chile. To many people at the time, it appeared as if the idealism of science had triumphed over politics.[8] This achievement was all the more remarkable for bringing the two superpowers together at the height of the cold war. Historians have tended to follow this idealistic interpretation of the connection between the IGY and the Antarctic Treaty, and the southern continent tends to be held up as an all-too-rare example of scientific cooperation leading to political harmony.[9]

This essay focuses on IGY research in the contested Antarctic Peninsula region—an area defined as being within the territories disputed by Great Britain, Argentina, and Chile, and therefore stretching into the Weddell Sea and including the surrounding sub-Antarctic islands. It will argue that the IGY did indeed play an important role in the resolution of the Antarctic sovereignty dispute, but not in quite the idealistic way the traditional narrative has suggested. The actual science of the IGY, and the improved understanding of the Antarctic environment that it facilitated, played an important role in the partial resolution of the question of sovereignty. As politicians and officials in Great Britain and the United States learned more about the reality of the Antarctica environment through the work of IGY—in particular the realization that it contained little or nothing of immediate economic value— they came to favor a limited internationalization of the continent. In order to bring this about, they then exploited the "goodwill" generated by the IGY. Argentina and Chile found themselves swept along, somewhat reluctantly, by the process of internationalization. The end result, however, was generally favorable to all participating countries, as they found themselves members of the Antarctic Treaty system's "exclusive club," which continues to govern the continent to this day.[10]

## Historical Background

Until the early nineteenth century, Antarctica remained a purely hypothetical continent, untouched by wider historical processes. It was not until the early 1820s that expeditions from Great Britain, the United States, and Russia claimed to be the first to set eyes on the Antarctic continent.[11] It would be a further 80 years before the so-called heroic era of Antarctic exploration stirred any real popular interest in the new lands.[12] The heroic era began with the provocative resolution of the Sixth International Geographical Congress, held in London in 1895, which stated that "the exploration of the Antarctica regions is the greatest piece of geographical exploration yet to be undertaken."[13] Lack of scientific knowledge about Antarctica appeared to be an affront to the civilized world and propelled a series of expeditions southward to "conquer" the Antarctic environment. The heroic era climaxed in the famous race to the South Pole between Norwegian explorer Roald Amundsen and British naval captain Robert Falcon Scott in the austral summer of 1911–12.[14] After Amundsen's famous victory and Scott's failure to return, some figures among the British polar establishment muttered that the Norwegians' "dash to the pole" compared unfavorably with the "genuine" scientific work of the British party.[15] From the beginning of human presence in Antarctica, science, the environment, and politics have been inextricably connected.

The heroic era also served to remind the world of the marine wealth of the Southern Ocean.[16] From the end of the nineteenth century, with whale populations in the Northern Hemisphere devastated by over-fishing, companies from Norway, Great Britain, Argentina, and Chile began to establish whaling factories on South Georgia and sub-Antarctic islands around the Antarctic Peninsula. These operations proved remarkably successful and radically changed perceptions of the economic worth of the Antarctic environment. By the outbreak of World War I, Antarctica had become the undisputed center of the world's whaling industry. Because of this expansion in the Antarctic, the world's whale-oil production increased from 149,100 barrels in 1904 to 812,000 in 1914. Initially, the whaling companies were restricted to shore-based whaling, but by the 1920s large factory ships began to dominate the industry. Over time, the dominance of factory-ship-based pelagic whaling reduced the importance of shore-based stations in and around the Antarctic Peninsula.

In 1908 the British Empire responded to the boom in Antarctic whaling by making a formal sovereignty claim to the vast Antarctic Peninsula region, the center of the nascent whaling industry. The so-called Falkland Islands Dependencies, administered from Stanley by the governor of the Falkland Islands, would turn out to be the largest territory under the jurisdiction of the colonial office. The principal motivation for this claim was economic: the British wished to tax and regulate the whaling industry.[17] There was also a widespread belief, resulting from a lack of any real understanding of the environment, that Antarctica could contain a treasure trove of mineral wealth. In the aftermath of World War I, it became imperial policy to bring the whole of Antarctica, piece by piece, into the British Empire, a policy implemented via claims by New Zealand (in 1923) and Australia (in 1933). Although a sovereignty claim by France in 1924 thwarted this continental ambition, on the eve of World War II more than two thirds of Antarctica appeared pink (the standard color for the British Empire) on imperial maps.

One of several ways Great Britain legitimized its claims to Antarctica sovereignty was through recourse to "environmental authority." British officials claimed only they had the scientific and administrative skills to administer the Antarctic whaling industry in a sustainable fashion and prevent it from declining, as had whaling in the Northern Hemisphere. Sir Miles

Clifford, governor of the Falkland Islands from the mid 1940s, made the case explicitly:

> The only true wealth that this area contains, so far as we know today, is still as in the past its marine wealth—its whales and seals; these, as we have noted earlier, could readily be exterminated by indiscriminate killing and it was the recognition of this danger which decided His Majesty's Government to bring their industries under control and led to the establishment of British sovereignty over the area now known as the Falkland Islands Dependencies. *The motive was a purely unselfish one, to conserve the harvest of these seas for the benefit of mankind as a whole* [emphasis added].[18]

Such claims to environmental authority as a justification for the empire mirrored similar strategies throughout the British Empire.[19] In the years after World War I, the British established an oceanographic research project, with a particular focus on the marine life of the Southern Ocean, an endeavor that became known as the Discovery Expeditions.[20] The Discovery Expeditions initially used the ship from Captain Scott's first expedition, adding a layer of romantic association to the scientific case for British sovereignty.

On the eve of World War II, Norway called for a conference in Bergen to discuss Antarctic sovereignty. As the country with the greatest involvement in the Antarctic whaling industry, the Norwegian government had a significant interest in the question of who owned Antarctica. At the time, an expedition from Nazi Germany was exploring the region directly to the east of the British claim. In January 1939, fearing a Nazi presence in the southern continent and encouraged by the United States, the Norwegian government went ahead and made a claim of its own. Because the Norwegians disputed the so-called "sector principle" in their claims to sovereignty in the Arctic, their claim to Antarctica deliberately set no southern limit. As a consequence, the Norwegians were the only nation whose claim does not stretch to the South Pole. Because of the outbreak of World War II in September 1939, and the German invasion of Norway shortly afterward, the proposed polar conference in Bergen never took place.

In the wartime environment of 1939–40, U.S. Navy Admiral Richard E. Byrd embarked on his third expedition to Antarctica. In his first expedition (1928–30) Byrd had become the first person to fly over the South Pole. In his second expedition (1933–35), he famously attempted to winter "alone" at a station in the interior of the continent.[21] This third expedition, however, would be the first to have the official backing of the U.S. government. It had two bases: one at Little America III in the Pacific sector, and one at Marguerite Bay on the Antarctic Peninsula. With this base on the peninsula, this expedition would also be Byrd's first venture into the region already claimed by the British as the Falkland Islands Dependencies. The expedition clearly had the question of sovereignty high on its agenda, and President Franklin D. Roosevelt issued instructions for markers to be laid out that could form the basis of future U.S. claims. The traditional U.S. policy toward Antarctica had been to reserve its rights but to make no formal claim. This policy was not set in stone, however, and North American politicians and officials were certainly willing to consider making a claim if it could be shown that Antarctica contained anything of economic or strategic worth that would justify such a measure. During the expedition Byrd dutifully laid out sovereignty markers, then returned to the United States to report to Congress that he had found 147 different types of minerals in Antarctica.[22] As the quintessential Antarctic booster, Byrd knew exactly what his political masters wanted to hear.

In the early years of World War II, Argentina and Chile advanced competing claims to the Antarctic Peninsula region, which overlapped with the British Falkland Islands Dependencies. The direct catalyst for these claims was the 1939 Norwegian invitation to a polar conference in Bergen and the apparent need to assert to the world South American rights to the Antarctic continent. The Argentinean and Chilean governments argued that they were not making new claims, but rather delimiting the traditional rights they had inherited from the Spanish Empire, rights dating to the 1494 Treaty of Tordesillas. The South Americans shared the widespread belief that Antarctica might be a "frozen El Dorado" containing vast mineral wealth. Argentinean and Chilean publications, for example, frequently quoted Byrd's claim that he had discovered 147 different types of minerals in the continent.[23] But the principal motivation for South American claims was nationalism. Assertions of sovereignty in Antarctica—especially when set against the imperial pretensions of the British—appealed to the rising tide of nationalist sentiment in both Argentina and Chile. Antarctica offered a relatively safe place for nationalist posturing, because, unlike the British-occupied Islas Malvinas (as the Argentinians termed the Falklands), the region contained no permanent population, and any offense caused by Argentine landings would not be directly threatening to British communities.

Assertions of Argentinean and Chilean sovereignty began an active three-way dispute with British that would last for the next 20 years. At the heart of the dispute was the contest for environmental authority: the Argentineans and Chileans challenged Great Britain's claims to dominion over nature in Antarctica and argued instead that the natural environment demonstrated the validity of South American claims. The idea of geological continuity was central to their argument: the Andes Mountains, they argued, plunged under the sea and reemerged in the Antarctica peninsula as the "Antarcandes." They also used weather, fossils, wildlife migration, and even apparent similarities between the snow and ice of Patagonia and of Antarctica to demonstrate their "natural rights" to Antarctic sovereignty. Taken together, the South American challenge to British assertions of environmental authority might be thought of as a form of "environmental nationalism." Not only did Argentina and Chile argue against the idea that British scientific work in Antarctica gave Great Britain political rights to the region, but they also sought to incorporate the natural environment—or at least their perceptions of that environment—into their respective cases for Antarctic sovereignty.

The South American campaign for Antarctic sovereignty opened the possibility that the Antarctic Peninsula region could indeed be recognized internationally as an integral part of Argentinean and Chilean national territory.

In the later years of World War II, in an effort to respond to Argentinean and Chilean claims, Great Britain became the first country to establish permanent stations on the Antarctic Peninsula.[24] In addition to attempting to fulfill the legal requirement of "effective occupation"—one of the means of demonstrating sovereignty in an international court—these stations offered scientists an opportunity to continue their Antarctic research and strengthened the connection between science and politics. In the 1946–47 Antarctic season, a veritable "scramble for Antarctica" took place, as expeditions from Argentina, Chile, and the United States constructed bases within the contested region. During the next few years, these competing claims led to tense standoffs, which more than once came close to open hostility. A report from the U.S. State Department noted that "wars have broken out over lesser issues." From the late 1940s, with the annual voyage of the *Slava* whaling fleet to the Southern Ocean, the Soviet Union began to show increasing interest in the Antarctic continent. This Soviet presence in Antarctica threatened to bring the cold war to the Antarctic continent and added a further layer of complexity to an already fluid and confused political situation.[25]

### The IGY on the Antarctic Peninsula

The origins of the IGY had little direct connection with the politics of the Antarctic continent, and the IGY itself came as something of a surprise to many of the participants in the sovereignty dispute. The IGY had its genesis at a dinner party held at the house of James Van Allen in Silver Spring, Maryland, in honor of visiting British geophysicist Sydney Chapman.[26] At the dinner, the assembled scientists—led by the redoubtable Lloyd Berkner—proposed a third International Polar Year (IPY) to coincide with a period of high sunspot activity. During the next couple of years the idea won the support of the major international scientific unions and changed its name to the IGY. Although the polar regions would remain a major focus of global research, the enterprise was to have global scope. As the IGY won the backing of governments around the world, it became a classic example of the new era of big-science, with a complex intermingling of scientific, economic, and political concerns.[27]

The proposal to hold an IGY was built on two previous IPYs, which had taken place in 1882–83 and 1932–33. These had focused predominantly on the Arctic region, but they provided a model for international cooperation and coordinated research agendas for the IGY of 1957–58. The first IPY of 1882–83 sparked only small amounts of Antarctic research, with such nations as France and Germany still focused on the far north, and in particular the Northwest Passage. The second IPY of 1932–33 certainly intended to pursue Antarctic research as part of its planned investigations, but the global economic depression precipitated by the Wall Street crash of 1929 severely restricted its scope. Some research was done in the Antarctic region during this period: for example, the Argentineans contributed observations from their meteorological base on Laurie Island.[28] But in general the second IPY did not have the dual-polar focus that its planners had intended. The IGY of 1957–58 would therefore be the most sustained and coordinated effort to understand the Antarctic continent that had ever been made. The scale of its ambition, and, at least to some extent, its application of the latest technology for scientific research, marked the IGY as both quantitatively and qualitatively different from earlier polar years.

British scientists, led by Sydney Chapman, had been involved in the planning and organization of the IGY since its inception. This global scientific endeavor appeared to offer an excellent opportunity to demonstrate, on a worldwide stage, British assertions of environmental and scientific authority. Nevertheless, the Antarctic section of the IGY posed problems for British politicians and officials charged with promoting the empire's sovereignty in the Falkland Islands Dependencies. Disagreements as to what form British participation should take rocked the British polar establishment.[29] Some leading figures advocated the eye-catching Trans-Antarctic Expedition of Vivian Fuchs and Edmund Hillary, an attempt to complete Ernest Shackleton's aborted heroic-era expedition to the South Pole. Others argued for less spectacular, but more scientifically useful work within the boundaries of the Falkland Islands Dependencies. In the end, enough money was found to support three major projects associated with the IGY: the Fuchs-Hillary "Crossing of Antarctica," Peter Mott's aerial survey of the peninsula region, and the construction of the Royal Society's Halley Station on the Weddell Sea. Of these three projects, only the Royal Society's research was officially part of the IGY but barely hidden political objectives underlay all of these ostensibly scientific activities.

British officials also viewed the IGY in Antarctica as an opportunity to assess the economic potential of the Falkland Islands Dependencies. In the aftermath of the Suez Crisis of 1956 (when Harold Macmillan replaced Anthony Eden as prime minister), the British government conducted a strategic rethink of its colonial possessions. This came to be known as Macmillan's "Audit of Empire." The section of the audit dealing with the Antarctic Peninsula region explicitly noted the importance of the IGY for assessing its economic potential: "Our withdrawal from Antarctica would mean a loss of UK prestige and influence, especially in scientific circles. It might also involve the loss of strategic minerals, but this will be easier to evaluate when the results have been assessed of the work done during the International Geophysical Year. Argentina and/or Chile, which have claims (partly competing) to the Dependencies, would probably step in if the United Kingdom withdrew."[30]

If there were exploitable minerals of economic or strategic value to be found in the Falkland Islands Dependencies, British officials fully expected the IGY to reveal their presence. Great Britain's attachment to exclusive sovereignty in Antarctica had a financial bottom line: If the region could not pay for itself, then its value would be questioned.

    The British Commonwealth Trans-Antarctic Expedition of 1957–58—an attempt to traverse the entire continent overland—was Great Britain's most blatantly politicized activity associated with the IGY.[31] The surveying and geological activities of the expedition, along with its obvious nonscientific agenda, meant the expedition was not officially part of the IGY. During the planning stages, Prime Minister Anthony Eden boasted the expedition would cross the whole of Antarctica, never once setting foot outside the British Empire. The Fuchs-Hillary "Crossing of Antarctica" was a conscious attempt to complete the Shackleton expedition, which had failed after the sinking of his ship, *Endurance*. The British made every effort to infuse the expedition with the spirit of the heroic era. Around the world, even in Chile and Argentina, journalists followed the story with enthusiasm. In an age of lofty big-science that was often difficult to understand, the simple narrative of human against nature continued to grab people's attention. The undertone of rivalry between Fuchs and Hillary added an extra layer of sensationalism to the story: Edmund Hillary—whose role was to lay stores for the second half of Fuch's journey to the Ross Sea—broke with the plans when he made a "dash for the pole" to arrive before the leader of the expedition. With echoes of the Scott-Amundsen contest more than 50 years earlier, Fuchs implied that his work was genuinely scientific, and this is what had slowed him down on his way to the pole. As it turned out, the scientific contributions of the Trans-Antarctic Expedition were not quite as useful as some had hoped.[32]

    In a similar fashion to the Trans-Antarctic Expedition, Peter Mott's aerial survey of the Antarctic Peninsula was associated with the IGY, but not formally part of the enterprise. This was because IGY planners viewed cartography, survey work, and geology as areas that were potentially too contentious to promote political harmony. Nevertheless the aerial survey would be the most important endeavor associated with Great Britain's economic survey of the Falkland Islands Dependencies.[33] The Royal Society constructed Halley Base on the shores of the Weddell Sea as Great Britain's formal contribution to the IGY.[34] This enterprise had the greatest claim to being purely scientific, although it was no less politicized. Additionally all 14 of Great Britain's permanent stations in the Antarctic Peninsula, established across the region during the previous 15 years, contributed to the work of the IGY.

    In contrast to British attitudes, by the mid-1950s Argentinean and Chilean claims to the Antarctic Peninsula region had become somewhat detached from the environmental reality of the southern continent. The region mattered more as an "integral part" of their national territories than as a source of potential wealth. At the IGY Antarctic planning meeting in Paris in July 1955, representatives from Argentina and Chile successfully argued in favor of a "gentleman's agreement" in which the participants established that the scientific work of the IGY would in no way affect questions of sovereignty in Antarctica.[35] This agreement offered some insurance against the possibility that "scientific freedom" would be used to erode Argentinean and Chilean territorial claims. In the long term, however, there was little the two South American countries could do to stand in the way of precisely such a process.

In Argentina, the September 1955 overthrow of President Juan Domingo Perón plunged the country's Antarctic program into chaos on the eve of the IGY. Under President Perón, Argentina had become—by many indicators— the world's leading Antarctic power in the first half of the 1950s.[36] The purchase of an icebreaker *San Martín*, and the construction of Base Belgrano on the east coast of the Weddell Sea, both named after independence-era Argentinean heroes, attested to Perón's ambition to "saturate" Antarctica

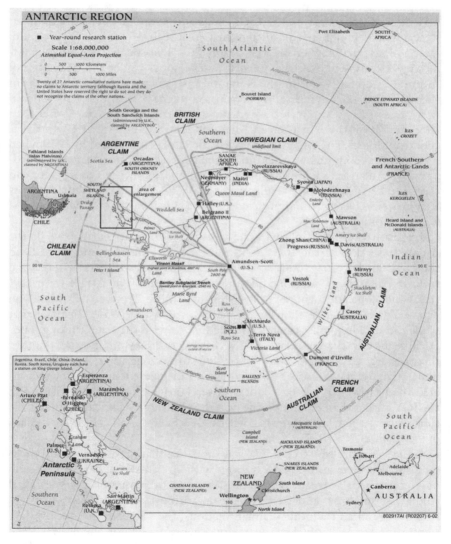

**Figure 13.1**  The territorial claims of Antarctica as defined in 2002. (Map produced by U.S. Central Intelligence Agency, Langley, Virginia.) (Courtesy of National Oceanic and Atmospheric Administration, NOAA Library Collection, Washington, D.C.)

with Argentinean bases. The military coup that toppled Perón destroyed Argentina's leading position. At Base Belgrano, Hernán Pujato, Perón's right-hand in Antarctic policy, survived for a short while as a curious vestige of Perónism until the new military government refused to supply him with the materials necessary to continue his scientific investigations, which might have included an expedition to the South Pole. As a consequence, Argentina went into the IGY with its leading position seriously compromised.

Nevertheless, despite the chaos precipitated by the Revolución Libertador, all eight of Argentina's permanent stations participated in some way with the IGY research, as did several research ships and temporary scientific bases.[37] Research was conducted in all ten areas of the IGY's Antarctic program: meteorology, geomagnetism, the aurora, the ionosphere, glaciology, seismology, oceanography, tides, gravity, and biology. Under the leadership of Admiral Rodolfo Panzarini, the Antarctic Institute represented Argentina in IGY planning conferences and participated in the formation of the Special (later Scientific) Committee on Antarctic Research (SCAR).[38] Between 1955 and 1959 the Argentinean Antarctic Institute published 47 scientific "contributions." The Argentinean scientific effort culminated in 1959 when a geophysical congress was held in Buenos Aires. The world's leading geophysicists, many on their way home from Antarctica, met in the Argentinean capital to discuss the early results of the IGY.[39] Argentina continued to play a significant role in Antarctic science, although it no longer occupied the leading position.

In relative terms, Chile was the poorest country to participate in the Antarctic section of the IGY. Politicians and officials in Santiago realized they did not have the funds to compete scientifically with their territorial rivals in Antarctica, and some questioned participation in a "race for bases" that Chile could not win. In May 1955 the Chilean Antarctic Commission debated whether Chile should even participate in the IGY.[40] The fact that Chile did decide to participate had much to do with the strength of geopolitical feeling in the country. Led by General Cañas Montalva, geopolitical theorists argued that the retention of Chilean sovereignty in Antarctica was imperative for the country's survival as an independent state. They considered the best way to protect Chilean sovereignty would be to participate in the IGY, rather than remain on the outside. Chilean IGY research, therefore, took place in the wider context of what Cañas liked to call the "New Science of Geopolitics."[41]

Chile's contribution to the IGY was similar to that of Argentina, although on a smaller scale: auroral observations with an "all-sky camera," meteorology reports sent to weather-central, glaciology, oceanography, gravitational studies, seismology, and participation in various "World Days." Just like Argentina and all the other claimant nations, the Chileans did not conduct significant research outside the region they claimed for themselves. The Chileans constructed a new base specifically for scientific research, which they named Base Risopatrón, after one of the originators of the idea of an "American Antarctic." Unfortunately for the Chilean IGY program, this base burned in March 1958, severely hampering Chilean scientific efforts.[42] As some officials had feared, the Chilean contribution to the IGY was, in relative terms, much less than the other countries with geopolitical interests in the region. In the new era of big-science in Antarctica, the Chileans were being forced to compete on a playing field that was far from level.

On a continental scale, research by the United States and the Soviet Union dwarfed the contributions of the nine other participating nations combined.[43] Thankfully for Great Britain, Argentina, and Chile, the Soviet Union did not establish any bases within the contested region: The focus of the Soviet effort was the relatively unexplored East Antarctica. The United States was the only country to challenge the contentious status quo by building a base in the region.[44] Initially, U.S. planners wanted to locate a base west of Weddell Sea, with the intention of filling a "gap" within scientific observations in addition to acting as a riposte to the three claimant countries. In any event, sea-ice conditions prevented the construction of a base in this area, and Ellsworth Station was built east of the Weddell Sea, close to Argentina's Base Belgrano and Great Britain's Shackleton and Halley bases. From a scientific perspective, therefore, the U.S. base was not quite as useful as it might have been, although it did conduct valuable work on ice thickness, meteorology, and upper-atmospheric research in the Weddell Sea region.[45]

Additionally, the U.S. South Pole station, deliberately sited at the conjunction of six of the seven territorial claims in Antarctica, represented a physical demonstration of the American reservation of territorial rights in any part of the continent (the exception being the Norwegian claim, which had no southern boundary). The U.S. construction of the South Pole station was also a classic example of technological and scientific bravado: demonstrating the U.S. ability to construct a base in such hostile surroundings. Such bravado—where the hostility of the environment actually became part of its attraction—was possibly trumped by the Soviet construction of a base at the so-called "Pole of Relative Inaccessibility," the place farthest from any Antarctic coastline, which lay in the sector claimed by Australia.[46] It was a sign of the changing times that when the British overland expeditions of Fuchs and Hillary staggered to the South Pole, they were greeted by U.S. scientists who had flown in and were living a life of relative luxury at the bottom of the world.

In common with the British, U.S. Antarctic planners also viewed the IGY as an opportunity to assess the continent's economic potential. Uncertainty and vacillation in U.S. Antarctic policy stemmed from the fact that the State Department and the Defense Department could not agree between themselves which parts of Antarctica to claim for the United States, or whether sovereignty claims should even be made at all. The Defense Department was keen on reserving U.S. rights to the whole of Antarctica until it was known which parts of the continent, if any, contained valuable minerals. Admiral Duncan, for example, the Navy's representative to the National Security Council agued that "we should carry out our plans to achieve our scientific purposes, but do nothing about establishing claims to Antarctica until we see what is actually revealed about the resources of the area in the course of carrying out our program for the IGY."[47] Both Great Britain and the United States, therefore, participated in the IGY with their official policies toward Antarctica dependent—at least to some extent—on the actual scientific results of the enterprise.

## Internationalization

The scientific results of the IGY began to emerge almost before the research had officially begun. With hundreds of scientists traveling to Antarctica, establishing research stations, and conducting research, scientific understanding of the Antarctic environment increased exponentially.[48] Among some of the most important scientific results of the IGY: a series of overland traverses confirmed the average depth of the ice was several kilometers thick; meteorologists made important advances in understanding the workings of the Antarctic polar front; and geologists demonstrated that Antarctica was in fact one continent.[49] Importantly the continental scale of IGY research enabled a much better understanding of the Antarctic Peninsula region to emerge. Geophysical research demonstrated interconnections among various geophysical processes, and it was becoming impossible to understand one part of the continent without reference to the Southern Hemisphere as a whole.

Before the IGY, there was no consensus on the depth of the ice in Antarctica. Many scientists speculated the average depth would be several hundred meters thick: In 1956, for example, British scientists wrote confidently that Antarctica contained significantly less ice than the Arctic. The Norwegian-British-Swedish Expedition to Queen Maud Land of 1951–52 conducted the first detailed seismic soundings of the ice depth in the interior of the continent.[50] The expedition discovered the ice in this region was actually several kilometers thick. It was not until the widespread traverses conducted as part of the IGY, however, that scientists could confirm that "thick ice" was the norm across Antarctica. The average thickness of Antarctic ice appeared to be at least 2,500 meters, meaning that the region in fact contained a much greater volume of ice than the Arctic. So deep was the ice that it actually appeared to be compressing the land underneath it, creating a pear-shaped continent. According to one author, as a result of these investigations estimates of Earth's water budget had to be revised upward by at least 20 percent.[51] These results threw into doubt the assumption that the south polar region was necessarily going through the same climatic process as the north polar region. More subtly, revised estimates of the depth of the ice challenged assumptions that any mineral deposits found beneath the ice would be possible to exploit with existing technology.

The location of scientific research stations across the Antarctic continent made possible significant increases in the understanding of the meteorology of Antarctica, as did a strategy of taking coordinated meteorological readings, especially by radiosonde, from across the Antarctic continent. Sixteen stations used weather balloons tracked by radar; these gave coordinated readings on designated world days. IGY meteorological research revealed the continent had a "coreless" winter in which temperatures were not as low as expected and stayed the same for around six months. The IGY established the basic outline of Antarctic weather that is known today: succession of cyclones around latitude 60° south, which rarely penetrate into the high central plateau, with nearly stationary continental anticyclones over Marie Byrd Land and the Pole of Inaccessibility. As expected, records for low temperatures were broken, and the coldest temperature ever measured to that date (−87.4°C) was recorded at the Soviet Union Vostok base in July 1958.[52]

Owing to its political sensitivity, geology was not formally part of the IGY program. Nevertheless, many geologists traveled to Antarctica between 1957–58 and conducted surveys alongside official IGY activities. With more than 98 percent of Antarctica's land covered in ice, geologists faced serious practical difficulties. Using seismic soundings, and information gleaned from the exposed rock formations, geologists were able to piece together the basic structure of the Antarctic continent. Work conducted during the IGY demonstrated that although east and west Antarctica were geologically distinct, the two parts were indeed connected—there was no deep trough as some had thought.[53] Geology during this period took place under the shadow of the continental drift controversy: it would not be until the early 1960s that the theory of plate tectonics was widely accepted. Geological research in Antarctica played a role in contributing data to this broader debate, but the southern continent was not central to the resolution of the controversy. Ultimately, from a political perspective, the most significant geological result of the IGY was negative: the participants found no major mineral deposits in Antarctica. Although neither before, during, nor after the IGY was there any single point when scientists declared that Antarctica contained little or nothing of immediate economic value, this dawning realization had a major impact on the politics of the Antarctic continent.

The scientific results of the IGY caused British politicians and officials to reassess the worth of maintaining exclusive political sovereignty in the Falkland Islands Dependencies. Politicians in all departments ceased to use the potential for mineral wealth as an argument for maintaining exclusive sovereignty in Antarctica. The concurrent decline of the Antarctic whaling industry in the late 1950s also detracted from the continent's economic luster. The ability of the Falkland Islands Dependencies to pay for themselves was central to British imperial policy. When the region no longer appeared to hold any significant economic potential, at least in the short-to-medium term, officials began to consider other options for the retention of exclusive sovereignty. The British did not, however, want to abandon their political influence entirely. Some form of limited internationalization, they reasoned, offered the best means of defusing tensions in Antarctica while retaining political influence in the region.

In a similar fashion, the results of the IGY convinced U.S. politicians and officials that internationalization would be the best course for Antarctic politics. U.S. officials wanted to prevent the cold war from spreading to the southern continent, and ideally achieving this goal by excluding the Soviet Union from the region. They feared, with some justification, that the Anglo-Argentinean-Chilean sovereignty dispute might be exploited by the Communist Bloc as a demonstration of divisions within the Western Alliance. When it became increasingly clear that Antarctica contained little in the short-to-medium term that could justify a formal U.S. claim to any part of the continent, the central U.S. policy objective became the defusing of political tensions. Just like their British counterparts, U.S. officials were unwilling to give up permanently their "historic rights" in Antarctica and so, as a consequence, did not want any sort of genuine internationalization, such as what might have happened had the Antarctic question been handed over to the United Nations.

Having decided that some form of limited internationalization would be the best option for their interests in Antarctica, officials in Great Britain and the United States set out to bring this about. The rhetoric of scientific internationalism would be one of the principal tools for achieving their political goals. Science in general, and the goodwill generated by the IGY in particular, offered a nonthreatening way to bring territorial rivals together to discuss political questions. A series of secret meetings involving Great Britain, United States, Australia, and New Zealand in the second half of 1957 and into 1958—excluding Argentina and Chile—set in motion the process that would lead to the signature of the 1959 Antarctic Treaty. Initially, the four countries disagreed on the question of Soviet involvement. U.S. officials, perhaps somewhat naively, believed they could create a treaty regime for Antarctica that would exclude the Soviet Union. British officials—who were especially keen on resolving the dispute—argued, more realistically, that for any internationalization of Antarctica to work the Communist superpower would have to be included. After some discussion, the British position prevailed.

In May 1958 the United States invited the other 11 nations that had participated in IGY research to Washington in order to discuss the political future of Antarctica. As a nonclaimant power, the United States felt itself best placed to host such a conference. The explicit connection between science and politics within the invitation offered a neat justification for the exclusion of potential "troublemakers," including Soviet satellites and newly independent states of the Third World. After months of preliminary negotiations, the Washington conference of October–December 1959 led to the signature of the Antarctic Treaty. Article 4 of this treaty suspended all existing sovereignty claims, neither recognizing them nor rejecting them. This provision effectively brought to an end the active phase of the sovereignty dispute among Great Britain, Argentina, and Chile. The Antarctic Treaty was also innovative in several of its other clauses: Article 3 established the free exchange of information; Article 5 prohibited nuclear explosions or the disposal of radioactive waste; and Article 7 created an open-inspection regime. The Antarctic Treaty was especially remarkable for being signed in the midst of cold-war tensions, and many of its clauses—especially the prohibition

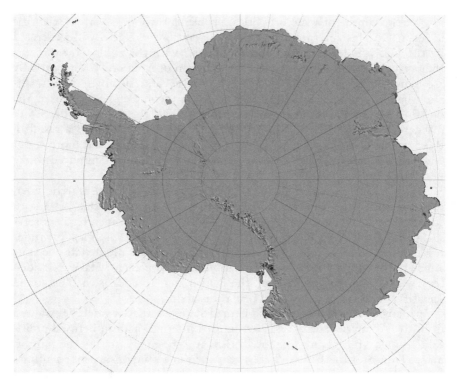

**Figure 13.2** Map of the continent of Antarctica, 2000. (Courtesy of National Oceanic and Atmospheric Administration, NOAA Library Collection, Washington, D.C.)

of nuclear explosions—would later be cited as precedents for international cooperation.

Argentina and Chile found themselves swept along by the process of internationalization. Their ratification of the Antarctic Treaty helped legitimize the southern continent's new international regime. They found themselves members of an "exclusive club," in which the vestiges of imperial influence not only survived but in some ways flourished. Rather than remaining on the outside to challenge the continued "Antarctic imperialism" of countries such as Great Britain, the United States, and the Soviet Union, the Argentineans and Chileans decided to collaborate with the new regime. This collaboration took away the possibility of a united front of postcolonial nations against the Antarctic Treaty system, thereby weakening one of the few potential challenges to the new status quo.

## Conclusion

The traditional narrative of the internationalization of Antarctica suggests that the IGY functioned as a deus ex machina, coming out of nowhere to resolve the continent's vexed political problems through a wave a scientific idealism. In this interpretation, the cooperation and goodwill generated by the IGY acted as a force "above politics," with the ability to overcome the petty squabbling that plagued the question of Antarctic sovereignty. By ascribing a full role to IGY science and to the Antarctic environment, this essay has shown such an interpretation to be incomplete. Speaking at the opening of the Washington conference of 1959, the head of the Soviet delegation stated: "It may be said without exaggeration that, as a result of this international scientific co-operation, mankind has learned more about Antarctica in the last three or four years than in all the one hundred thirty years since the day of discovery."[54]

IGY science fundamentally changed perceptions of the Antarctic environment and, at least in the short term, dispelled the myth that Antarctica could be a frozen El Dorado brimming with valuable minerals. This reassessment of the economic worth of Antarctica changed political attitudes toward the continent, and both Great Britain and the United States became amenable to a limited internationalization of the continent. Having decided they wanted internationalization, the British and Americans then exploited scientific goodwill as a means to bring this about.

In many ways, Great Britain and United States got exactly what they wanted from the Antarctic Treaty of 1959: limited internationalism defused political tensions, while claims (for the British) and the reservation of the right to make claims (for the United States) remained in a state of suspended animation, to be renewed if ever the occasion should demand. The Argentineans and Chileans viewed the Antarctic environment differently, seeing it as an "integral part" of their national territories. They opposed any form of internationalization and only participated in the Antarctic Treaty negotiations when they realized the weight of international opinion was against them. Nevertheless, despite this reluctance, their participation helped give credibility to the solution of limited internationalization. Science also offered the Antarctic Treaty signatories a useful tool for excluding unwanted countries from their new political club. Far from being a simple story of (good) science trumping (bad) politics, the history of the connection between the IGY and the Antarctic Treaty involved the political exploitation of scientific goodwill to achieve essentially political objectives.

By examining the dynamic interaction of environment, science, and sovereignty during the IGY in the Antarctic Peninsula, this essay offers a slightly more critical way of looking not only at the origins of the Antarctic Treaty, but also at the contemporary environmental politics of the Antarctic continent. This historical perspective raises important questions about who gets to speak for the Antarctic environment. The 12 original signatories deliberately exploited scientific goodwill in order to retain their political interests in the continent, to the continuing exclusion of many countries in the global south. During treaty negotiations, the signatories of the Antarctic Treaty were explicit in their desire to exclude potential "troublemakers" from the continent.[55] The requirements for a country to be conducting significant scientific research in Antarctica in order to become a member of the Antarctic Treaty system offered a convenient and relatively uncontroversial barrier to entry. Today, for example, South Africa remains the only African nation to be a member of the Antarctic Treaty system, despite the fact that the consequences of environmental change in Antarctica—in particular global warming—will disproportionately affect the world's poorest nations. The exclusion of these economically marginalized countries leaves them without a voice at the political negotiating table and reliant on the paternalistic goodwill of the member states to serve their best interests. This situation perpetuates a colonial mentality in Antarctica.

Rather than bringing imperial interests in Antarctica to an end, as the traditional interpretation would suggest, the Antarctic Treaty reformulated and retained those interests. This observation opens Antarctica to study from within a postcolonial framework. Postcolonial scholarship seeks to highlight and challenge continued imperial practices of exclusion and unequal power-relationships after the "decolonization" of most of the colonized world in the mid-twentieth century.[56] Despite the numerous achievements of the Antarctic Treaty system in protecting the environment and maintaining peace, it remains firmly rooted in the power structures of Western imperialism and the cold war. The Antarctic Treaty itself is a distinctly postcolonial treaty, because the retention of imperial influence is written into its text. Article 4, in particular, which effectively suspends—rather than refutes—all preexisting rights and claims, highlights the continuation of imperial influence in Antarctica. If the treaty were ever to lapse, then the legal status of Antarctica could revert to its pre-1959 situation. Hints of this have already been seen in debates about a minerals regime for Antarctica and assertions of maritime "exclusive economic zones" as part of the United Nations Convention on the Law of the Sea.[57] In these debates, Antarctic claimant countries have maneuvered to seek advantages for themselves based largely on imperial claims and activity that took place before the signing of the Antarctic Treaty. Not only does this situation discriminate against member states that have no claims of their own, but it leaves the nonmember states firmly on the outside of any potential minerals bonanza in Antarctica.

The Antarctic Treaty created a continent "dedicated to peace and science." In addition to having an intrinsic value of its own—especially at a time of growing awareness of the centrality of the southern continent to the global environmental system—science has also done much to keep the peace in Antarctica. Scientific cooperation has laid the basis for half a century of peaceful coexistence in a region that was becoming increasingly contentious in the 1940s and 1950s. As a consequence of this success, scientific activity plays an important role in legitimizing the Antarctic Treaty system, in both its positive and negative dimensions. The IPY of 2007–08 demonstrates the ongoing dominance of science in the south polar region. The historical approach adopted by this essay allows for comparison across time. Just as the British Empire once justified its sovereignty to the Falkland Islands Dependencies through assertions of environmental authority "for the good of humanity," the Antarctic Treaty system continues to justify its dominant role in Antarctic politics through assertions of environmental authority, "for the good of humanity."

## Notes

1. Richard S. Lewis, *A Continent for Science: The Antarctic Adventure* (New York: Viking Press, 1965), 95–97.
2. G. E. Fogg, *A History of Antarctic Science*, Studies in Polar Research (New York: Cambridge University Press, 1992), 290.
3. Laurence Gould, head of the U.S. IGY committee, quoted in Philip W. Quigg, *A Pole Apart: The Emerging Issue of Antarctica* (New York: McGraw-Hill, 1983), 46.
4. Peter Beck, *The International Politics of Antarctica* (London: Croom Helm, 1986); Stephen Martin, *A History of Antarctica* (Sydney, New South Wales: State Library of New South Wales Press, 1996).
5. Oscar Pinochet de la Barra, *La Antártica Chilena* (Santiago de Chile: Editorial del Pacífico, 1948), 164.
6. Klaus Dodds, *Pink Ice: Britain and the South Atlantic Empire* (New York: I.B. Tauris, 2002).
7. Adolfo Scilingo, *El Tratado Antártico; Defensa De La Soberanía Y La Proscripción Nuclear* (Buenos Aires: Librería Hachette, 1963).
8. Walter Sullivan, *Assault on the Unknown; the International Geophysical Year* (New York: McGraw-Hill, 1961).
9. Examples of authors who make this argument include Sullivan, *Assault on the Unknown;* Fogg, *A History of Antarctic Science;* Martin, *A History of Antarctica;* and Stephen J. Pyne, *The Ice* (London: Weidenfeld & Nicolson, 2003).
10 The Antarctic Treaty does not contain a built-in expiration date. The Madrid Environmental Protocol to the Antarctic Treaty, which entered into force in 1998, has a specified duration of 50 years, at which point revisions might be considered.
11. Robert Headland, *Chronological List of Antarctic Expeditions and Related Historical Events*, Studies in Polar Research (New York: Cambridge University Press, 1989), 111–118.
12. Beau Riffenburgh, *The Myth of the Explorer: The Press, Sensationalism, and Geographical Discovery* (New York: Oxford University Press, 1994).
13. Headland, *Chronological List of Antarctic Expeditions*, 219.
14. Roland Huntford, *Scott and Amundsen* (New York: Atheneum, 1984).
15. Max Jones, *The Last Great Quest: Captain Scott's Antarctic Sacrifice* (New York: Oxford University Press, 2003).
16. J.N. Tønnessen and Arne Odd Johnsen, *The History of Modern Whaling* (Berkeley: University of California Press, 1982), 147–201.
17. Tønnessen and Johnsen, *The History of Modern Whaling*, 178–182.
18. Miles Clifford, "Broadcast Address by His Excellency the Governor," in *Sir Miles Clifford Papers* (Oxford: Rhodes House Library, 1948).
19. Examples include John M. MacKenzie, *The Empire of Nature: Hunting, Conservation, and British Imperialism* (Manchester, UK: Manchester University Press, 1988); David Arnold, *The Problem of Nature: Environment, Culture, and European Expansion* (Cambridge, Massachusetts: Blackwell, 1996); and Richard Harry Drayton, *Nature's Government: Science, Imperial Britain, and the "Improvement" of the World* (New Haven, Connecticut: Yale University Press, 2000).
20. Ann Savours and Margaret Slythe, *The Voyages of the Discovery: The Illustrated History of Scott's Ship* (London: Chatham, 2001).
21. Richard Evelyn Byrd, *Alone* (New York: G. Putnam's Sons, 1938).

22. Pinochet de la Barra, *La Antártica Chilena*, 164. Hearings before the Subcommittee on Appropriations, 76th Congress, 1st and 3rd sessions.
23. Pinochet de la Barra, *La Antártica Chilena*, 164. Hearings before the Subcommittee on Appropriations, 76th Congress, 1st and 3rd sessions.
24. Dodds, *Pink Ice: Britain and the South Atlantic Empire*, 14–18.
25. Frank Klotz, *America on the Ice: Antarctic Policy Issues* (Washington, D.C.: National Defense University Press, 1990), 14–26.
26. Lewis, *A Continent for Science*, 62.
27. Allan A. Needell, *Science, Cold War, and the American State: Lloyd V. Berkner and the Balance of Professional Ideals* (Amsterdam: Harwood Academic, 2000).
28. Fogg, *A History of Antarctic Science*, 169.
29. Michael Smith, *Sir James Wordie, Polar Crusader: Exploring the Arctic and Antarctic* (Edinburgh: Birlinn, 2004).
30. The Cabinet Office, "Future Constitutional Development in the Colonies," (London: TNA, 1957).
31. Vivian Fuchs and Edmund Hillary, *The Crossing of Antarctica: The Commonwealth Trans-Antarctic Expedition, 1955–1958* (London: Cassell, 1958).
32. Klaus Dodds, "The Great Trek: New Zealand and the British/Commonwealth 1955–58 Trans-Antarctic Expedition," *Journal of Imperial and Commonwealth History* 33, no. 1 (2005).
33. G. Mott, *Wings over Ice: An Account of the Falkland Islands and Dependencies Aerial Survey Expedition 1955–57* (Long Sutton: Mott, 1986).
34. David Brunt and Royal Society (Great Britain), *The Royal Society International Geophysical Year Antarctic Expedition: Halley Bay, Coats Land, Falkland Islands Dependencies, 1955–1959*, 4 vols. (London: Royal Society, 1960).
35. Oscar Pinochet de la Barra, *Medio Siglo De Recuerdos Antárticos: Memorias* (Santiago de Chile: Editorial Universitaria, 1994).
36. Eugenio A. Genest, *Pujato Y La Antártida Argentina En La Década Del Cincuenta* (Buenos Aires, Argentina: H. Senado de la Nación, Secretaría Parlamentaria, Dirección Publicaciones, 1998); Susana Rigoz, Hernán Pujato: El Conquistador Del Desierto Blanco (Buenos Aires: Editorial María Ghirlanda, 2002).
37. Instituto Antártico Argentino, *Report to SCAR on Antarctic Scientific Activities during the International Geophysical Year, 1957–1958* (Buenos Aires: 1960).
38. S.M. Comerci, *La Ciencia Argentina En El Antártico: El Instituto Antártico Argentino* (Buenos Aires: 1979).
39. Headland, *Chronological List of Antarctic Expeditions*, 386.
40. Comisión Antártica Chilena, 1955.
41. Biblioteca del Congreso Nacional, *Donación General Cañas Montalva: Catálogo Del Fondo Bibliográfico* (Santiago: Biblioteca del Congreso Nacional, 1972).
42. Headland, *Chronological List of Antarctic Expeditions*, 374.
43. Sullivan, *Assault on the Unknown*.
44. Dian Olson Belanger, *Deep Freeze: The United States, the International Geophysical Year, and the Origins of Antarctica's Age of Science* (Boulder, Colorado: University Press of Colorado, 2006), 203–215.
45. John C. Behrendt, *Innocents on the Ice: A Memoir of Antarctic Exploration, 1957* (Niwot: University Press of Colorado, 1998).
46. Headland, *Chronological List of Antarctic Expeditions*, 376.
47. The State Department, "Memorandum of Discussion at the 272nd Meeting of the NSC," (Washington: Foreign Relations of the United States, 1955–57, 1965).
48. Quigg, *A Pole Apart: The Emerging Issue of Antarctica*, 40.
49. Lewis, *A Continent for Science*.
50. Fogg, *A History of Antarctic Science*, 265.
51. Lewis, *A Continent for Science*, 72–93.
52. Fogg, *A History of Antarctic Science*, 304.

53. Lewis, *A Continent for Science*, 125–145.

54. Quigg, *A Pole Apart: The Emerging Issue of Antarctica*, 40.

55. British Embassy Washington, letter to Foreign Office, January 14, 1958. The National Archive (London), FO 371/131905.

56. Barbara Bush, *Imperialism and Postcolonialism*, first ed., History: Concepts, Theories, and Practice (New York: Pearson Longman, 2006).

57. See, for example, Owen Bowcott, "Argentina Ready to Challenge Britain's Antarctic Claims" *The Guardian* (October 19, 2007).

## Chapter 14

# The International Geophysical Year in Antarctica: A Triumph of "Apolitical" Science, Politics, and Peace[1]

*Dian Olson Belanger*

Half a century ago, during the depths of the cold war, nations and personalities from around the globe managed to link their respective self-interests, rivalries, suspicions, and fears to a noble common purpose—the cooperative pursuit of physical knowledge of Earth. Their scientific endeavor, called the International Geophysical Year (IGY, 1957–58), began as a proposal for intensified study of the polar regions. And in Antarctica, the IGY represented 18 months[2] of breakthrough scientific-pioneering, the first systematic, continent-wide attempt to understand the limitless ice and the atmosphere enveloping it. Enabling the science were logistical and operational victories hard won by the U.S. Navy. Its extraordinary success, with that of 11 other participant nations, inspired the Antarctic Treaty of 1959, under which the IGY never ended in the polar south. Science, politics, and humanity have all been well served. All three major players in the Antarctic program—military, diplomatic, and scientific—brought to the IGY era prior south-polar experience.

First, scientists, the personification of the IGY, had commonly accompanied early Antarctic exploring expeditions. Their observations and investigations, however puny overall, exploded the knowledge base. But so much remained unknown: How much ice was there? How vast, how deep? How did Antarctic cold influence global weather patterns? How did solar and atmospheric phenomena interplay with the southern magnetic and geomagnetic poles? By the 1950s, advanced technologies greatly extended scientists' reach: radar-tracked and rocket-propelled weather balloons; new and improved aurora-revealing spectroscopes, ionosphere sounders, and cosmic ray recorders; and electronic computers. Scientists hungered to apply these tools to the secrets of the ice.[3]

Second, the navy, which would assume all logistical support for the U.S. IGY in Antarctica, had twice penetrated the icy Southern Ocean. Lieutenant Charles Wilkes, leading the 1838–42 U.S. Exploring Expedition on long cruises off the coast of East Antarctica, recognized as a continent what he was seeing distantly; it was he who named it Antarctica.[4] A century later, in 1946–47, the navy sent Operation Highjump—its largest cold-weather training exercise ever—to the southern high latitudes to enhance readiness for a feared Soviet attack over the *North* Pole and, surreptitiously, to strengthen America's basis for a territorial claim. Scientists sailed with both expeditions and contributed scientific output that holds up well to scrutiny today. But the navy's objectives were strategic and would remain so, at least covertly, in Operation Deep Freeze, to come.[5]

Third, the U.S. government, led by the Department of State, had agonized for decades about the political question of Antarctica's "ownership." Seven friendly countries—Argentina, Australia, Chile, France, Great Britain, New Zealand, Norway, and the United Kingdom—already claimed wedges of territory. Except for Norway's undefined northern and southern boundaries, these claims terminated at the South Pole. Some sectors overlapped, with occasional conflict. The United States asserted its basis for a claim, surely a "right" given its domination of polar exploration in the interwar period. Indeed, at President Franklin Roosevelt's quiet urging, explorers such as Richard Byrd, Lincoln Ellsworth, and Finn Ronne deposited copious claim sheets. But these were never officially acted on, so for years, while government officials dithered, they pressed no claims and explicitly recognized no one else's. In 1948, seeking a solution, the United States proposed joint sovereignty over the polar continent to be exercised by the claimants plus itself. The idea went nowhere—with either the claimants or the uninvited Soviet Union. The issue continued to fester. Some of the seeds planted would, however, bear fruit a decade later in landmark treaty talks.[6]

Two profound outcomes of World War II also affected the unfolding of the IGY. The first was the swift postwar descent into a tense and terrifying cold war: for example, Europe divided by the Iron Curtain; Communist expansion in Asia; and, especially, the nuclear arms race. Indeed, in seizing a singular opportunity to step back from the nuclear abyss, the IGY community served humanity, writ large, in addition to science and politics.

The second result was that science, through such pivotal developments as radar, the proximity fuse, and the atomic bomb, was widely credited with winning the war. Scientists emerged from the conflict with enormous prestige and political influence. Several—such as Lloyd Berkner, ionospheric physics leader, radio engineer, and veteran of Byrd's first Antarctic expedition (1928–29)—were making highly visible careers of bringing science to bear on issues of national policy and security.[7]

In this historic context came the now-famous dinner party on April 5, 1950, at the Silver Spring, Maryland, home of space scientist James Van Allen, where the IGY began. It was in character, if allegedly spontaneous, for energetic, "ideafull" Berkner[8] to suggest a new International Polar Year (the third IPY since the first in 1882–83; a second had followed in 1932–33). Perhaps it was also no surprise that the honored guest, renowned British geophysicist Sydney Chapman, returned an enthusiastic "Good idea, Lloyd!"

along with the pertinent reminder that 1957–58 would be a period of maximum solar activity—a perfect moment for concentrated cooperative study of Earth's coldest regions and an excuse to cut by half the previous 50-year polar-year interval.[9]

The charm of this now iconic conversation was that a few science luminaries had the political clout, professional connections, and personal charisma to casually initiate what they knew would be a long process leading to an ambitious, expensive polar (soon global) science endeavor, with full confidence that it would happen and that someone would pay for it. They all knew that only governments could.

Their proposal won the enthusiastic embrace of the International Council of Scientific Unions (ICSU). When leaders of the otherwise keen World Meteorological Organization considered the polar focus too narrow, Chapman proposed an International Geophysical Year, and the third IPY morphed into IGY, a cooperative worldwide effort to probe the secrets of Earth entire.[10] Sixty-six nations would eventually participate. Thus, if its IPY birth made the IGY concept not strictly new, its scope, scale, and results would be unprecedented. Antarctica, along with outer space, one of Earth's two great unknowns of the time, remained a centerpiece. Twelve countries would pursue science on the polar continent at several dozen stations.[11]

**Figure 14.1**  Aerial view of the first permanent station built at the South Pole, taken on December 4, 1956. The U.S. Navy built seven stations, including one at the South Pole, in support of the International Geophysical Year. (Photograph by Dick Prescott. Courtesy of National Science Foundation, Washington, D.C.)

As politically attuned, influential, and chauvinistic as they were, the science leaders kept IGY planning determinedly apolitical—at least the public face of it. To that end, when ICSU invited its member national academies of sciences to participate, it did so without regard to national political philosophy. Even the Soviet Union's academy was invited, although it did not then belong to ICSU.[12] Argentina sent an ambassador, not a scientist, to the first Antarctic meeting in Paris in July 1955, and when this political appointee protested a wall map (used only for siting stations) showing Chilean and British claims that were in conflict with Argentina's, France's Georges Laclavère, the savvy presiding officer, who had insisted that only "technical" (scientific) matters be discussed, dramatically tore it down.[13]

When the science planners finalized the list of official IGY sciences to be studied in Antarctica, they deliberately excluded the quintessentially geophysical discipline of geology. It was thought too political: someone might discover a valuable mineral resource, which could set off a "gold rush" and inflame adversarial relationships or the claims issue. Unsurprisingly, enthusiastic geologists sought, picked away at, and quietly documented every attainable rock outcropping anyway.[14]

The IGY planners promoted international cooperation by agreeing to simultaneous, identical observations, measurements, and record keeping; coordinated radio communications, logistical support, and search-and-rescue operations; and intensified information-gathering during so-called World Days and occasional ten-day World Intervals. They also set up still-used World Data Centers, where all qualified researchers from anywhere could later access all findings.[15]

The ornament of international cooperation, though, was Antarctic Weather Central, the brainchild of Harry Wexler, U.S. IGY chief scientist for Antarctica. Scientists at this Little America facility would collect, analyze, research, and share weather data via a complicated "mother-daughter" radio relay system from every science station as well as from ships, planes, and field parties. Vladimir Beloussov, the Soviet leader, endorsed the concept, but only if it would be truly international—that is, if Soviet and other meteorologists could work there. Wexler shrewdly agreed, but only if it would be a true exchange—if a U.S. "met" could live and work at Mirny, the main Soviet station. At length, numerous political hurdles on both sides were overcome, and productive, mutually beneficial reciprocal assignments proceeded. Walter Sullivan wrote, "Nothing could have done more to allay [cold war] fears and suspicions than this exchange."[16]

Of course, politics was ever-present. Criteria for locating stations were one example. Byrd's science leader, George Toney, later opined that his station was planted "spang in the middle of a huge unclaimed wedge of Antarctica where the United States might well launch a claim later on, if it came to that." When, years later, an interviewer asked U.S. IGY vice-chair Alan Shapley if Wilkes, one of the so-called science "gap" stations, was sited comparatively near the Soviet Mirny for political reasons, his response was a conspiratorial "Sshhhh."[17]

**Figure 14.2** Wilkes Station was built by the United States for the International Geophysical Year. This photograph was taken in February 1957. The United States turned the station over to Australia in 1959. Several years later Australia built a new facility 2 km away and renamed it Casey Station. (Photograph by Olav Loken. Courtesy of National Science Foundation, Washington, D.C.)

The question of who would take the geographic South Pole, a coveted prize despite the daunting logistical challenges, was especially politically charged. When the late-arriving Beloussov announced his country's intent to build at the pole, early U.S. interest had not yet become a firm commitment, according to Paul Siple, Byrd's protégé, who would lead the station through its first winter. Whatever took place behind the scenes, the smooth Laclavère said with an "air of regret" that the Americans' prior offer had been accepted, but he noted a great gap in science coverage in East Antarctica. The conference urged, and the Soviet Union accepted, the South Geomagnetic Pole and, later, the Pole of Inaccessibility. Only the language of science was spoken (aloud).[18]

The planners carefully minimized politics organizationally. The ICSU set up a special international coordinating committee (Comité Spéciale de l'Année Géophysique Internationale) for the IGY, but each participating country planned and implemented its own program, according to its own means and interests. So there was no international program-management body or complex finance, thus ensuring minimal direct contact among political adversaries and low overhead, too. The IGY secretariat in Brussels never employed more than eight people; its annual budget over five years ran around $50,000 (1958 dollars). Laurence Gould, Byrd expedition veteran and chair of the U.S. IGY Antarctic committee, recalled "a simplicity, a flexibility and freedom from political consideration hitherto unknown."[19]

Government funding introduced domestic politics, of course. But U.S. IGY planners astutely asked Congress for the money only after their programs were substantially in place, at which point it would be harder for the lawmakers to say no—even harder when the planners testified that the Soviet Union had committed to a large and costly program. (The Soviets reportedly used the same ploy in Moscow.)[20]

At home, the IGY came to fruition under the guidance of the U.S. National Committee for the IGY (USNC–IGY), led and staffed by the National Academy of Sciences (NAS). But only a government agency, not the prestigious private academy, could request, receive, and dispense federal dollars, so the committee asked the fledgling National Science Foundation (NSF) to assume the role of IGY banker. Taking on the IGY in effect doubled NSF's modest budget. Despite their shared interests and goals, the NAS and NSF differed in personality, philosophy, and modus operandi. Their interactions would become increasingly laced with controlled tension and mutual frustration as they jockeyed for position within a multiagency effort, each with its own agenda.[21]

Logistical support was, as it still is, the most expensive and challenging component of doing science in Antarctica. For the United States, only the navy (assisted by other military services) had the expertise, experience, and specialized equipment to provide that support. Although the science-navy relationship would continue for 44 years, naval leaders early on questioned Antarctica's strategic value, their only defensible justification for involvement. Still, they could use likely outcomes of IGY studies in Antarctica such as better understanding of radio-wave propagation and improved weather prediction. And they were loath to see any other power gain either political or economic advantage—whether controlling a claim or being the first to find a mineable Mother Lode, which surely existed.[22]

Despite this ambivalence, Admiral George Dufek, commander of Operation Deep Freeze, did all he could to meet the scientists' objectives. That meant siting, building, and operating two foothold bases in Deep Freeze I (1955–56). One was a logistics base and air-operations facility at McMurdo Sound, essential to support the establishment and maintenance of a U.S. station at the South Pole. The other, Little America V, on the Ross Ice Shelf, was to become the flagship science station for the IGY. In Deep Freeze II the following brief austral summer (1956–57), the navy would have to site and establish five widely scattered stations, all of whose feasibility, even accessibility, was still unknown and iffy at best.[23]

All the proposed science stations presented staggering challenges of access, terrain, and weather—especially the two deep, interior sites. Byrd Station, 650 miles inland, had to be built and supplied by tractor train from Little America, through a seven-mile-wide belt of wild crevasses—terrifying for all and deadly for one Seabee driver who, while pushing snow into one blasted-open icy chasm, unknowingly backed his 35-ton D-8 Caterpillar tractor into

**Figure 14.3**   The South Pole Station in 1957. Taken during the Austral summer of 1956–57. (Photograph by Cliff Dickey. Courtesy of National Science Foundation, Washington, D.C.)

another.[24] Only personnel and the most delicate scientific equipment could be practicably landed by aircraft at the South Pole, if in fact anything could. The first successful test landing and—even more critical—takeoff took place on October 31, 1956, just weeks before pole construction had to begin. Some 850 miles distant, and nearly two miles above sea level, the station could only be equipped and supplied in the time available by airdropping everything else—from diesel-power generators to canned potatoes.[25]

The final three widely scattered coastal stations, ostensibly filling geographic "gaps" for science but also responding to political realities, were slated for the Cape Adare, Knox Coast, and Weddell Sea areas. The ships supporting these efforts, icebreakers and cargo vessels alike, suffered severe ice damage, some life-threatening. Impeded approaches forced leaders to settle for alternative locations, while ice-delayed arrivals pushed the Seabees to exhaustion in their base-construction efforts. All three stations were renamed—becoming Hallett (operated bilaterally with New Zealand), Wilkes, and Ellsworth stations, respectively.[26] The triumphs and tragedies of the mostly anonymous mid-century pioneers who accomplished these feats take no back seat to those of earlier heroes we know by name.

The unusual navy-IGY relationship, a clash of goals and cultures, was challenging, often testy. Navy captains facing hurdles and risks no one not

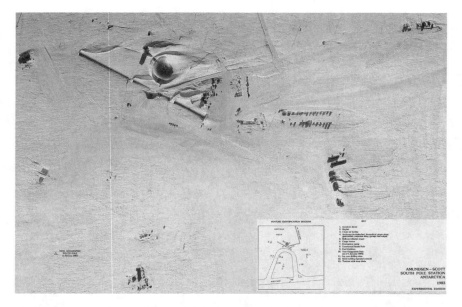

**Figure 14.4** Overhead aerial view of the Amundsen-Scott South Pole Station, Antarctica, taken in 1983. (Courtesy of U.S. Central Intelligence Agency, Langley, Virginia)

on the scene could appreciate would suggest more easily attained station sites, short of the scientists' targets. Science leaders, desperate to be to be in place on time to fulfill their scientific promises, would accuse the navy of indifference to the IGY, of not trying hard enough against the ice. The navy in turn would see the scientists' impatience as unseemly ingratitude and foolish disregard of the perils. A similar exchange of sentiments followed Dufek's order that flying IGY equipment to Byrd and the pole was a last priority until both had been stocked with food and fuel sufficient for a year.[27]

The most serious IGY–U.S. Navy dispute arose over the awkward issue of the command structure in Antarctica. Because the IGY was a scientific endeavor, science leaders insisted that a scientist be in charge of each station. The navy countered that military personnel could not take orders from a civilian, so a naval officer must exercise the station's leadership. After passionate argument, they compromised on a dual-command system, whereby a leader of each side would see to his own people. These two would be expected to work together for the good of the program. Ironically, this mutually deplored system worked remarkably well on the ice; the most conspicuous leadership failure during the IGY was at Ellsworth, the one station where one leader wore both hats.[28]

Science in Antarctica was thrilling for the scientists who pursued it-mostly young graduate students. One of the most exciting discoveries was pre-IGY, in early 1957, as the first ice-science traverse team drove Sno-Cats along the newly blazed trail, dubbed Army-Navy Drive, from Little America to Byrd Station. Taking seismic measurements en route, they learned that contrary to historic hypothesis the West Antarctic ice sheet beneath them was thick and grounded up to 1,500 meters below sea level. That is, if the ice were somehow removed, open ocean dotted with islands would be revealed. Traverse leader Charles Bentley said later, after a lifelong career in polar glaciology, that "it took quite a lot of experimentation before I actually came to believe my own results." IGY officers in Washington at first did not.[29]

About half the Antarctic scientific effort went to meteorology, the weather being of immediate importance to polar survival and also as a source of understanding continental, hemispheric, and even global climatic patterns. Other scientists measured variations in Earth's magnetic field, monitored incoming cosmic rays, observed and photorecorded the aurora australis in motion, and bounced radio waves off the outer layers of the ionosphere to probe the relationship between solar activity and radio propagation. Some spent the darkest, coldest months outdoors digging deep pits in the snow, which gradually turned to ice, to read the history of glacial formation in its characteristics and annual layers.[30]

Extensively covered by the press, all these activities, each revealing some secret of the mysterious frozen continent, found their way back home, where the IGY garnered enormous public interest. *The New York Times* alone published hundreds of news articles, updated maps, and feature stories on Antarctica. Walter Sullivan, a *Times* senior reporter assigned to cover the IGY fulltime, wrote two popular books and a scholarly monograph. The navy brought two artists to document Deep Freeze I (1955–56), and two Disney photographers wintered over that inaugural year. *Time, LIFE,* and other magazines did cover stories.[31]

The scientific community was enthralled. Even before the IGY officially began, excited U.S. science leaders were urging that the work in Antarctica be continued after the "year" expired at the end of 1958—even though they had promised Congress a one-shot expenditure. By February 1958 the international science community had optimistically established SCAR, the Scientific [at first, Special] Committee on Antarctic Research, to formulate and coordinate expanded post-IGY polar scientific investigation.[32]

The participating governments, daunted by costs, and some deeply anxious to reestablish their claimed territorial sovereignty after IGY intrusions, initially gave mixed responses on continuing, but when the Soviets announced, at an ad hoc ICSU meeting on Antarctica in September 1957, that they would be staying on, Wexler confided to his diary, "The last bit of opposition, if there was any, collapsed." A one-year add-on called the International Geophysical Cooperation–1959 afforded time to formulate a more permanent arrangement.[33]

Anxiety remained, however, over territorial claims. U.S. State Department experts, still asserting their country's undeniably strong basis for a claim, and not wanting to yield any potential advantage, worried anew about: what to claim that might prove profitable in one way or another; what to claim that would neither offend the existing friendly claimant states nor incite the distrusted Soviets to press their own claim; and what would any claim cost, not only in dollars or arms, but also in criticism for violating the IGY freeze on political activity, even for violating the IGY spirit?[34]

At length, American policy makers began to see greater advantages in free access throughout Antarctica than in specific ownership of a limited area. An internationalized continent, secured by treaty, might best serve everyone's interests. They began to see in the IGY an opportune moment, and a possible path, to institutionalize the scientific cooperation and avoid political confrontation on claims.[35]

Ambassador Paul Daniels, with "very old-fashioned but very sound" diplomacy, opened and guided the prolonged, delicate negotiations that would become the Antarctic Treaty. In order to encourage candid discussion, free of political posturing or second-guessing by the press, Daniels kept the preliminary talks secret. He rotated, by country, the chairing of these sessions to promote shared approval of draft treaty articles. When the Soviet representatives sat in glum silence, Daniels observed to worried allies that they kept coming (they wanted to hear what everyone else was saying); when the Soviets spoke obstructively, he heard them out with polite thanks and went on. With patient repetition, he assured the nervous claimholders that no country would acquire, alter, or relinquish claimant rights.[36]

The treaty, signed on December 1, 1959, enshrined peace during the depths of the cold war. It dedicated the polar continent to scientific pursuit, specifically on the IGY model, and specifically among the 12 IGY participating nations, both decisions basing the agreement on science, not politics—a unique and farsighted approach. It emphasized international cooperation. It set aside (legally "froze" in place) territorial claims. It prohibited nuclear explosions and military exercises and audaciously instituted unlimited, unannounced inspection for enforcement. It provided for orderly evolution through regular meetings of the treaty parties, and it limited its purpose and duration to the doable. Fragile and imperfect, the Antarctic Treaty has grown and lives on.[37]

Thus, the IGY in Antarctica never ended—a minor miracle from a tense and scary time. Those pursuing the current (fourth) IPY can thank the political and military officials of 50 years ago, who had the wisdom to allow scientists to lead where they could not. They can thank those IGY science leaders, who well knew the political stakes but conducted their planning in science terms alone. They can thank the anonymous pioneers on the ice who built a safe and workable infrastructure and began a systematic pursuit of knowledge too important and exciting to let go. They can thank the diplomats of 12 nations, who put science and peace ahead of confrontational politics. Today's IPY leaders can expand, intensify, and focus without having to invent and create new political systems and structural mechanisms. They have inherited a remarkable legacy.

## Abbreviations and Short Forms

All numbered record groups (RG) in the Notes section below are federal records held at the National Archives and Records Administration, Archives II, in College Park, Maryland.

RG 313      Records of U.S. Naval Operating Forces, Operations Plans and Reports 1955–72, Operation Deep Freeze, I–IV.

RG 307      Records of the National Science Foundation (NSF), Office of Antarctic Programs, IGY: NAS–USNC IGY Documents 1955–59.

RG 59      Records of the U.S. State Department, CDF (Central Decimal File) 1955–59. Boxes 1651–1656: International Organizations/Conferences VIII (399.829). Boxes 2772–2777: Regional/Antarctica (702.022).

NAS–IGY      Records pertaining to the IGY, Archives of the National Academy of Sciences.

BPRC      Archives of the Byrd Polar Research Center, Ohio State University, Columbus, Ohio.

CTF, TF      Commander Task Force, Task Force (with number).

DF      Operation Deep Freeze (with Roman numeral for year).

MCB (Spec)      Mobile Construction Battalion (Special), Seabees unit.

NAS      National Academy of Sciences.

NSF      National Science Foundation.

USNC–IGY      U.S. National Committee for the International Geophysical Year

## Notes

1. This essay is based on the author's recent book, *Deep Freeze: The United States, the International Geophysical Year, and the Origins of Antarctica's Age of Science* (Boulder: University Press of Colorado, 2006). The larger study was supported by the National Science Foundation under Grant No. OPP-9810431, which is gratefully acknowledged. Interpretations are the author's and do not necessarily reflect the views of NSF. "The International Geophysical Year in Antarctica: Uncommon Collaborations, Unprecedented Results," a previous paper of somewhat different scope and purpose appeared in the *Journal of Government Information* [subsequently *Government Information Quarterly*] 30 (2004): 482–489.

2. The IGY officially ran from July 1, 1957, through December 31, 1958. The longer "year" took advantage of the full span of maximum solar activity and facilitated completion of scientific programs in the less-accessible polar regions.

3. Walter Sullivan, "The International Geophysical Year," *International Conciliation* 521 (January 1959): 267–268.

4. Walker Chapman, *Antarctic Conquest: The Great Explorers in Their Own Words* (Indianapolis: Bobbs-Merrill, 1965), 114.

5. RG 313, Task Force 68, General Reports 1946–1947, Box 1, Report of Operation Highjump, vol. I, 2. Highjump's objectives were classified at the time.

6. Roosevelt to Byrd, November 25, 1939, cited in full in Kenneth Bertrand, *Americans in Antarctica, 1775–1948* (New York: American Geographical Society, 1971), 473. M. J. Peterson, *Managing the Frozen South: The Creation and Evolution of the Antarctic Treaty System* (Berkeley: University of California Press, 1988), 37–38.

7. See Allan A. Needell, *Science, Cold War, and the American State: Lloyd V. Berkner and the Balance of Professional Ideals* (Amsterdam: Harwood Academic Publishers, 2000), 2–8.

8. Needell, *Science, Cold War*, 189, quoting William T. Golden, Truman's unofficial science advisor.

9. Lloyd Berkner, Columbia University oral history, interview with Jay Holmes, June 4, 1959, 2–3, Library of Congress, Manuscript Division, L.V. Berkner papers, Box 19; James Van Allen, interview with Brian Shoemaker, November 18, 1997, 17–21, BPRC. See also James Van Allen, "Genesis of the International Geophysical Year," *Polar Times* (Spring–Summer 1998): 5.

10. For details of the organizational evolution of the IGY, see Harold Spencer Jones, "The Inception and Development of the International Geophysical Year," International Council of Scientific Unions, Comité Spéciale de l'Année Géophysique Internationale, *Annals of the International Geophysical Year, 1957–1958*, vol. 1 (London: Pergamon, 1959), 383–413; Berkner, interview, 9.

11. The Soviet launch of *Sputnik* on October 4, 1957, marked the achievement of a shared IGY goal, but it so overwhelmed all else that the IGY relationship is virtually unremembered.

12. Sullivan, "IGY," 270.

13. Philip Mange, by phone with author, December 10, 2001, and "The IGY of 1957–58: An Unprecedented Surge in International Scientific Achievement," paper delivered to American Association for the Advancement of Science, AMSIE 1997, Seattle, February 17, 1997; Minutes, First Antarctic Conference, Paris, 6–10 July 1955 (unofficial translation March 5, 1956), 7, 22, NAS–IGY: 1st Antarctic Conference, Delegates and Resolutions.

14. Little was committed to paper on the politics of excluding geology during the IGY (military interests favored it for strategic reasons), but the attitude becomes clear when, immediately following the IGY, the program "broadened" to all sciences, specifically, almost pointedly, including geology. See, for example, Odishaw to Waterman, July 10, 1957, NAS–IGY: IGY Termination, SCAR 1957–1959; NAS, A Report on the Status of Plans for an International Program of Antarctic Research for the Year 1959, March 14, 1958, RG 307, Box 2.

15. Minutes, First Meeting Technical Panel on World Days and Communications USNC–IGY, August 18, 1955, NAS–IGY: Science Program World Days; IGY *Bulletin* 2; "CSAGI and the International Geophysical Year," *Transactions, American Geophysical Union* 38, no. 4 (August 1957): 638–639; Stanley Ruttenberg (IGY program officer), by phone with author, December 12, 2001.

16. Wexler to Odishaw, Some Impressions…IGY–Antarctic Conference, July 4–10, 1955, July 11, 1955, NAS–IGY: 1st CSAGI/Antarctic Conference correspondence; Wexler to Odishaw, Report on Third CSAGI Antarctic Conference, August 7, 1956, NAS–IGY: 3rd Antarctic Conference, correspondence; Harry Wexler, Antarctic Diary, July 8, September 14, 15, 1955, Library of Congress, Manuscript Division, Harry Wexler papers, Box 27; Sullivan, "IGY," 323.

17. George Toney, interview with author, October 28, 1998, 5, BPRC. Alan Shapley, interview with Brian Shoemaker, January 31, 1997, untranscribed, BPRC.

18. Paul Siple, *90° South* (New York: G.P. Putnam's Sons, 1959), 99; Minutes, First Antarctic Conference; Wexler, diary, July 8, 1955; see also G.E. Fogg, *A History of Antarctic Science* (Cambridge, UK: Cambridge University Press, 1992), 173.

19. Harold Bullis, "The Political Legacy of the International Geophysical Year," Congressional Research Service, Library of Congress, November 1973, 2, 12–13; Mange, "IGY of 1957–58"; Laurence McKinley Gould, *Cold: The Record of an Antarctic Sledge Journey* (Carleton College limited edition, 1984; first edition 1931), 200.

20. See NAS–IGY: Budget Hearings and Budget Testimony files.

21. Congress established the NSF in 1950. Kaplan to members and reporters of USNC–IGY, December 10, 1953, NAS–IGY: Organization USNC 1953; [unsigned, probably

John Hanessian Jr.], Some Illustrations Suggestive of Pattern Being Established between NAS–USNC–IGY and NSF, June 1955, and Hanessian to Odishaw, NSF representation on the OCB [Operations Coordinating Board] Antarctic Working Group, February 15, 1956, NAS–IGY: OCB Antarctic Group. Albert P. Crary detailed interorganizational political and fiscal relationships in "History of the International Geophysical Year," 31–36, draft manuscript, privately held.

22. Quarles to Dodge, March 19, 1954, NAS–IGY: Organization USNC 1954; CTF 43 Staff Instruction I-55, 1955, 1–1, RG 313, Box 3. See also NSC 5421/1, July 19, 1954, and NSC 5528, December 12, 1955, RG 59, Box 2773.

23. George Dufek, *Operation Deepfreeze* (New York, Harcourt, Brace, 1957), 39, 41, and others. Philip M. Smith, later with both NAS and NSF, spoke at length about Dufek's leadership and commitment; see his interview with author, December 8, 1998, BPRC. See also CTF 43, DF I, Operation Plan no. 1–55, E-I-1, RG 313, Box 4.

24. Army-Navy Trail Party Report, RG 313, Box 12; Victor Young, interview with author, May 10, 1999, 21–22, BPRC; Smith, interview, 4–12.

25. See Richard Bowers, South Pole Station Daily Narrative, privately held and retyped by Robert L. Chaudoin, also in RG 313; see also Bowers and other pole construction party interviews, BPRC.

26. See MCB (Spec), Reports, Deep Freeze II and III, 1956–1958 (for Hallett, Wilkes, and Ellsworth stations), RG 313.

27. Among others, John C. Behrendt, in *Innocents on the Ice: A Memoir of Antarctic Exploration 1957* (Boulder: University Press of Colorado, 1998), reproduces illustrative exchanges of messages on Ellsworth Station's siting: see 25–45. Wexler's diary, January–February 1957, reveals the conflicting viewpoints on establishing Byrd Station where planned. Bowers, South Pole Narrative, January 4–7, 1957.

28. Minutes, first meeting, panel on Antarctic policies of the USNC–IGY Antarctic committee, February 25, 1955, NAS–IGY: USNC Antarctic committee—confidential; Paul Siple, position paper on command structure, July 2, 1956, and Antarctic Operational Command Relationships: NAS and DoD, July 11, 1956, NAS–IGY: IGY–DoD Command Structure Antarctic Stations.

29. Charles Bentley, interview with author, December 2, 1998, 8, BPRC; Hanessian to Files (528.6), Weekly Antarctic Message Report, 21 March 1957, NAS–IGY: Daily and Weekly Antarctic Message Summaries, Jan–Jun 1957.

30. USNC–IGY, Addendum to Antarctic Status Report No. 26, January 1958, IGY Summary of Technical Programs by Station Location, May 26, 1958, NAS–IGY: Antarctic. See Scientific Station Leaders Reports, by station and IGY year, RG 307, Box 5. The respective scientific leaders outlined the goals and achievements of each discipline.

31. Sullivan's major publications on the period, both by McGraw Hill, are *Quest for a Continent*, 1957, and *Assault on the Unknown* (on the IGY generally), 1961, and "The International Geophysical Year."

32. Wexler, diary, December 8, 1956; "US Asks Extension of Antarctic Study," *New York Times* (June 30, 1957): 28; Odishaw to Waterman, July 10, 1957, NAS–IGY: IGY Termination, SCAR 1957–1959; Odishaw to Waterman, January 16, 1958, NAS–IGY: IGY: ICSU/(SCAR), 1957–59; Minutes, International Council of Scientific Unions, Special Committee on Antarctic Research, The Hague, February 3–5, 1958, February 17, 1958, NAS–IGY: ICSU/(SCAR) 1957–1959.

33. Emmerson to State, Fourth Antarctic Conference, June 17, 1957, RG 59, Box 1652; Wexler, diary, June 16, September 12, 1957.

34. Wilson to Dreier, Antarctic Policy Review, January 14, 1957; Watrous to Files, Antarctic, April 23, 1957; Jones to Snow, Clearance of Draft NSC Policy Paper on Antarctica, June 3, 1957, with attached draft policy, and a plethora other internal memoranda, RG 59, Box 2773.

35. Showing the evolution of official thinking is Daniels to The Secretary, Antarctica, December 9, 1957, RG 59, Box 2774.

36. Allan Needell, interview with Brian Shoemaker, July 19, 2000, 30, BPRC. Memoranda [minutes] of the 60 informal meetings, spanning a year and a half, on Antarctic conference preparations recap the progress toward a treaty, RG 59, Boxes 1653–1654, 2775–2776.

37. See Antarctic Treaty, December 1, 1959, in U.S. Treaties and Other International Acts Series 4780 or www.scar.org/Treaty/Treaty_Text.htm. R. Tucker Scully, interview with author, June 10, 2002, untranscribed, BPRC.

# Chapter 15

# Tracking Diplomacy: The International Geophysical Year and American Scientific and Technical Exchange with East Asia

*Teasel Muir-Harmony*

## Introduction

Sir Patrick Moore, the highly visible host of the television series *The Sky at Night* and a fellow of the Royal Society, wrote the book *Earth Satellites* in 1956 for the popular audience on the exciting potential of satellite technology. Published one year before the launch of *Sputnik*, Moore's book included a drawing of satellite technology's possible ramifications for global culture. The drawing depicts people in both Africa and Latin America watching the same Jackie Gleason television program, which was being broadcast around the world thanks to satellite technology. The viewers have created sculptures and pottery mimicking Gleason's famous pose. Moore's picture suggests that people from all parts of the world would translate the ideas and images from television programs into their own visual culture. To Moore "whether this would be a blessing or not is a moot point."[1] What is relevant to Moore is that the globalizing effects of satellite technology could standardize the way people live and work.

Six years later Hal Cozzens, an American astronomer at the Smithsonian Astrophysical Observatory (SAO), wrote a scathing report about his visit to an optical satellite-tracking station in northeastern India. Perched atop a Himalayan foothill, the station was staffed by Indian astronomers who would scan the sky nightly for satellites, using the SAO telescopic camera. Cozzens's report offers another perspective on satellite technology and related programs' potential cultural and scientific influences. Cozzens wrote:

> I now believe I have been [here] a sufficient length of time to give you an accurate report on the meat of the situation that exists and the lack of efficiency that is shown at the station. The staff is as good as can be found in India [but] they simply do not meet the standards of the majority of other satellite observers and never will. What is normally expected of the other stations simply

cannot be expected of India because of their outlook, mental capacity and general ability and ambition. Carl Hagge (an SAO administrator) described the situation very well when he said, "It is like teaching a three-year-old how to drive a car." Am sorry I can't report some good news but the above is a true picture of how things are here.

As his report conveys, Cozzens did not believe that rank-and-file Indians had the potential to assimilate a Western work ethic and to practice experimental science. His solution was to move the station to another country "where it could at least contribute to the Program."[2] Cozzens's ground level view clearly differs from Moore's visionary, yet popular, view of the future effects of satellite technology.

What then is the potential impact of a global scientific project such as the SAO optical satellite-tracking program that began as a result of the International Geophysical Year (IGY)? If Cozzens is correct, then there are limits to scientific and cultural dissemination. If, on the other hand, Moore's view is correct, surely the Smithsonian's program brought political and cultural influence to other countries along with its satellite-tracking telescopic cameras. To what extent can scientific programs and technologies impact culture and politics? Does the incorporation of foreign scientific programs and technologies alter or transform the political, cultural, and scientific landscape of a country? This essay addresses these questions by examining the experience of American observers working at satellite-tracking stations in Japan and India and by exploring whether the program affected the scientific, political, or cultural views of its participants.

## The IGY and the Establishment of the SAO Satellite-Tracking Program

At the beginning of the space age, under the auspices of the IGY, the SAO was chosen to develop a network of optical satellite-tracking stations in a number of countries around the world. Located in such diverse places as Japan, Australia, India, Hawaii, Peru, Iran, and Ethiopia, the observatories were staffed with a combination of local scientists and Americans. The SAO undertook a number of measures to promote the standardization and unification of its tracking system. Each observatory was equipped with the same tracking camera, blueprints to construct the same tracking buildings, and the same supplies. SAO administrators stressed the importance of observer exchanges and frequent formal and informal communication in an effort to normalize observing techniques and establish a cohesive scientific community. I analyze these efforts to standardize and unify the tracking system by investigating the reception of standardized equipment, communication, and observer exchanges at the Indian and Japanese tracking stations. As Gabriel Hecht and Thomas Hughes, among others, have suggested, a cross-national comparison of scientific practices can delineate cultural and national differences.[3] In order to tease out the distinct ways in which the satellite-tracking program was incorporated into the scientific, political, and cultural landscape of Japan and India, I will chronicle the histories of the tracking stations in India and Japan, compare how an American-organized scientific program was received in each country, and consider the ways in which the collaborative relationship among the United States, India, and Japan served the interests of all the countries involved. In *Trackers of the Skies,* E. Nelson Hayes has already discussed the scientific fruit of the program.[4] This essay is not a detailed chronology of the Baker-Nunn program or a technical account of early optical satellite tracking. It is, alternatively, an investigation of international scientific cooperation as it played out at two observatories in East Asia in the late 1950s and 1960s.

In the 1950s Japan was in the midst of reconstruction after World War II, and India was a newly independent nation. Japan and India's interest in modernization and commitment to building up their countries' scientific infrastructure created a context that was different from the other countries that collaborated with the SAO. This essay will investigate the numerous ways these countries negotiated with an American observatory to advance their scientific institutions and participate in an international scientific program while still maintaining their authority and autonomy. Ultimately, this essay argues that in order to understand the nature and impact of global scientific programs, the day-to-day workings of the scientific outposts (the backbone of the programs) have to be examined.

On July 7, 1955, the United States announced that it would launch an Earth satellite as part of its commitment to the IGY. Many of the major early scientific benefits of satellites, including precision mapping of Earth's surface, detection of variations in Earth's gravitational field, and information about the density and temperature of the upper atmosphere, require satellite tracking.[5] In late 1955, the U.S. National Committee (USNC) for the IGY selected the SAO to run the optical tracking and scientific analysis program, while the Naval Research Laboratory (NRL) was given charge of developing a radio tracking system. Hugh Odishaw, executive secretary of the USNC–IGY, pointed out to the executive committee that "since the Smithsonian Institution and Dr. [Fred] Whipple, the new Director of its Astrophysical Observatory, have independently operated observing stations at isolated points on the Earth, this experience can be of marked assistance to the success of the satellite observing program, which requires the location of some stations outside the limits of the continental United States."[6] Whipple believed that optical satellite tracking would contribute to the observatory's long-term goals of making astronomical observations and atmospheric studies, and that it would foster new techniques that could improve the SAO research program.[7]

## The SAO

At the time of Odishaw's announcement, the SAO existed largely on paper at Harvard, and Whipple was rapidly building his new infrastructure. In the early 1950s, as Ron Doel and David DeVorkin have noted, the Astrophysical Observatory of the Smithsonian Institution, located in Washington, D.C., since the 1890s, was moribund, and limited by the vision of Charles Greeley Abbot, an emeritus astronomer who had failed to change with the times.[8] In 1953, Leonard Carmichael, the new secretary of the Smithsonian Institution, set in motion a series of changes that created, as he told Vannevar Bush, "a modern research program that will be to the maximum degree possible scientifically significant."[9] Carmichael's plan was to find a suitable academic base for SAO. He ultimately chose Harvard. The SAO still needed a new director interested in modern astrophysics and administration who could organize a research team.[10]

Carmichael and Donald Menzel, the Harvard College Observatory director, believed Whipple, a professor of astronomy at Harvard, was the person for the job. Whipple had his own reasons for accepting the directorship of the observatory. He had close ties with the military, as an advisor to the Office of Naval Research and the Air Force. At Harvard, military funding was severely limited and classified research was prohibited. As the director of the SAO, Whipple could carry out a well-funded, optical satellite-tracking program with greater freedom than if he remained under the institutional constraints at Harvard.[11]

In 1955 Whipple was named the new director, and the SAO moved to the grounds of the Harvard College Observatory in Cambridge, Massachusetts. This move created a wholly new form of hybrid astronomical institution that was nimble enough to take quick advantage of the latest funding opportunities and prestigious enough at first to attract those funds on a major scale.[12]

## The IGY and the Satellite Program

IGY administrators in the United States were well aware that the success of a U.S. satellite program depended on the participation of other countries.[13] The U.S. satellite program's association with the IGY facilitated its cooperative tracking and data collection programs with foreign countries. As Allan Needell has shown, the Central Intelligence Agency and the Department of Defense agreed on the importance of associating the U.S. satellite program with the IGY and the benefit of disseminating information to the international scientific community.[14]

Between April and September of 1956, a working group on tracking and computation within the USNC–IGY considered and approved Whipple's proposals for establishing 12 to 14 optical satellite-tracking stations. After agreeing on constructing a north–south fence and distributing the stations as widely as possible within the latitude range of 30–35° north and south for optimal coverage of the orbits of the IGY satellites, the group considered issues concerning international collaboration.

Because satellite tracking could be an issue of national policy, the Department of State was involved in the planning stages of the program.[15] Walter Rudolph, the Department of State science advisor, noted that a successful satellite would depend on the cooperation of scientists around the world.[16] To ensure this cooperation, special efforts were made to sell the optical satellite-tracking program. During the planning for the international IGY assembly in Barcelona in September 1956, J.G. Reid, Department of State, told Whipple to make sure his report on the SAO optical tracking program was "suave" and had "diplomatic appeal for international participation so as to kill the idea that this was the U.S. program, that we had all the answers and had done all that there was to be done." In particular he stressed the importance of inviting countries to participate in the program to ensure that it did not come across as a strictly American program.[17]

Before he officially announced the Smithsonian's satellite-tracking program at the Barcelona IGY meeting, Whipple wrote to many of his colleagues abroad to drum up support and interest. He received almost immediate support from astronomers and administrators in Japan and India.[18] In early August 1956, Whipple wrote to Dr. A.C. Banerji, Jadavpur University, Calcutta, proposing that India participate in the optical satellite-tracking program. Banerji, enthusiastic about the prospects, wrote to Prime Minister Jawaharlal Nehru, suggesting that India play a part.[19] On the same day Whipple also wrote to Dr. P.L. Bhatnagar, Department of Applied Mathematics at the Indian Institute of Science, "If, by any chance, you could encourage this relationship [cooperative station operation between the United States and India] I would be very happy as I think that the scientific advantages would be marked for both of our great countries." In his letter, Whipple acknowledged the institutional, national, and international incentives of a cooperative station for both "great countries."[20] Japanese scientists and officials were also eager to participate in the optical satellite-tracking program. Masasi Miyadi, the director of the Tokyo Astronomical Observatory (TAO), sent his endorsement for the program to the Japanese National Committee for the IGY a month before Whipple formally announced the program to the world.[21]

The September 1956 Barcelona IGY meeting started off with a bang. On the first day of the meeting the Soviet Union announced that it would carry out a satellite program for the IGY.[22] Whipple's presentation of the SAO optical satellite-tracking program followed a few days later, amid widespread enthusiasm for the burgeoning field of artificial satellites. He outlined the volunteer and professional sides of the tracking program, the potential station locations, the design and function of the new satellite-tracking telescopic camera, plans for data acquisition and dissemination, the layout and operation of the stations, and the importance of the collaboration of professional astronomers around the globe to carry out the program.[23] The NRL radio satellite-tracking program, which would be less precise than the SAO system but could function under variable weather conditions, was also outlined at the meeting. The general report for the meeting noted that President Dwight Eisenhower was enthusiastic that it was an American program that would provide the opportunity for scientists from other nations to participate.[24]

## Political Interests

SAO administrators, State Department officials, and Japanese and Indian officials recognized the political implications of the satellite-tracking program and other large-scale cooperative programs in the IGY.[25] Both India and Japan had a vested interest in developing their country's scientific infrastructure. At the same time, the United States had a vested interest in being a part of this development to ensure its influence in East Asia in the midst of the cold war.

When Higashikuni Naruhiko was named prime minister of Japan on August 16, 1945, two days after Japan's surrender, he claimed that Japan's defeat was due to its inferior science and technology.[26] After World War II, Japanese government officials stressed the importance of science and science education in the country's reconstruction. Japan and India both underwent significant political change in the middle of the twentieth century and both looked to scientific development to advance their countries. India had gained independence in 1947. Shortly thereafter the central government began establishing national laboratories, like the UPSO, to expand its scientific and engineering infrastructure. Japan's education minister, Maeda Tamon, argued that "the cultivation of scientific thinking ability" was key to "the construction of a Japan of culture."[27] The IGY and its related programs, like the SAO satellite-tracking program, offered a venue to build up scientific institutions and contribute to a program that was in the international spotlight.

India and Japan's participation in the satellite program also offered notable advantages to the United States. In the mid-1950s U.S. officials looked for ways to maintain a presence in East Asia. Walter Kotschnig, director of the Office of International Economic and Social Affairs at the U.S. Department of State, summed up India's current situation in 1956. He noted that India was key to the future socioeconomic development and the political ideology of Asia. Although India was technically neutral, Kotschnig believed that it was more sympathetic to the Soviet Union than to the United States, and internally it was clear that India was swiftly becoming a socialist nation. Kotschnig's report warned that "realizing that India is fascinated by scientific progress Russia is making every effort to send Russian scholars and scientists to [Indian scientific] meetings."[28]

From the American point of view, Japan was also seen as an essential component in the future of Asia in the struggle between communism and democracy. Alfred Eckes and Thomas Zeiler have argued that U.S. Secretary of State John Foster Dulles believed the future of the free world rested on the reconstruction, recovery, and incorporation of Japan into the West, and that "Washington deemed Japan the linchpin of the future Asian economy."[29] A state department report prepared for President Eisenhower on the Japanese situation after the Bikini accident on March 1, 1954, noted, "We feel that in the long run scientific interchange is the best remedy for Japanese emotion and ignorance and we intend to push such projects. The most important thing that we can do to help is to treat Japan as a full, free world partner and bring her as much as possible into our own and free world counsels. This is essential if we are to count upon the use of Japanese bases and other cooperation in any future conflict."[30]

In the wake of anti-American street demonstrations, Eisenhower and Dulles tried to ease tension, as the Soviet Union was simultaneously wooing Japan. As Japan and India became increasingly neutral, U.S. officials searched for ways to maintain a firm alliance.[31]

The initial planning stages of the optical-tracking program reflected these bilateral interests. In January 1957, Lloyd Berkner, after meeting Nehru, wrote a letter to the U.S. office for the IGY alerting them to the keen interest the prime minister had shown: "India is most eager to participate in the satellite program. (There is political dynamite in this so India better get an early camera.)" Berkner was closely tuned to the political implications of the program and told the IGY office that India was also. During a lecture Berkner gave on the subject at the National Physics Laboratory in New Delhi on January 9, 1954, "[K. S.] Krishnan got up and remarked that it was wonderful that the U.S. was taking the world into its confidence on the satellite program so that all nations could cooperate. He said further, that it was a shame that the satellite program of the USSR was so secret, and he hoped they might follow the example of the U.S. I thought I saw the USSR representatives present squirm a little."[32] By associating the United States with the open structure of one of its scientific programs, Krishnan, director of the National Physical Laboratory and president of the Indian National Committee to the IGY, drew attention to the significant political potential of SAO collaboration with India.[33]

### Initial Stages of the Tracking Program: After the Barcelona IGY Meeting

Negotiation about the authority, equipment, and level of collaboration with the Cambridge headquarters played a central role in the satellite-tracking program from its inception until its closure. Unlike the other stations in the network, the stations in India, Japan, and Australia were under the jurisdiction of their host countries. The IGY national committees in each country were in charge of selecting satellite-tracking station coordinators and suggesting station locations.

By the middle of November the Japanese national committee gave Miyadi the position of satellite-tracking station coordinator. Even though Miyadi, a Hiroshima native, was cooperating with the SAO a little more than 10 years after the United States dropped an atomic bomb on his hometown, I have only found a roster of the Tokyo observatory staff, with two red exclamation points next to Miyadi's hometown in Whipple's hand, that acknowledged this point.[34] Shortly after he became the coordinator, Miyadi selected the Tokyo observatory in Mitaka as the site for the station. Miyadi wrote that newspapers were "in competition to report the news about the program, which was making a sensation in Japan."[35] The correspondence from

Japanese administrators to the SAO was continually marked by expressions of enthusiasm and assertions to maintain their station's autonomy.[36]

In April 1957 the Indian National Committee for the IGY appointed M.K. Vainu Bappu as India's satellite-tracking coordinator. Bappu, director of the UPSO, left India in 1949 to study astronomy and astrophysics at Harvard. Like many of his predecessors and colleagues, Bappu saw science as a tool to improve his country and looked for ways to establish the UPSO as a center for astronomical research in India.[37]

## J. Allen Hynek's World Tour of Tracking Stations

Throughout the course of 18 months, from 1956 through 1957, J. Allen Hynek, SAO associate director and head of the satellite-tracking program, took two trips around the world to visit station sites. On his first trip Hynek visited potential station locations and Moonwatch teams in Argentina, Peru, Uruguay, Chile, and the Netherlands Antilles. On his second trip he traveled to Iran, India, Japan, and Hawaii. In a report he explained that his approach was always that the "Smithsonian Institution was making representations, on a strictly cooperative basis" and he "scrupulously avoided creating any impression of 'American Imperialism' taking over a foreign site for our own purposes." Even with the lack of astronomical infrastructure in some countries he visited, Hynek found that people were eager to take part in the satellite-tracking program. Hynek noted, "It occurred to me that, in the establishment of satellite observing stations abroad, the Smithsonian Institution was fostering far more than operating astronomical units. Its long-established distinction and, indeed, its experience in international scientific cooperation appears to be thrusting upon Smithsonian the conception of just so many semi-permanent nuclei for even greater scientific cooperation—especially at the astronomical level."

Hynek was quick to point out that in order for the SAO to contribute to the growth of astronomical communities in each of these countries the attitude "in operating the station must be one of mutual cooperation rather than the mere operation of a branch station on a selfish basis." As with most documents from the planning stages of the optical satellite-tracking program, Hynek's report stressed the importance of cooperation in fulfilling the scientific aims of the program and potentially impacting scientific development in each country.[38]

After spending two weeks in Iran, giving talks and selecting a station site, Hynek flew to India to meet with Bappu and see Nainital on June 2, 1957. When he was in Delhi, before he made his way to Nainital, Hynek spoke with members of the Indian National Committee for the IGY in addition to journalists. Although Hynek only spent 12 hours in Nainital, he was able to select a station site with Bappu that could be accessed by a paved road and could be easily equipped with power and communications lines. Hynek thought Nainital was pleasant, but that American observers would have to have a sense of adventure to thrive there.

Dr. H. Hirose, an astronomer at the TAO, greeted Hynek when he arrived in Japan on June 9, 1957. Hynek observed that in Japan, more than at any other station, the astronomers took complete charge of the station. At the time of Hynek's visit the station location was already selected and the foundation for the building had been erected. In addition to the astronomers' enthusiasm for the satellite-tracking program, Hynek commented "I must say that I was surprised at the extent to which the papers covered my visit." Roughly 30 reporters, who brought an "utter maze of cameras and flash-bulbs" with them, attended a press conference at the observatory. A Tokyo radio station set up a special shortwave sound truck to broadcast the news from the press conference. After the press conference Hynek gave a technical talk to the 100 astronomers at the Tokyo Observatory. He was "tremendously impressed with the organization of astronomy in Japan and the efficient manner in which the Tokyo Observatory was run by Dr. Miyadi."

Hynek's report revealed markedly different situations in Japan and India. When writing about India he emphasized the contrast between the country's developing scientific institutions and the "poverty and a very backward way of life." When writing about Japan Hynek emphasized the well-established scientific community, the receptiveness of the press, and the general enthusiasm toward the program. In June 1957, the satellite-tracking stations in Japan and India were at different stages in their development. The fact that a building foundation had already been laid in Japan, whereas in India the station site was just being selected, points to an important contrast between these locations, which would later be manifested to a greater extent. This is not to suggest that the Japanese station would be more advanced or have a greater output than the Indian station. In fact, in future correspondence Whipple pointed to the Indian station as an example of successful satellite tracking when he tried to encourage the Japanese station to instate observer exchanges.[39] The difference between the early developments of these two stations is significant because it illuminates important dimensions in the relationship between each station and the SAO. Throughout the course of the tracking program the Japanese station worked toward complete autonomy. The Indian station's relationship, on the other hand, was defined and confined by pressures from a newly formed government and pressure to establish a nationalistic scientific infrastructure.

## The Design and Development of the Optical-Tracking Stations

During the initial planning stages of the program, SAO administrators decided that each tracking station should be isolated and self-sufficient. Each station was to be housed in a T-shaped building with a retractable roof, electronics lab, communications room, a darkroom, and an administration room. The building would have air conditioning, heating systems, and a backup power generator. Early SAO reports stressed again and again that the station plans should not be considered as rigid requirements, but instead they should be seen as a guide. In a preliminary planning document from October 1956, SAO administrators noted "in those cases where most of the auxiliary equipment and the buildings are to be supplied by cooperating agencies or governments, the layout of the station grounds, the design of the buildings, and the facilities provided for the comfort and convenience of observers will be left to the judgment of local collaborators."[40] Some stations incorporated local materials in their construction and took environmental factors into consideration. The station building in Nainital, for instance, was built from stones quarried locally and chipped into blocks by hand.[41] By allowing for flexibility in station designs and leaving decisions up to cooperating governments and agencies, the SAO sought to ensure the cooperative character of the optical satellite-tracking program.

The Smithsonian equipped each station with an identical Baker-Nunn, wide-field camera-telescope to guarantee both precise and standardized data collection. Twelve cameras around the world had to function in exactly the same way if the resulting data were going to be valuable.[42] Although camera-telescopes had been in use for years, the SAO program required a completely new optical tracking technology. Whipple and Hynek concluded that to accurately determine the position and time of a satellite's orbit, the camera-telescope had to use continuous rolls of film, a dual shutter system, a crystal clock, and preliminary observational data. Instead of hiring a single contractor, the SAO outsourced the production of the camera-telescope to a number of people and firms. In 1955, Whipple met with James Baker, a consultant to the Perkin-Elmer Corporation and an engineer who had previously worked with the Harvard College Observatory, to discuss the development of a new telescopic camera. Baker designed an optical system that merged a camera with a wide-field Schmidt telescope. Whipple also consulted with Joseph Nunn, an engineer living in South Pasadena, California, to design a gimbal ring and drive mechanism for the camera-telescope, which would allow it to alternate between fixed positions and tracking motion. The camera-telescope was also designed with a Norrman crystal clock, an optical system that allowed for the maximum passage of ultraviolet rays, and a 32-inch Pyrex mirror designed by Corning Glass Works.

The 12-foot-tall Baker-Nunn camera was a complex and innovative scientific instrument. Sunlight that was reflected off the metallic surface of a satellite would travel through the Baker-Nunn's three 20-inch corrector lenses before hitting a 32-inch mirror, which would reflect the light onto Cinemascope film. The Norrman precision clock was caught on the film along with the image of the satellite to record the time down to a millisecond. The staff at each station developed the film and then sent the film to SAO headquarters, where data was processed and studied. The Baker-Nunn could photograph satellites as small as 20 inches in diameter at a distance up to 2,000 miles away or roughly up to the 13th magnitude. It could take as many as 60 photographs per minute. Each telescopic camera cost roughly $100,000 to produce. Although the Baker-Nunn optical system could only track satellites at night with clear weather, the system provided accurate position determination, long-range tracking capability, and multiple target resolution.[43]

## Life at a Satellite-Tracking Station: Samuel Whidden and the Early Years at Nainital

In late January 1958, Samuel Whidden, an astronomer from SAO headquarters and Bappu's former classmate at Harvard, arrived in India with his wife, Mary Holt, to help Bappu and his staff set up the satellite-tracking station.[44] Founded in 1954, as part of the central government initiative to establish national laboratories, the UPSO was intended to be a center of modern astronomical and astrophysical research in India, but the fledgling observatory was hindered by its small collection of instruments, variable weather, and its relocation to Nainital in 1955.[45] The advent of the IGY offered new opportunities for the developing observatory. An article in the *New York Times* commented that the observatory "isn't much yet, just a bungalow and a ten-inch-mirror telescope housed in a tin shed. But the small staff of Indian astronomers feels that the IGY has made it part of a great world scientific movement."[46] The most ambitious of the projects undertaken by the observatory for the IGY was optical satellite tracking.[47]

In Nainital Sam Whidden and Mary Holt found the satellite-tracking station still at the beginning stages of construction. Unlike the station in Japan, which had been completed in September 1957, ahead of schedule, the station in India was nowhere near completion, slated for March 1958. The station building was not the only thing behind schedule. The Baker-Nunn cameras were taking longer to make than SAO had originally thought they would. Whidden started growing a beard when the telescope finally arrived in New Delhi in March 1958. He joked in a report that he "intended to shave it off upon the occasion of our satellite tracking camera's first photograph (considering that it [the beard] would first have to be braided and bound)."[48] The station staff was prepared to observe by July 5, but the summer monsoon delayed testing, which in turn delayed observing until the end of August.[49]

## Problems with Jurisdiction of the Indian Station

In addition to problematic construction delays, disagreements about whether the station was under American or Indian jurisdiction almost prompted closure of the station. E.C. Balch, an SAO engineering consultant sent to help set up the station, noted that jurisdiction of the station was a "touchy subject" and it had "political implications as well as a personal one."[50] Whidden wrote to Cambridge, warning that if the SAO did not act immediately they were in risk of losing the station in India. Bappu made it clear to Whidden that unless the station was under complete Indian jurisdiction, it would not be supported by the Indian government. Whidden commented in a letter to Cambridge that even the prestige value of having one of the twelve optical tracking stations was not enough to justify any Indian government expenditure. Bound by political obligations, Bappu noted further that an American would only be accepted at the station as a technical advisor who had no authority. Kenneth Drummond, the executive officer of the satellite-tracking program, let Whidden know that the SAO would "certainly bend over backwards if necessary to insure [*sic*] smooth operation and international cooperation with India and the Uttar Pradesh State Observatory."[51]

On top of construction delays and the problem of jurisdiction, progress at the station was hindered by the presence of two U.S. Army map service personnel who were sent to help set up the station. The Indian national government had classified all maps of the region, because the observatory was located in a restricted border area. According to Whidden, because of the presence of American military personnel, Bappu felt he was forced to withhold full cooperation with the Baker-Nunn tracking program.

Whidden tried to relay Bappu's multiple frustrations with the SAO in his reports to Cambridge. He noted that Bappu felt the "Smithsonian had taken actions affecting his observatory without consulting him. He also feels, however unjustly from our point of view, that Smithsonian, in retaining control of the station, is pulling a fast one."[52] The question of jurisdiction coupled with the presence of military personnel put Bappu in a difficult position. With the help of an official letter from Whipple and more explicit correspondence about the organization of the Nainital tracking station, the American and Indian observers came to an agreement. The Indian tracking station would be under the complete jurisdiction of the Uttar Pradesh State Observatory.[53]

## Communication at the Indian Station

Whidden, with the help of his wife, sent frequent reports to Karl Henize at SAO headquarters, detailing everything from the progress of the station, administrative matters, mechanical and technical problems, cooperative issues, and adjustment to life in northeastern India. In addition to strongly encouraging the stations to send reports, the SAO informed its observers that it would not only send "formal communications and memoranda, but also frequent informal narrative letters from Cambridge regarding the progress of the program as a whole and at each of the twelve stations"[54] This policy helped to ensure the uniformity of the system and promote a bond among the stations. Whidden was grateful for the frequent letters he received from the SAO. In an early report he thanked SAO headquarters for the information it supplied him on the progress at other stations and wrote "our hearts go out to our compatriots in the far-flung satellite realm. We commiserate wholeheartedly, with understanding and sympathy, with everyone who can't find a double-edged backhanded framitz in the local market place; we can't either."[55]

In early May 1958, when Whidden showed Bappu a letter from SAO about the progress at other stations, Bappu took it as an affront to the UPSO instead of encouragement to get the station up and running. Although Whidden meant to convey the urgency of getting the observatory ready for tracking, Bappu took offense and wrote to Cambridge. According to Whidden, Bappu implied that the SAO motivations for getting the network running were political while the UPSO motivations were scientific.[56] Whidden's account illuminates some of the sensitivities present at the initial stages of the program. These sensitivities could have been heightened by the long hours (seven days a week, 11 hours a day, by all staff members), the equipment delays, and that they were working during the hottest season in 35 years. Even so, the tension about the SAO role at the station that Whidden encountered when he first arrived in India was still present a few months after the issue of jurisdiction of the station was settled.[57]

Whidden frequently expressed frustration about his lack of authority at the Nainital station. Even though on paper Whidden was only a technical advisor, he was also supposed to ensure that the SAO interests were met. In early July 1958, Whidden reported that the "Smithsonian is at present represented here by four things: its plans, its actions, its equipment, and me." According to Whidden, although the UPSO plans were in line with the SAO objectives, it was difficult for him to encourage respect for the Smithsonian's program. In a letter from late July, Whidden explained that even though the station was up and running, relations between the SAO and the UPSO remained "a tug of war." He explained that "Bappu still complains that Smithsonian doesn't trust UPSO" and that the Smithsonian representative in Nainital refuses to "understand local problems in the light of local conditions." Whidden suggested that, "Bappu is like a dedicated piano student performing his first concert after years of study with a teacher of whom he is quite in awe. The teacher is in the front row and his opinion matters very much. When Smithsonian seems to criticize, the awe turns to dislike. He says he doesn't care what Smithsonian thinks, but his actions show that Smithsonian's opinion is a serious matter to Bappu's self respect."[58]

Whidden left the station for good in the middle of December 1958. After spending a number of months in India, Whidden came to believe that the problems he encountered setting up the station were rooted in what he called the "Indian attitude." He did not want to give the impression that Indian people were incompetent, instead he emphasized the importance placed on science and not engineering in India. Whidden observed that India was "not lacking in intelligent people; the social structure in India, however, is oriented away from producing a desire for personal technical ability among its educated people. India has science but not technology." He believed that the remnants of the caste system accounted for the emphasis put on science above technology.[59]

Whidden was in agreement with a number of his Indian colleagues about the lack of engineering in India when he wrote: "Science cannot grow on books and theory alone. It requires the support of the technician, the instrument maker, the mechanic, and the man who makes the tools of science and provides the equipment for experiment. This breed is almost nonexistent in India and India is more in need of such people than of economic aid, invitations to the summit, or anything else if she is to assume a place of importance among nations."[60]

Even with setbacks, including a misalignment of the film transport system, the Nainital station was making a respectable number of observations

by the end of 1958.[61] The station in India, along with the 11 other stations, increased their quarterly observations exponentially during the next
few months. In April and May 1959, the worldwide network made 2,902
observations, 300 of which came from the station in India.[62]

### After the IGY

Because of the success of the Baker-Nunn tracking network during the IGY,
the SAO and the National Aeronautics and Space Administration (NASA)
decided to extend the program beyond the official end of the IGY on
December 31, 1958. Whipple, realizing that the networks formed through
scientific collaboration could contribute to better political relationships,
encouraged positive interactions among American observers and locals.[63]
Throughout the 1960s, Baker-Nunn stations became local centers for information on space exploration in their host countries. The staff of the stations
gave lectures, led tours, hosted open houses, and also became involved in the
local communities, by loaning tools, teaching English, and attending social
functions. Like their sister stations, the optical satellite-tracking stations in
India and Japan became an integral part of the developing social, political,
and scientific landscapes. Open houses were held regularly for the public,
providing locals with an opportunity to observe the sky, learn about the
equipment, and investigate the newest addition to the community.[64]

Like Whipple, NASA administrators, State Department officials, and
the president's science advisors noted that cooperative international scientific programs, like the SAO optical satellite-tracking network, were having
a "profound influence on the future strength and position of [the] country
in world affairs."[65] In a statement before the U.S. House Committee on
Science and Astronautics on January 27, 1960, T. Keith Glennan, NASA
administrator, discussed the optical satellite-tracking program and remarked
that the policy of openness and disclosure of data was "building good will
throughout the world."[66] The president's Science Advisory Committee noted
in 1960, "scientific activities can serve to strengthen political ties through
demonstration of mutual interest and common purpose."[67] The committee
suggested that cooperative scientific activities were an ideal way to encourage
"identification of interests and attitudes with the U.S.," and to "demonstrate
U.S. scientific leadership to those nations that look to science and its applications as the keys to their future." At the U.S. State Department the U.S.–
Japanese collaboration in satellite related science was viewed as a means to
advance U.S. objectives toward Japan and Free Asia. A memo from the State
Department noted "the Japanese do find the latest gadgets fascinating, and
genuine collaboration in the outer space field, if tactfully managed, could
become an important further bond between our two countries."[68] Space
science collaboration with India was also viewed as an important way to
advance U.S. objectives in India. A 1961 State Department memo to Arnold
Frutkin, NASA administrator, noted that it would be politically desirable to
have more collaborative space science programs with India.[69]

## Hal Cozzens's Reports on Tracking Stations

Throughout the course of the satellite-tracking program, SAO administrators and station chiefs would visit stations periodically to ensure that they were operating properly. From October to December 1961, Hal Cozzens, SAO astronomer and former station chief, visited the tracking stations in Japan and India, and on Mt. Haleakala in Hawaii. Cozzens reported that the station in Hawaii was well run and that morale was high, however, the stations in Japan and India needed improvement. Cozzens was disappointed in the Japanese station, noting that the "station facilities were grossly inadequate. Mechanically and optically speaking the camera seemed to be in good condition but in my four years of being around Baker-Nunn Cameras I have never seen a camera as wet and damp."

In order to explain to his colleagues in the United States what India was like, Cozzens suggested imagining "living in Egypt 2,000 years ago and you will have an exact picture of India as it is today." Nainital, although slummy by U.S. standards, according to Cozzens, was bearable. In his report, he stressed that it was more difficult for American observers to work at the Nainital station because it was "further in the wilds" and more disconnected from Western civilization than the rest of the other stations.[70]

Cozzens's impression of the Indian tracking station was higher than his impression of India. He reported that the UPSO new facility was impressive and the observers' morale was high. According to Cozzens, the prestige of the observers' positions in India encouraged them to be dedicated to their work. At first, although the Baker-Nunn's mirror was scratched by dirt and the timing system was still problematic, Cozzens was optimistic about the Indian station. After a month in Nainital, however, Cozzens believed that the only way to improve the Baker-Nunn station in India was to move it to a different country.[71]

Cozzens's reports, especially the one quoted at the beginning of this essay, are striking, not just for their depiction of cultural tension, but also because they were not out of the ordinary. His reports reflected the tensions that existed among American and local observers at a number of the stations around the world.[72] Many observers in the field communicated with R.C. Brock, an SAO administrator in Cambridge, about cultural tensions at the stations. For instance, in 1962, Fred McCallum, an observer in Curacao, wrote, "I presume you will agree that I should prefer to have an American living at the station. If this is prejudice, I am guilty."[73] Although the SAO administrators before and after Cozzens's visit referred to the Indian station as one of the best stations in the network, the tone and difficulty expressed by Cozzens was not uncommon among American observers and their families who were having difficulty assimilating to, "living among prehistoric people," as Mary Holt had complained.[74] In February 1962, Cozzens noted that the highlight of the past two weeks was a visit from SAO Administrator Carl Tillinghast because he was "the first white man we've spoken to for over three months."[75]

Cozzens' complaints about the Indian tracking station were not solely based on cultural differences. Bob Citron, another SAO observer, visited India in December 1962 and reported that the productivity of the Indian station was the lowest among all the SAO stations during the previous four years; that most of the Indian observers were incompetent satellite trackers; and that "real observing skills do not exist at this station."[76] Citron had been a station chief in South Africa, but he was sent to India on an emergency basis to ensure that the SAO did not lose the right to have a visiting American observer at the station. Administrators at the SAO noted that it was extremely difficult to find "Peace Corps–oriented" American observers to work at the station.[77] S.D. Sinvhal, the new director of the UPSO, wrote to the SAO explaining that he was also concerned about the station's efficiency and proposed a number of initiatives that the UPSO and SAO could take to improve the station. Unlike Cozzens, Citron was optimistic and proposed that if the station improved maintenance of its equipment, modernized tracking techniques, and fine-tuned observing skills and efficiency, it would successfully contribute to the network.[78]

In 1962 administrators at the SAO were worried about both the Indian and Japanese stations. Because these stations were under local jurisdiction, the Smithsonian had less influence and control over the standardization of satellite-tracking procedures than at its other stations. A visiting SAO observer at the Japanese station noted that its film processing was not as uniform as it was at the other stations, because the TAO did not update its processing techniques along with the other stations.[79] Citron suggested that the satellite trackers in India needed extensive training to bring the station up to the standard of the other SAO tracking stations. According to Citron, "strong leadership and real observing skill did not exist at this station, and it will be up to Cambridge headquarters to supply support personnel to fill these needs."[80]

## Observer Exchanges

In the early 1960s administrators at the SAO saw observer exchanges as the key to improving the Japanese and Indian satellite-tracking stations' proficiency and morale. The SAO looked to these exchanges, among stations and among the headquarters and stations, to promote the spread of ideas, foster the station's relationship with the Cambridge headquarters, refine the methods of using the Baker-Nunn camera, and establish a cohesive global scientific community. In January 1961, Arvind Bhatnagar, an observer in Nainital, commented that his visits to the tracking stations in Hawaii and Japan "infused [him] with new ideas to make the satellite tracking business [in India] much more pleasant."[81] By 1964, just two years after Cozzens's scathing report quoted at the beginning of this essay and Bob Citron calling it one of the least productive stations in the network, Whipple noted that, "I think the fact that India had the best station in the network last year is evidence of the value of observer exchanges."[82] In 1966, four years after Cozzens's report, Carlton Tillinghast, assistant director of the SAO, on a visit to India reported, "Nainital is one of the best Baker-Nunn stations: SAO views observer exchanges as one means of keeping it that way."

Although these observer exchanges were vital to the improvement of many stations, Sinvhal, current director of the UPSO, opposed any extended American presence in India, writing to Tillinghast, "Please appreciate that to most Indians still, a Westerner coming here to work, is a superiorly qualified person and comes only to train, guide and supervise. The concepts of 'participation' and 'exchange of views' are only beginning to get acceptance."[83] In Japan, administrators at the station also resisted a continued American presence but their resistance was rooted primarily in an interest in maintaining the autonomy of the station.

## James Latimer and the Tokyo Station

When James Latimer arrived in Japan in early 1965, he hoped that he could get the observers caught up to the average Baker-Nunn station speed and "by osmosis, or just my presence, somehow try to influence them into a more receptive attitude."[84] After spending about a month in Japan, Latimer began to find it easier to speak with the Japanese observers and observed that they were "impressive as individuals—intelligent, capable, personable, etc."[85] According to Latimer, the only reason the Japanese station was not top notch was because the observers lacked motivation, believing that they were doing the SAO a service. Latimer proposed "to help out in this big unity campaign," that the SAO should send all routine correspondence through the observatory director to emphasize the TAO authority over that station.[86]

In a report that he wrote for future American observers visiting the station, Latimer suggested that the biggest shock they would face was not material inconveniences but "the nature of the people" and their way of thinking.[87] For instance, in Japan the "woman is kind of secondary to the man; therefore, she does not take part in his conversation and the most basic rule is she never expresses an opinion."[88] In India the "nature of the people" was also commented on by a number of visiting American observers and their families. Hal Cozzens's wife reported that "everything from religion, speech, customs, dress, living conditions, and especially the people, conspires to make this country more different than could be imagined."[89]

Throughout his correspondence with the SAO, Latimer stressed the unwillingness of the Japanese observers to incorporate his suggestions and customary SAO procedures. Latimer noted that some of the tension he encountered stemmed "from their desire to be known as TOKYO ASTRONOMICAL OBSERVATORY (capital letters, if you please) rather than SC-5, or Smithsonian satellite-tracking station."[90] He relayed that when he suggested a standard Smithsonian procedure, the Japanese observers would respond by reminding him that he was at TAO, not SAO. In addition, Latimer believed that TAO was not sending the SAO rosters, because they did not like the implication that they were SAO employees. Carl Hagge from the SAO headquarters responded that he thought the SAO should make an effort to acknowledge the Japanese observers' assertions of identity and autonomy.[91]

The SAO and TAO were in constant negotiations about the day-to-day workings of the tracking station. When the SAO suggested that they send an engineer to Japan to help with the timing system, Latimer responded, "it is very likely that his 'salesmanship' skills will be needed far more than his technical skills."[92] Latimer's comment reflected the frustration he felt when he tried to convince the TAO astronomers to do something the SAO way and not the TAO way. When John Hsia, an SAO administrator, visited Japan in December 1965 he noted that the Japanese observers were accustomed to doing things their way and not "the SAO way on faith as the other stations do." In India the tracking station made a policy, according to Hsia, to operate just like any other SAO station. "I am sure" Hsia wrote "a lot of problems with Japan in the past have gone unresolved because of the extra burden of selling that is necessary."[93]

Throughout their collaborative relationship, communication was a point of tension with both the SAO and the TAO. The SAO was often frustrated with the lack of communication between the Japanese station and headquarters. Latimer expressed difficulty getting Japanese observers to send weekly reports, material requests, and letters to SAO. The lack of communication was not one-sided however. The SAO did not tell the TAO how long Latimer would stay in Japan, with the expectation that by evading the issue they could keep him at the Tokyo station longer.[94] Hirose and other TAO astronomers frequently asked Latimer how long he intended to stay, and Latimer did not offer a direct answer to their questioning. Latimer was surprised that the Japanese observers did not write the SAO directly and voice their frustrations with the vagueness surrounding the length of his visit. He wrote SAO for advice, remarking "I am highly uncertain just how much I should compromise SAO intentions in the interest of cooperating with them."[95] At the end of March, the SAO finally sent a letter to the observatory director asking that Latimer stay for six months or more.[96]

Shortly after he arrived, Latimer began sending what he called "clandestine memos" to the SAO. In these memos Latimer voiced his frustrations, contemplated cultural differences, and was more direct about the situation at the TAO than he was in the correspondence sent through TAO channels. He made arrangements with the U.S. Embassy to handle his mail because he believed that sending these memos from the TAO was too risky for the SAO.[97] Over time Latimer's clandestine memos changed tone from cynical or suspicious to optimistic and understanding. Latimer suggested that most of the difficulty that the SAO had with the Japanese station rested on the TAO lack of funding and unwillingness to ask for SAO support. Although the SAO provided the funds to carry out the tracking program, the TAO was under other financial constraints.[98] According to Carl Hagge the TAO did not ask for funding because that might "represent to them loss of autonomy."[99] Eventually Latimer wrote to the SAO that he understood the Japanese observers' "reluctance to having an American always here ... there is a bit of a strain on them at times, even with our best intentions. They are really doing a marvelous and kind thing for us."[100]

## Conclusion

During the roughly 15 years that it ran an optical satellite-tracking network, the SAO was in constant negotiation with its field stations in Japan and India. The SAO continually adjusted its relationship with these stations in order to ensure uniform data collection, the cohesiveness of the program, and the individual authority of the TAO and UPSO.

The Baker-Nunn satellite-tracking network had a substantial impact on the growth and administrative system of the SAO. By 1959, after one year in operation, the optical satellite-tracking system had produced more than 6,000 photographs and had 38 astronomers running 12 stations. In this year the SAO staff had grown to 179 people, a significantly large number compared to the staff of roughly 15 in 1955. Already, by 1959, the *Annual Report to the Smithsonian Institution* notes, "The Baker-Nunn camera has produced results of inestimable scientific value."[101] In 1961 data gathered from the satellite-tracking stations helped SAO scientists determine that Earth is egg-shaped at the equator, and that it is roughly 14,000 feet wider at some diameters of the equator than at others.[102] In the early 1960s, the observatory's largest source of revenue came from contracts with the military and NASA for satellite tracking, which were worth millions of dollars annually. This substantial funding enabled Whipple to develop the observatory's research in space studies, astrophysics, planetary science, meteoritic, and cometary studies.[103] In 1963, Whipple described the progress of the SAO when he wrote "during the decade, the Smithsonian Astrophysical Observatory experienced greater change and generated more scientific data than in any other comparable period since its establishment." The optical satellite-tracking cameras had made more than 81,750 observations of 73 satellites, which were used to study Earth's gravitation potential, fluctuating densities of the high atmosphere, and refine geodetic data.[104] From data collected by the Baker-Nunns, the location of each tracking station was refined to within 10 meters. Along with the other optical satellite-tracking stations, the data collected at the stations in India and Japan contributed to improved determinations of the size of the Earth, intercontinental distances, and distribution of Earth's atmosphere.[105] By the early 1970s, optical satellite-tracking technology was superseded by a more precise satellite-ranging laser system that the Smithsonian developed. After its successful run, the SAO closed most of its Baker-Nunn stations.[106]

In Japan and India the question of how to incorporate the Baker-Nunn tracking stations into existing institutions was at the forefront of ongoing negotiations at the observatories since the inception of the program. By the mid 1960s, artificial satellites and their associated programs had fulfilled some of the expectations foreseen by Sir Patrick Moore. The global network of satellite-tracking stations set up by the SAO promoted cultural exchanges and political cooperation. But what was standardized through this global scientific cooperation? Did this standardization spread beyond data collection? Does Moore's suggestion of widespread cultural influences hold true for the satellite-tracking stations?

The experiences of American and local observers at the Japanese and Indian tracking stations suggest that negotiation, more than cultural exchange or standardization, played a pivotal role in the success of the program. The archetypal satellite-tracking observatory did not exist in Japan or India. Even though each station was equipped with the same telescope, roughly the same buildings, and collected the same kind of data, the stations' social, political, and cultural contexts impacted how they were run. Observer exchanges, although tension-filled at times, ultimately contributed to system cohesiveness, as Whipple had hoped. From the frequent reports of visiting observers, it is clear that their habits, tastes, and political views were not drastically altered by their experiences in other countries, nor were the habits, tastes, and political views of the local observers drastically altered by the influence of visiting observers. The visiting American observers were mediators not missionaries. Acting as liaison among headquarters and the stations, visiting observers could recommend the best course of action to serve all the parties' needs and maintain the cooperative character of the program without undermining the jurisdiction of the tracking stations.

The important role of negotiation between the SAO and its field stations is a reflection of the true cooperative character of the optical satellite-tracking program. The SAO was able to run an international scientific program that produced standardized data without having to standardize each station. It was essential that the tracking system did not require a homogeneous network because in order for the TAO and UPSO stations to serve the interests of their host countries, a requirement for their installation, they had to be under Japanese and Indian jurisdiction. This collaboration is significant because American, Indian, and Japanese astronomers, working under institutional, national, and international pressures, had to design, and continually refine, their relationships to serve complex interests. Their relationships were synergistic. The SAO needed observatories around the world, like the UPSO and TAO, to carry out a successful satellite-tracking program that strengthened the observatory's presence on the international scientific stage and reinforced positive impressions of America. The TAO and UPSO needed to participate in a large-scale scientific program, like the SAO optical satellite-tracking program, to build their countries' scientific infrastructure and participate in large-scale space science projects. The SAO optical satellite-tracking program is not an example of scientific and cultural standardization, but instead, is a case study of what can be maintained and gained through international scientific cooperation.

## Notes

1. Patrick Moore, *Earth Satellites* (New York: W.W. Norton & Company, 1956), 116.
2. Hal Cozzens to R.C. Brock, January 5, 1962, Record Unit 263, Box 64, Smithsonian Astrophysical Observatory (SAO), Satellite-Tracking Program (STP), Satellite-Tracking

Station Records (STSR), 1956–1968, Smithsonian Institution Archives (SI Archives), Washington, D.C.

3. Gabrielle Hecht, *The Radiance of France: Nuclear Power and National Identity after World War II* (Cambridge, Massachusetts: The MIT Press, 1998); Thomas Hughes, *Networks of Power: Electrification in Western Society, 1880–1930* (Baltimore, Maryland: Johns Hopkins University Press, 1983).

4. E. Nelson Hayes, *Trackers of the Skies* (Washington, D.C.: Smithsonian Institution, 1967), 10–15; Shirley Thomas, *Satellite Tracking Facilities: Their History and Operation* (New York: Holt, Rinehart, and Winston, 1963).

5. W. Patrick McCray, *Keep Watching the Skies! The Story of Operation Moonwatch and the Dawn of the Space Age* (Princeton, New Jersey: Princeton University Press, 2008), 6–7.

6. Hugh Odishaw, executive secretary, USNC–IGY to the USNC Executive Committee, November 30, 1955, 6.11 Earth Satellite Program, Project 30.3: Smithsonian Institution: Initial Development of Optical Tracking System, 1955–1959, National Academy of Sciences Archives (NAS Archives), Washington, D.C.

7. *Annual Report of the Smithsonian Institution, 1957* (Washington, D.C.: Smithsonian Institution Press, 1957), 69.

8. David DeVorkin, "Defending a Dream: Charles Greeley Abbot's Years at the Smithsonian," *Journal for the History of Astronomy* 21 (1990): 121–136; Ron Doel, "Redefining a Mission: The Smithsonian Astrophysical Observatory on the Move," *Journal for the History of Astronomy* 21 (1990): 137–153.

9. Leonard Carmichael to Vannevar Bush, September 9, 1954, Record Unit 50, Box 15, Folder "A.P.O. Future," SI Archives.

10. See Leonard Carmichael memorandums: December 7, 1953; June 29, 1954; and October 18, 1954, Record Unit 50, Box 15, Folder "A.P.O. Future," SI Archives.

11. W. Patrick McCray, "Amateur Scientists and the Ambitions of Fred Whipple," *Isis* 97 (December 2006): 639.

12. DeVorkin, "Defending a Dream," 121–136; Leonard Carmichael to Donald Menzel, June 21, 1955, Record Unit 40, Box 15, Folder "A.P.O Future," SI Archives.

13. Sydney Chapman to Joseph Kaplan, January 27, 1956, 10.2 Comité Spéciale de l'Année Géophysique Internationale, CSAGI: Correspondence: Earth Satellites, 1955–1957, NAS Archives, Washington, D.C.

14. "Technological Capabilities Panel Recommendations Report to the National Security Council," 56, Tab A, attached to Allen W. Dulles to executive secretary, NSC, June 6, 1955, "NSC Action 1355," Folder NSC 5522 Technological Capabilities Panel, Box 16, NSC Policy Papers, Office of the Special Assistant for National Security Affairs: Records, 1952–1961, White House office, DDE Library, quoted in Allan A. Needell, *Science, Cold War, and the American State: Lloyd V. Berkner and the Balance of Professional Ideals* (Amsterdam: Harwood Academic Publishers, 2000), 340.

15. Odishaw asked Rudolph if the SAO should establish a satellite-tracking station in either Egypt or Iran because, as he noted, the Middle East was an area of government concern. Either location was suitable to the SAO, but Egypt would be more advantageous scientifically. Rudolph called Odishaw and let him know that Iran was preferable to the Department of State. Odishaw to Rudolph, February 16, 1957, Record Unit 12, Box 10.1, Folder "CSAGI: Participating Countries: Iran, 1956–1959," NAS Archives.

16. Walter Rudolph, meeting notes, December 5, 1956, Record Unit 12, Box 1.8, Folder "USNC Meetings: Executive Session: Special Satellite, Dec. 1956," NAS Archives.

17. J.G. Reid Jr. notes on a phone call with Whipple, August 24, 1956, Record Unit 12; J.G. Reid to Fred Whipple, September 4, 1956, Record Unit 12, Box 10.4, Folder "CSAGI Assemblies: Fourth: Barcelona: Correspondence, 1956," both in NAS Archives.

18. Hugh Odishaw to Walter Rudolph, June 10, 1957, Record Unit 12, Box 10.1, Folder "CSAGI: Participating Countries: Japan, 1957"; Summary of Negotiations for the

Optical Satellite-tracking station, India, May 10, 1957, Record Unit 12, Box 10.1, Folder "CSAGI: Participating Countries: India, 1956–1959," NAS Archives.

19. *Summary of Negotiations for the Optical Satellite-tracking station, India,* May 10, 1957, Record Unit, 12, Box 10.1, Folder "CSAGI: Participating Countries: India, 1956–1959, NAS Archives.

20. P.L. Bhatnagar became a visiting scientist at the SAO in the early 1960s. See Fred Whipple to P.L. Bhatnagar, April 7, 1964, Record Unit 7431, Box 7, Folder "B"; Fred Whipple to P.L. Bhatnagar, August 2, 1956, Record Unit 7431, Box 7, Folder "B," both in SI Archives.

21. Hugh Odishaw to Walter Rudolph, June 10, 1957, Record Unit 12, Box 10.1, Folder "CSAGI: Participating Countries: Japan, 1957," NAS Archives.

22. R.C. Peavey, memo for files, September 19, 1956, Record Unit 12, Box 10.4, "CSAGI: Assemblies: Fourth: Barcelona: Correspondence, 1956," NAS Archives.

23. Packaged document on the Smithsonian Institution (Smithsonian Astrophysical Observatory) Program for Optical Tracking of the Satellite for Release at Barcelona, September 10–15, 1956, Record Unit 188, Box 118, "Old STP," SI Archives.

24. *General Report,* 1954, Record Unit 12, Box 10.4, Folder "CSAGI: Assemblies: Fourth: Barcelona: General Report, 1956," 304, NAS Archives.

25. In addition to writing to his colleagues around the world about participation in the satellite-tracking program, Whipple wrote to the Department of State about the government's preference for station locations. A September 1956 issue of *The Daily Telegraph* noted that Americans were "trying to 'hog' the limelight" and give the "appearance of trying to make political capital out of their contributions to the program." U.S. Information Services, London to U.S. Information Agency, September 27, 1956, on an article in the British newspaper *The Daily Telegraph* (September 14, 1956), Record Unit 12, Box 11, "CSAGI Barcelona Meeting, September 10–15, 1956," NAS Archives.

26. John D. Dower, *Embracing Defeat: Japan in the Wake of World War II* (New York: W.W. Norton & Company, 1999), 494.

27. See S.D. Sinvhal, "The Uttar Pradesh State Observatory—Some Recollections and Some History (1954–1982)," *Bulletin of the Astronomical Society of India* 34 (2006): 73; S.D. Sinvhal, *Twenty-Five Years of Uttar Pradesh State Observatory* (Naini Tal, India: The Observatory, 1979), 5; Stuart Leslie and Robert Kargon, "Exporting MIT: Science, Technology, and Nation-Building in India and Iran," *Osiris* 21 (2006): 113.

28. "The Eighth Session of the Industry and Trade Committee at the Twelfth Session of ECAFE, held in Bangalore between January 25 and February 14, 1956, were more than run of the mill annual conferences of subsidiary bodies of ECOSOC. They gained exceptional importance due to the fact that at Bangalore the USSR and the United States, for the first time after the Bulganin–Khrushchev visit to India, confronted each other across a conference table in the very country which the Soviets have chosen as the major target in their new type of ideological and economic warfare." Walter Kotschnig to Wilcox, February 25, 1956, General Records of the Department of State Record Group 59, Box 413, Folder "21.44 Country File: India, a. Agreements, 1955–59," Archives II, National Archives and Records Administration, College Park, Maryland.

29. Alfred Eckes and Thomas Zeiler, *Globalization and the American Century* (New York: Cambridge University Press, 2003), 137.

30. Brumright to Scott, May 27, 1954, General Records of the Department of State, Record Group 59, Box 422, Folder "21.52 Country File: Japan, e. General, 1954," National Archives.

31. Walter LaFeber, *The Clash: U.S.–Japanese Relations throughout History* (W.W. Norton & Company, 1997), 310–312; Leslie and Kargon, "Exporting MIT."

32. Lloyd Berkner excerpted from J.W. Joyce to Hugh Odishaw, January 22, 1957, Record Unit 12, Box 10.1, Folder "CSAGI: Participating Countries: India, 1956–1959," NAS Archives.

33. In 1953 the Indian National Committee (INC) for the IGY announced that India would participate in a number of cooperative programs, including a longitude survey, solar observations, and ionosphere studies. As programs were added to the IGY, the Indian national committee expanded its research plans. By 1956 a number of scientific stations had been set up and numerous observatories were outfitted with equipment to carry out research for the IGY.

34. I have only found one acknowledgement of this connection. Odishaw to Nicolet, November 15, 1956, Record Unit 12, Box 10.2, Folder "Correspondence: Earth Satellites, 1955–1957," NAS Archives; Whipple to Masaki Miyadi, March 13, 1957, Record Unit 263, Box 23, Folder "Japan: 200—Personnel, 1957–1964," SI Archives.

35. Miyadi to Henize, December 18, 1956, Record Unit 263, Box 23, Folder "Japan: 300—Administration, 1956–64," SI Archives.

36. Miyadi to Henize, December 18, 1956, Record Unit 263, Box 23, Folder "Japan: 300—Administration, 1956–64," SI Archives. See also Shigeru Naakayama and Morris Low, "The Research Function of Universities in Japan" *Higher Education*, 34 (September 1997): 253; Yamamoto to Whipple, August 17, 1956, Record Unit 263, Box 23, Folder "Japan: 300—Administration, 1956–64," SI Archives.

37. Bappu received his PhD in astronomy from Harvard University in 1952. His dissertation discussed Wolf-Rayet atmospheres. While at Harvard, Bappu along with Bart Bok and Gordon Newkirk discovered the "Bappu-Bok-Newkirk Comet." Bok, a professor at Harvard, was working at the observatory when negotiations were made between the UPSO and SAO. Donald Menzel, director of the Harvard College Observatory, was Bappu's dissertation advisor. Bappu was well acquainted with countless astronomers at Harvard and the SAO. Hugh Odishaw to Mitre, April 16, 1957, Record Unit 12, Box 10.1, Folder "CSAGI: Participating Countries: India, 1956–1959," NAS Archives.

38. All quotes on this trip are from J. Allen Hynek, Report, July 29, 1957, J. Allen Hynek Papers, Box 10, "STP—Stations: JAH Trip Reports on Travel to Possible STP Station Locations, 1957" Northwestern University Archives, Evanston, Illinois.

39. Fred Whipple to H. Hirose, March 19, 1964, Record Unit 7431, Fred Lawrence Papers, circa 1927–1983, Box 7, SI Archives.

40. Preliminary Plans for Satellite-tracking stations, October 23, 1956, J. Allen Hynek Papers, Box 10, Folder "STP—Stations: Preliminary Planning and Designs for Stations, 1956," Northwestern University Archives.

41. Whidden to Karl Henize, February 7, 1958, Record Unit 263, Box 18, Folder "India: Incoming Correspondence, 1956–June 1958," SI Archives.

42. Record Unit 188, Box 118, Folder "Old STP," "Packed Document on the Smithsonian Institution Program for Optical Tracking of the Satellite for Release at Barcelona, September 10–15, 1956," 3.

43. Hayes, *Trackers of the Skies*, 10–15; Thomas, *Satellite Tracking Facilities*, 31–39; Robert Citron, *Report on the Operational Capabilities of a Baker–Nunn Tracking System for Satellite Surveillance*, March 28, 1964, Record Unit 263, Box 65, Folder "Oslo Administratively Confidential, 1963–65," SI Archives; McCray, *Keep Watching the Skies*, 76–77.

44. Smithsonian Institution to the American Embassy, New Delhi, India, May 6, 1958, Record Unit 12, Box 10.1, Folder "CSAGI: Participating Countries: India, 1956–1959," NAS Archives; A. M Rosenthal, "Indian Town Gets a Satellite Role," *New York Times* (April 6, 1958): 31.

45. The Uttar Pradesh Observatory was renamed the Uttar Pradesh State Observatory in 1957. Even though the region is awash during the monsoon season, the SAO and Indian National Committee to the IGY selected the site for the optical satellite-tracking station. All the potential station locations in India were not ideal. Although observing conditions were one of the major variables for location selection, a number of the stations were set up in areas affected by the frequent rains, including the stations in Iran, Argentina, and Japan. See Air Weather Service Report, January 10,

1957, Record Unit 188, SI Archives; Sinvhal, "The Uttar Pradesh State Observatory," 73; Sinvhal, *Twenty-Five Years of Uttar Pradesh State Observatory,* 5.

46. Located 150 miles northwest of New Delhi, the hill town of Nainital had a population, in the late 1950s, of approximately 40,000 people in the summer and 7,000 in the off season. See A.M. Rosenthal, "Indian Town Gets a Satellite Role," *The New York Times* (April 6, 1958): 31.

47. During the IGY the INC–IGY assigned the UPSO the (1) observation of aurora, (2) photography of the Moon, and (3) optical satellite tracking. The UPSO low latitude limited any possible observations of aurora. Working with the U.S. Naval Observatory (USNO), the UPSO took 284 photographic plates of the Moon to refine the positions of the lunar center and observatory. The photographs of the Moon were taken for a program, run by USNO, organized to determine the center of the Moon, which in turn could be used to determine the position of the observatory. The USNO provided the UPO with a Markowitz dual motion camera in December 1957. See Sinvhal, "Uttar Pradesh State Observatory," 71. On Moonwatch, also see chapter 16.

48. Whidden to Henize, report no. 13, June 14, 1958, Record Unit 263, Box 18, Folder "India: Incoming Correspondence, 1956–June 1958," SI Archives.

49. "We received with great interest your photograph of the completed camera house at your observatory. It is one of the first such buildings to be completed." Henize to Miyadi, September 24, 1957, Record Unit 262, Box 23, Folder "Box 23 Folder: Japan: 319 Weekly Reports, 1957–1964" SI Archives; Whidden to Karl Henize, February 7, 1958, Record Unit 263, Box 18, Folder "India: Incoming Correspondence, 1956–June 1958," SI Archives; *Smithsonian Institution Astrophysical Observatory Quarterly Progress Report No. 5 for the Establishment of Optical Tracking Stations for the IGY Satellite-tracking program,* 31, February 1958, 6.11 Earth Satellite Program: Project 30.6: Smithsonian Institution: Establishment of Optical Tracking Stations: 1956–1961, NAS Archives; Smithsonian Institution; *Astrophysical Observatory Quarterly Progress Report No. 7 for the Operation of Optical Tracking Stations for the IGY Earth Satellite Program,* September 30, 1958, 6.1 Earth Satellite Program: Project 30:8: Smithsonian Institution: Operation of Optical Tracking Stations: 1956–1963, NAS Archives.

50. E.C. Balch to Karl Henize, January 7, 1958, Record Unit 263, Box 62, Folder "India Agreement," SI Archives.

51. K. Drummond to S. Whidden, February 20, 1958, Record Unit 263, Box 18, Folder "India: Outgoing Correspondence, 1957–59," SI Archives.

52. Whidden to J.A. Hynek, February 11, 1958, Record Unit 263, Box 18, Folder "India: Incoming Correspondence, 1956– June 1958," SI Archives.

53. Whidden to Drummond, March 7, 1958, Record Unit 262, Box 18, "India: Incoming Correspondence, 1956–June 1958," SI Archives.

54. Henize to Miyadi, April 14, 1958, Record Unit 263, Box 23, Folder "Japan: 319 Weekly Reports, 1957–1964," SI Archives.

55. Samuel Whidden, Report 11, May 31, 1958, Record Unit 263, Box 18, Folder "India: Incoming Correspondence, 1956–June, 1958," SI Archives.

56. V. Bappu to Karl Henize, May 22, 1958, Record Unit 263, Box 18, Folder "India: Incoming Correspondence, 1956–June 1958," SI Archives.

57. S. Whidden to Karl Henize, May 24, 1958, Record Unit 263, Box 18, Folder "India: Incoming Correspondence, 1956–June 1958," SI Archives; S. Whidden to Karl Henize, May 31, 1958, Report No. 11, Record Unit 263, Box 18, Folder "India: Incoming Correspondence, 1956–June 1958," SI Archives.

58. Whidden to Hynek, July 30, 1958, Record Unit 263, Box 18, Folder "India: Incoming Correspondence, July–Dec. 1958," SI Archives.

59. Whidden, Report 22, August 9, 1958, Record Unit 263, Box 18, Folder "India: Incoming Correspondence, July–Dec. 1958," SI Archives.

60. Whidden, Report 24, August 21, 1958, Record Unit 263, Box 18, Folder "India: Incoming Correspondence, July–Dec. 1958," SI Archives.

61. *Smithsonian Institution Astrophysical Observatory Quarterly Progress Report No. 8 for the Operation of Optical Tracking Stations for the IGY Earth Satellite Program*, December 31, 1958, Record Unit 12, Box 6.1 Earth Satellite Program: Project 30:8: Smithsonian Institution: Operation of Optical Tracking Stations: 1956–1963, NAS Archives.

62. *Smithsonian Institution Astrophysical Observatory Quarterly Progress Report No. 10 for the Operation of Optical Tracking Stations for the IGY Earth Satellite Program*, June 30, 1959, Record Unit 12, Box 6.1 Earth Satellite Program: Project 30:8: Smithsonian Institution: Operation of Optical Tracking Stations: 1956–1963, NAS Archives.

63. Fred Whipple to Charles Tougas, March 29, 1960, Record Unit 188, Box 114, "NASA 1 (Stations)," SI Archives.

64. Fred Whipple to SAO Scientists and Administrators, June 2, 1964, Record Unit 7431, Box 8, Folder "NASA Station Operns. 1964–1965," SI Archives; Hayes, *Trackers of the Skies*, 72.

65. E. B. Skolnikoff memo to the Science and Foreign Affairs panel members of the President's Science Advisory Committee, December 15, 1960, Record Group 58, Box 298, Folder "14.G.42 President's Science Advisory Committee–General, 1959–61," National Archives.

66. T. Keith Glennan, Statement to the House Committee on Science and Astronautics, January 27, 1960, Record Unit 50, Box 259, "National Aeronautics and Space Administration Press Releases," SI Archives.

67. E.B. Skolnikoff memo to the Science and Foreign Affairs panel members of the President's Science Advisory Committee, December 15, 1960, Record Group 58, Box 298, Folder "14.G.42 President's Science Advisory Committee–General, 1959–61," National Archives.

68. George Morgan, Department of State, draft letter to MacArthur, October 30, 1958, Record Group 59, Box 259, Folder "Outer Space, 14.B.8, International Cooperation," National Archives.

69. Gathright to Frutkin, October 19, 1961, Record Group 59, Box 250, Folder "Cooperative Space Program. 9. India, 1961–62," National Archives.

70. Hal Cozzens to R. C. Brock, December 3, 1961, Record Unit 263, Box 19, Folder "India: 319 Weekly Reports, 1960–62," SI Archives.

71. Hal Cozzens to R. C. Brock, January 5, 1962, Record Unit 263, Box 64, SAO, STP, STSR, 1956–1968, SI Archives.

72. Shortly after reporting on the Baker-Nunn station in Nainital, Cozzens became a station chief in Iran. While Cozzens was working in Iran, Fred Whipple wrote a letter to H.K. Afshar of the Universite de Teheran about Cozzens's inability to "understand the people and the environment" of the country. See Whipple to H.K. Afshar, January 8, 1963, Record Unit 7431, Fred Lawrence Whipple Papers, circa 1927–1983, Box 6, SI Archives.

73. Fred McCallum to R.C. Brock, March 20, 1962, Record Unit 263, Box 64, SAO, STP, STSR, 1956–1968, SI Archives.

74. Mary Holt, *Newsletter to Wives of Observers*, July 7, 1958; Fred McCallum to R.C. Brock, March 20, 1962, Record Unit 263, Box 64, Folder "Curacao: Administratively Confidential Memoranda, 1961–1966," SI Archives.

75. Cozzens biweekly report to Brock, February 12, 1962, Box 19, Folder "India: 319— Weekly Reports, 1960–62," SI Archives.

76. Bob Citron to Jan Rolff, December 10, 1962, Record Unit 263, Box 19, SAO, STP, STSR, 1956–1968, SI Archives.

77. The SAO had difficulty convincing American observers to visit India. As one observer reported, there was an absence of other American personnel within a radius of 150 miles, poor housing, lack of cultural and recreational facilities, inadequate transportation; the cost of living and living conditions were the main reasons observers did not want to work in India. See Brock to Tillinghast, November 13, 1962; Rolff to Taylor, December 5, 1962, both in Record Unit 263, Box 37, Folder "Station Operations; Station Records, India, 1962–1963," SAO, STP, STSR, 1956–1968, SI Archives.

78. Sinvhal to Tillinghast, June 25, 1962, Record Unit 263, Box 19, "India: 300—Administration, 1959–64," SAO, STP, STSR, 1956–1968, SI Archives.

79. Hagge to Brock, April 9, 1962, Record Unit 263, Box 23, Folder "Japan: 400—Equipments, Supplies, and Services, 1957–64," SAO, STP, STSR, 1956–1968, SI Archives.

80. Citron to Rolff, December 10, 1962, Record Unit 263, Box 19, SAO, STP, STSR, 1956–1968, SI Archives.

81. Arvind Bhatnagar to SAO, January 12, 1961, Record Unit 263, Box 19, Folder "India: 300—Administration, 1959–64," SAO, STP, STSR, 1956–1968, SI Archives.

82. Fred Whipple to H. Hirose, March 19, 1964, Record Unit 7431, Fred Lawrence Papers, circa 1927–1983, Box 7, SI Archives.

83. Sinvhal to Charles Tillinghast, February 15, 1965, Record Unit 263, SAO, STP, STSR, 1956–1968, SI Archives.

84. Jim Latimer to Carl Hagge, January 18, 1965, Record Unit 263, Box 65, Folder "Japan: Administratively Confidential, 1965," SAO, STP, STSR, 1956–1968, SI Archives.

85. Latimer to Hagge, March 6, 1965, Record Unit 263, Box 48, Folder "STADAD Japan 320—Organization, 1965–66," SAO, STP, STSR, 1956–1968, SI Archives.

86. March 10, 1965, memo from Latimer to Hagge, Record Unit 263, Box 48, "STADAD Japan 320—Organization, 1965–66," SAO, STP, STSR, 1956–1968, SI Archives.

87. Latimer to La Count, October 23, 1965, Record Unit 263, Box 48, Folder "STADAD Japan 620—Recurring Reports, 1965–1966," SAO, STP, STSR, 1956–1968, SI Archives.

88. Latimer to La Count, October 23, 1965, Record Unit 263, Box 48, Folder "STADAD Japan 620—Recurring Reports, 1965–1966," SAO, STP, STSR, 1956–1968, SI Archives.

89. Helen Cozzens to the SAO, undated, Record Unit 263, Box 19, "India: 300—Administration, 1959–64," SAO, STP, STSR, 1956–1968, SI Archives.

90. Latimer to La Count, January 18, 1965, Record Unit 263, Box 65, Folder "Japan: Administratively Confidential, 1965" SAO, STP, STSR, 1956–1968, SI Archives.

91. Hagge to Latimer, January 27, 1965, Record Unit 263, Box 65, Folder "Japan: Administratively Confidential, 1965," SAO, STP, STSR, 1956–1968, SI Archives.

92. Latimer to La Count, July 10, 1965, Record Unit 263, Box 48, Folder "STADAD Japan 620—Recurring Reports, 1965–1966," SAO, STP, STSR, 1956–1968, SI Archives.

93. John Hsia to Carl Hagge, December 8, 1965, Record Unit 263, Box 80, Folder "DAE: Japan 420—Operations, 1963–1967," SAO, STP, STSR, 1956–1968, SI Archives.

94. Latimer believed that if the SAO was upfront with how long it wanted him to stay he could "continue trying to get them to modernize their system—although (he was) pretty luckless thus far." See Latimer to Hagge, March 10, 1965, Record Unit 263, Box 48, Folder "STADAD Japan 320—Organization, 1965–66," SAO, STP, STSR, 1956–1968, SI Archives.

95. Latimer to Hagge, March 27, 1965, Record Unit 263, Box 48, Folder "STADAD Japan 320—Organization, 1965–66," SAO, STP, STSR, 1956–1968, SI Archives.

96. Hsia draft letter to Hirose, April 29, 1965, Record Unit 263, Box 48, Folder "STADAD—Japan Organization, 1965–66," SAO, STP, STSR, 1956–1968, SI Archives.

97. In June Latimer had received a memo that if Hirose had seen it, it would have been clear that he would be aware of the private channel of communication between Latimer and Cambridge. Administrators at SAO "advised that, since the impression Mr. Latimer had created at the Japanese station was of an honest, straightforward person, he should admit that this channel existed and explain that there are certain problems that arise that aren't really problems." See John Hsia memo on a telephone call with Latimer, June 28, 1965; Latimer to Hagge, February 11, 1965, both in Record Unit 263, Box 48, Folder "STADAD—Japan Organization, 1965–66" SAO, STP, STSR, 1956–1968, SI Archives.

98. Latimer to Hagge, March 10, 1965, Record Unit 263, Box 48, Folder "STADAD Japan 320—Organization, 1965–66" SAO, STP, STSR, 1956–1968, SI Archives.

99. Hsia to Hagge, December 8, 1965, Record Unit 263, Box 80, Folder "DAE: Japan 420—Operations, 1963–1967," SAO, STP, STSR, 1956–1968, SI Archives.

100. Jim Latimer to Ron La Count, September 28, 1965; Jim Latimer to Carl Hagge, January 31, 1965, both in Record Unit 263, Box 48, Folder "STADAD—Japan Organization, 1965–66," SAO, STP, STSR, 1956–1968, SI Archives.

101. See *Annual Report of the Smithsonian Institution, 1959* (Washington, D.C.: Smithsonian Institution Press, 1959), 100–108; Doel, "Redefining a Mission," 137–153.

102. Science Notes, *New York Times* (February 12, 1961): E7.

103. W. Patrick McCray, "Amateur Scientists, the International Geophysical Year, and the Ambitions of Fred Whipple," *Isis* 97, no. 4 (2006): 657.

104. Fred Whipple to Leonard Carmichael, September 10, 1963, Record Unit 7431, Fred Lawrence Whipple Papers, circa 1927–1983, SI Archives.

105. Sinvhal, *Twenty-Five Years of Uttar Pradesh State Observatory*, 11–27.

106. M.R. Pearlman, N.W. Lanham, C.G. Lehr, J. Wohn, and J.A. Weightman, "Smithsonian Astrophysical Observatory Laser Tracking Systems [and Discussion]," *Philosophical Transactions of the Royal Society of London. Series A, Mathematical and Physical Sciences* 284, no. 1,326, A Discussion on Methods and Applications of Ranging to Artificial Satellites and the Moon (1977): 431–442.

# Chapter 16

# Geodesy, Time, and the Markowitz Moon Camera Program: An Interwoven International Geophysical Year Story

*Steven J. Dick*

## Introduction and Context

The Markowitz Moon camera program, originated at the U.S. Naval Observatory in Washington, D.C. in 1952 for the determination of "ephemeris" time, was transformed a few years later into one of the many programs of the International Geophysical Year (IGY). The Moon camera's stated IGY goal was to improve geodesy—the study of the gravity field, shape and size of the Earth; geodetic positions are related to time through longitude (360 degrees of longitude is 24 hours of time, so 1 degree at the equator is equal to 4 minutes of time, 69 statute miles or 60 nautical miles). Although in the end the Moon camera program failed in its ambitious geodetic goals, it is in keeping with the theme of this volume because geodesy is a global endeavor by definition, and because this particular program emphasizes that not all IGY programs were successful. Moreover, the Moon camera highlights an important aspect of "making science global" that historians usually ignore— the determination and dissemination of time worldwide. Without accurate timing and synchronization of observations on a global scale, much of the data from the IGY would have been compromised or useless. The Moon camera program was undertaken in the midst of rapid changes in timekeeping and time dissemination, events in which the eponymous William Markowitz played an important role.

Given the lack of historical treatment of this program compared to the Smithsonian's Moonwatch/Baker-Nunn program with its similar aims, in this essay I will answer several questions: Who was William Markowitz and what was the context in which the Markowitz Moon camera program emerged? What were the results of the Markowitz Moon camera program? Where does the program fit in the context and history of the IGY? The answers to these questions will help us assess the broader social and intellectual significance of the program. And along the way we will come to

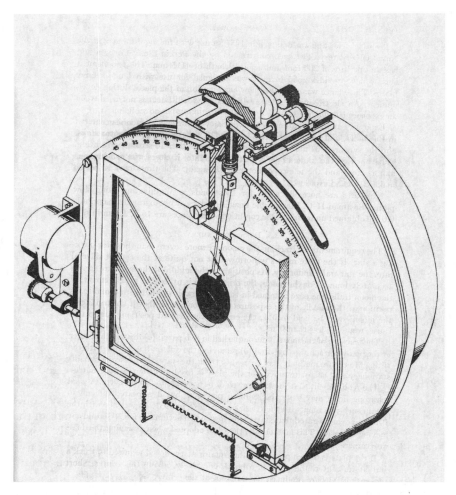

**Figure 16.1** Schematic view of the dual-rate camera built by William Markowitz. (Courtesy of U.S. Naval Observatory Library, Washington, D.C.)

appreciate the importance of time measurement and synchronization for IGY global observations.

## Context

Geodesy is characterized for most of its history by triangulation networks such as pioneered by Karl Friedrich Gauss beginning in 1821.[1] These networks were gradually improved and expanded, and provided increasingly accurate geographic positions over increasingly wide areas; as geodetic maps over time show, it was truly a literal example of a science that had to be made global by definition. Chronometer trips during the 19th century, and the advent of powerful radio transmitters had even enabled some intercontinental longitude determinations to be made, notably during the World Longitude Operations of 1926 and 1933, but not enough for geodetic accuracy.[2] By the 1950s, driven by the needs of intercontinental ballistic missile systems, satellite tracking, and global navigation, the continents needed to be geodetically tied together by a World Geodetic System. Triangulation of satellites against the starry background was the revolutionary new way toward that old problem of geodesy, and it proved very successful. This paper examines an innovative ground-based method for geodesy that was carried out as part of the IGY just before the Space Age, and that illustrates how difficult the problem of geodesy would have been in the absence of the Earth-orbiting satellites.

NOAA (National Oceanic and Atmospheric Administration) historian John Cloud has delineated three eras of geodetic positional accuracy: From 1957–1970 ground-based systems observed satellites visually or by Doppler shifts, providing 10-meter accuracy and the revelation of a "pear-shaped" Earth in 1958. During the years 1970–1990, laser-ranging satellites such as LAGEOS, and Very Long Baseline Interferometry (VLBI), as well as satellite radar altimetry from GEOS-3 (1975), Seasat (1978) and TOPEX-Poseidon (1992) provided accuracies to 1 meter, allowing instantaneous determination of dynamic sea level heights. The current era began around 1990, when an array of ground and space-based systems yielded positional accuracies of a centimeter or less, allowing observations of plate tectonics.[3] VLBI, satellites such as CHAMP, and pairs of satellites such as GRACE (Gravity Recovery and Climate Experiment), were used to determine the structure of the Earth's gravity field with more resolution.[4]

The first World Geodetic System (WGS), known as the Department of Defense World Geodetic System, was achieved in 1960; Deborah Warner has detailed how it was the result of an intense rivalry between the Army and the Air Force.[5] Many WGS's have been developed since then, culminating in WGS 84, which has kept that name despite numerous updates since it was developed in 1984.[6]

Within Cloud's tripartite scheme, the IGY is situated firmly at the very beginning of what he defines as the first era. The International Council of Scientific Unions (ICSU) proposed the IGY in 1952, with the Comité Spécial de l'Année Géophysique Internationale (CSAGI) as the governing body for all IGY activities. "Longitude and Latitude" was one of eleven subject areas, and the location of "World Data Center A" for this area was the U.S. Naval Observatory in Washington, D.C. The Longitude and Latitude program for the IGY was twofold, using different methods: to obtain *astronomical* longitude and latitude (a process affected by the deflection of the vertical defined by a plumb line), and to obtain *geodetic* (or geocentric) longitude and latitude.[7] It is the latter we concentrate on here in the context of the Markowitz Moon camera program. In addition to its geodetic ambitions, the stated goals of the Latitude and Longitude program were to determine accurate coordinates of participating observatories, to improve terrestrial time determination, knowledge of the irregularities of the Earth's rotation, and the accuracy of star catalogues.

Two programs for obtaining geocentric coordinates on the Earth for determining the shape of the Earth (the geoid), were planned for the IGY.[8] One, the Smithsonian Astrophysical Observatory's (SAO) optical Moonwatch program organized by Fred Whipple, has received considerable attention from historians.[9] It, and the SAO's complementary Baker-Nunn program, involved optical and photographic observations of artificial satellites. While the primary goal of the SAO program was satellite tracking, such tracking was the means to several scientific ends, including geodetic data.[10] The other program, the Markowitz Moon camera, originated and administered from the U.S. Naval Observatory in the early 1950s as part of its mission to determine time, is less well known but just as interesting from an international point of view, even though it had better success in the determination of time than for any direct geodetic results. These timing achievements of the Markowitz Moon camera, were, however, in themselves important for geodesy, not to mention broader aspects of science. Conceived and implemented five years before the launch of the first artificial satellite, the Markowitz program for time determination was based on observations of the real Moon, not an artificial one. It was made possible in the 1950s by improved star catalogs against which the Moon's position could be measured, an improved lunar ephemeris, and, above all, a new photographic technique invented by William Markowitz, whose career we describe in the next section. Only in 1954 was geodesy made one of its goals, when Markowitz sent a proposal to the U.S. National Committee for the IGY for "a program to measure the geodetic distance between continents, the size and shape of the Earth, and the variations in its speed of rotation."[11]

The expectations for the Markowitz Moon camera were high. Almost all accounts of the IGY, both before and after it took place, devote a few paragraphs to this program. As early as 1954 Joseph Kaplan, first Chair of the U.S. National Committee for the IGY, wrote that "Highly precise longitude and latitude measurements, important in surveying, mapping, and navigation, can now be made by direct photography of the Moon. The new technique, developed by William Markowitz, makes it possible also to measure changes in the speed of rotation of the earth to ten times the existing precision, and the observational material obtained will shed new light on the inner constitution of the earth. It is planned to photograph the moon and stars nearby every possible night, using the newly designed cameras attached to existing telescopes at astronomical observatories. Analysis of the data obtained will yield longitude and latitude at all stations, and, using existing geodetic nets, the exact distance in miles between all stations can be determined with a precision of 90 feet. During the International Geophysical Year it will be possible for the first time to triangulate the whole earth, because observations will be taken simultaneously all over the earth. Observations will be made at more than fifteen stations distributed over the Earth. England and France, for example, contemplate studies at Greenwich, the Cape of Good Hope, and Paris. The United States expects to make observations at Washington, D.C.; San Diego, California; and Hawaii."[12]

The year following the IGY, Sydney Chapman, then President of the Special Committee for the IGY, described geodetic methods and then the Moon camera. Chapman wrote:

> Seen from different observatories at the same instant, the moon occupies different positions relative to the stars. Each observatory makes two or more such observations during the night. Calculations from such observations will determine the geometry of the observatory distribution over the earth, independent of gravity. Intercontinental distances will be measured with errors of the order of no more than 100 to 200 feet. At present the uncertainty is greater. The positions of some ocean islands are uncertain by a mile. This program will also give an improved orbit of the moon, and a more accurate standard of time measurement.[13]

After the IGY was completed, the Moon camera program is still mentioned, but with a notable lack of discussion of results, at least relating to geodesy. Writing in the *Annual Report of the Smithsonian Institution* for 1959, before pointing out the value of artificial satellite observations (such as Vanguard) for geodesy, Elliott B. Roberts of the Coast and Geodetic Survey wrote that "The framework of international cooperation established for the IGY was seen at the outset to favor establishing a new and better measure of the longitude differences between continents and major isolated island

**Figure 16.2** William Markowitz and his Moon Camera. (Courtesy of U.S. Naval Observatory Library, Washington, D.C.)

groups, such as Hawaii. This was realized through use of new instruments and techniques, including the American dual-rate [Markowitz] moon camera, which provided new precision in the relation of terrestrial positions to the celestial firmament. Better absolute knowledge of geographic locations of the earth's landmasses was obtained, with advantages in mapping, operation of earth satellites, scientific studies of the earth, and the mechanics of its rotation, including problems of timekeeping."[14]

In his well-known popular account of the IGY, *Assault on the Unknown* (1961), *New York Times* science writer Walter Sullivan juxtaposed the Markowitz Moon camera program with the Smithsonian's Baker-Nunn program, and remarked "In addition to the prototype at the Naval Observatory, cameras of this design were delivered to the observatories at Greenwich, Paris, and seventeen other points around the world."[15] But again, no results were reported. We are thus confronted with the mystery of the significance of the Markowitz Moon camera given its high expectations.

## William Markowitz and the Origins of the Moon Camera Program

William Markowitz (1907–1998) obtained his PhD in astronomy from the University of Chicago in 1931, working under W. D. MacMillan and writing his dissertation on the statistics of binary stars. During this period he also worked at Yerkes Observatory under the noted instrument maker Frank Ross. He joined the U.S. Naval Observatory in Washington, D.C. in 1936 and remained there until his retirement in 1966, working under Paul Sollenberger and with Gerald Clemence, spending almost his entire career in the Time Service Department. One of Markowitz's early duties was operating the Photographic Zenith Tube (PZT), designed by Ross and used at the Naval Observatory beginning in 1915 to observe the variation of latitude, and since 1934 to determine Universal time.[16] One of Markowitz's early articles was on latitude and longitude determinations.[17]

The variation of latitude was one of the chief interests of Markowitz throughout his career; it was the analysis of this data that led to his contributions to the study of polar motion. Markowitz directed the Time Service Department from 1953–1966, a period during which methods for more uniform and accurate time were being devised. In addition to time determination, Markowitz was fundamentally concerned with improvements to the dissemination of time. At the International Astronomical Union meeting in Dublin in 1955 he proposed the system of UT0, UT1 and UT2, which corrected observed Universal time successively for polar motion and seasonal variations in the Earth's rotation (see below). The system went into effect within months, with UT2 being disseminated internationally. Markowitz later participated in experiments in synchronizing time around the world using artificial satellites and portable atomic clocks transported by airplanes; relativistic corrections were not yet determined until a famous experiment conducted by his Naval Observatory colleagues and published in 1972. He contributed to the Navy system of time and frequency transmissions, and the control of precise time via Loran-C and the Transit navigation satellite.

Markowitz was heavily engaged in international cooperation in time determination and dissemination. He served as President of the commission on Time of the IAU from 1955–61, and was also active in the International Union of Geodesy and Geophysics (IUGG), the American Geophysical Union (AGU) and the International Consultative Committee for the Definition of the Second. In short, he was well-connected with the American and international scientific communities, connections that helped him implement the Moon camera program.

In examining the reasons for the origins of the Moon camera program, one must keep in mind the varieties of time evolving in the 1950s. It was a time of extraordinary advances in time determination, timekeeping and time dissemination, and in setting standards that are largely still with us today. Universal time was based on the rotation of the Earth, and came in the 3 flavors (UT0, UT1 and UT2) that had been defined by Markowitz and the IAU in 1955. As the accuracy of quartz crystal clocks, and then atomic clocks, improved, the rotation of the Earth was shown to be irregular at the level of several tens of milliseconds, thus the need for a more uniform time.[18] That more uniform time came to be known as "Ephemeris time," because it was based on the orbit of the Earth as reflected in the ephemerides of the Sun and Moon. The difference between Ephemeris time and Universal time (specifically UT1) gave "delta T". Why was delta T important? Because it gave the most uniform time possible, and for historians and scientists alike, an accurate and uniform timescale was highly desirable. If an historian or astronomer wanted to study ancient eclipses, for example, or any phenomenon that required accurate timing over long periods, uniform time was indispensable for accurate results.[19] Aside from ET and UT, atomic clocks were being developed in the 1950s; the relation between Ephemeris time and atomic clocks proves essential to our story.

Given Markowitz's background and his central role in the determination and dissemination of time, the origin of the Markowitz Moon camera is not surprising. Because since the 17th century official time had been based on the rotation of the Earth (for UT and its later refinements) or the orbit of the Earth (since 1960 for ET), astronomers had controlled it for centuries. The Moon camera now became part of a larger struggle in the 1950s to determine whether physicists or astronomers would be the arbiters of time. In two oral history interviews, Markowitz recalled clearly the origins of the Moon camera well before the IGY came on the scene. At the June 1951 meeting of the American Astronomical Society (AAS) at the National Bureau of Standards (NBS) in Washington, D.C., papers were presented related to the subject of Atomic time. As Markowitz recalled, "this was shortly after Harold Lyons and coworkers at the NBS had constructed an atomic clock, which was operated by the principle of absorption of ammonium microwaves. At the meeting of the AAS there were some glowing accounts of the accuracy that the atomic clocks would achieve, and that it would eliminate the necessity for astronomical time."

"Shortly thereafter," Markowitz further recalled,

[Gerald] Clemence, who was then the Director of the Nautical Almanac Office, called me in to discuss the question of time. Ephemeris time was obtained from lunar occultations, but with a long delay of probably two or three years or more from observing occultations and getting the resulting Ephemeris time, say a value of Delta T. Clemence thought the claims were a little overblown—there was still a necessity for astronomical time of the nature of what is called Ephemeris time. What he was afraid of was that the physicists on their own might adopt some value for the frequency of an atomic clock that did not correspond to the Ephemeris time. That is to say, the two kinds of seconds would differ in value, and this would cause considerable confusion and other problems. What he wanted was a method of obtaining Ephemeris time rapidly. If this could be done then we could give physicists the value needed for obtaining the frequency of cesium [which had taken over from ammonium as the favored atomic standard], so that an atomic second and an ephemeris second would be the same. The result was that he asked me if there were some way of getting Ephemeris time rapidly.[20]

It did not take Markowitz long to take up the challenge: "I returned the next day to Clemence and said that, yes, I had a way that I thought would work."

He then laid out the problem and a relatively simple solution:

The problem in getting accurate positions of the Moon by photography was that if one took a short exposure, then the seeing or the scintillation would cause displacements of the stars and Moon. Whereas, if one took a long exposure, say of about 20 or 30 seconds, then there would be a blurred image of either the stars or the Moon. So I told Clemence that I had a method in mind, and I proposed to track both the Moon and the stars for about 20 seconds, each at their own rate. He was astounded at this proposal, and asked how I intended to accomplish it. The explanation was very simple. One would use a glass filter to cut down the light of the Moon, and one would tilt it during the exposure. The tilting of the glass would move the image of the object being photographed and it was a simple matter to calculate the rate so as to keep it stationary. He was enthusiastic, and proposed immediately that we set about with the construction of a camera.[21]

Thus, because the Sun itself could not be accurately observed, and because the planets moved too slowly in their orbits for accurate time determination, Markowitz proposed to use the Moon, and it was for this purpose the dual-rate Moon camera was developed. Markowitz designed the instrument himself, and it was built in the USNO machine shop under the direction of Mr. George Steinacker. The dual-rate camera was attached to the

12-inch refractor at the Naval Observatory, and a regular observing program began June 2, 1952. The 12-inch refractor was soon dismounted and the Moon camera was attached to an 8-inch telescope in the same dome atop the Observatory's main building.[22]

As with the navigational method of lunar distances, the Moon was to serve as the hands on the face of a clock. However, these much more accurate observations were to be made from the Observatory, not from shipboard. The tilting dark-glass filter effectively held the Moon fixed relative to the background stars during a simultaneous exposure of 20 seconds. The result was an image of the Moon that could be accurately measured with respect to the background stars. The Moon camera did indeed succeed in determining ET, delta T, and the calibration of the atomic second. The latter, 9,192,631,770 vibrations of radiation in cesium (resulting from the hyperfine transition due to the electron-spin flip), is the number still used for the official definition of the second.[23] Those contributions are documented and undisputed, and would themselves have justified the camera. But our question in this paper is about the Moon camera's contributions to geodesy, linked to time through longitude.

### The Moon Camera, the IGY, and Geodesy

In 1952 the International Council of Scientific Unions (ICSU) appointed a committee for the International Polar Year (IPY), renamed the effort the International Geophysical Year (IGY) to cover the whole Earth, and fixed the IGY to cover the period from July 1, 1957 through Dec 31, 1958. Eventually it would embrace 67 nations. Detailed national proposals were due by May, 1954, and so it was that in late 1953 or early 1954 the Naval Observatory sent a proposal to the US National Committee for the IGY outlining a program to measure the geodetic distance between continents, the size and shape of the Earth, and the variations in its speed of rotation. The proposal, undoubtedly written by Markowitz, pointed out the difficulties of extending triangulation nets across the oceans. Although in recent years advances in precision of observation and in knowledge of the motion of the Moon had made possible the use of the Moon as a triangulation point, those methods used total solar eclipses and occultations of stars, rare phenomena that did not allow much progress and had their own accuracy problems.[24]

The proposal pointed out that during the past 20 years the Moon had also been used as a clock to measure variations in speed of the rotation of the Earth. Now Markowitz outlined a new soon-to-be published technique to photograph the Moon, a technique that he claimed, with a well-distributed system of observatories, could accomplish in a single year what might take a century by older methods. The probable error of a single observation would be 0.15 seconds of arc, about 900 feet on the surface of the Earth, but many observations could reduce it to 90 feet, with undetermined systematic errors.[25]

The proposal noted that the equipment required was a refractor telescope with minimum aperture of 10 inches and focal length 10–20 feet, standard

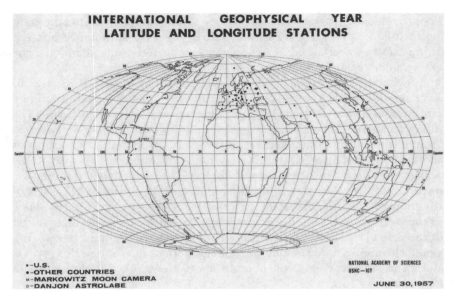

**Figure 16.3**   Locations of the Markowitz Moon Camera during the International Geophysical Year. (Courtesy of U.S. Naval Observatory Library, Washington, D.C.)

equipment at many observatories, plus a special camera that would cost about $700 at site. Also required was a more precise ephemeris of the Moon which the proposal promised was to come with support of the Office of Naval Research, the Army Map Service, the U.S. Naval Observatory, and the British Nautical Almanac Office. Ten to 20 observatories, preferably at the extremities of existing continental triangulation nets, would be required. Each observatory would measure and reduce its own observations, but the Naval Observatory would render assistance. The total cost for the Moon cameras was estimated at $12,000 for 20 copies, $40,000 for four measuring machines, plus personnel costs. Beyond the IGY, the Naval Observatory also foresaw a permanent Moon position program for determining delta T.[26]

By way of preparation, the first meeting of the CSAGI in Brussels in July, 1953 had already adopted a resolution recommending "that for longitude determinations observations shall be made by the method of Dr. Markowitz at a number of fundamental stations, and that the International Association of Geodesy shall, through its appropriate section, study the corrections required in order to refer the determinations to a common system of reference."[27] No details were given, and Markowitz's first paper to describe the technique and its application to geodesy appeared in the *Astronomical Journal* for March, 1954.[28]

1954 was a crucial year for latitude and longitude program preparations, as well as for the other programs of the IGY. As with those programs, the progress of the Moon camera program can be followed in the volumes of the

*Annals* of the IGY, at the USNO archives, and in correspondence now in the IGY archives at the U.S. National Academy of Sciences in Washington, D.C. It was at the second meeting of the CSAGI in Rome (30 Sept-4 October, 1954), that detailed programs were laid out, including determination of longitudes and latitudes. As mentioned above, it is important to remember that there were two parts to the IGY program, the first to determine *astronomical* latitude and longitude with the Danjon astrolabe and the PZT, the second to determine *geodetic* latitude and longitude with the Moon camera.[29] The first (for which the central agency was the Bureau Internationale de l'Heure (BIH) in Paris) was to be undertaken with standard programs planned at as many as 39 stations, while the Moon camera was "an auxiliary lunar observation campaign...envisaged (by the Markowitz method), which has the double aim: a) To improve certain tabular data on lunar movement as well as the definition of uniform time; b) To determine, for the observatories participating in these observations, the variation between the vertical and the normal to a conventional ellipsoidical surface." In two separate annexes to the report, the plans for the Markowitz Moon camera were laid out, specifically mentioning the desirability of extending triangulation nets across oceans. The US National Committee commented on this need "There are not enough islands to use as stepping stones, nor has it been possible to connect the separate networks of longitudes and latitudes with each other more accurately than to 200 or 300 ft. The location of some islands is uncertain by as much as a mile.[30]

The Moon camera, the report stated, could overcome this deficiency. The plan was as follows: "The United States proposes to employ the new technique for photographing the Moon, and also to determine astronomical longitudes and latitudes using the new Danjon astrolabe, at three stations: Washington, San Diego and Hawaii. It will supply dual-rate cameras suitable for attachment to existing long-focus telescopes at about seventeen astronomical observatories strategically located in other countries. It will also undertake to measure approximately one-fourth of the photographic plates with the twenty cameras, and will supply three measuring engines of special design to observatories in other countries willing to undertake the remaining measures." Lists were given of participating observatories, and the four with measuring engines were specified as Washington, the Cape (S. Africa), Paris and Greenwich. The values of Delta time (ET minus UT) were to be published by the measuring centers; the reduction of the plates was to be completed by 1959, and the general solution by 1960.

Annex B to the same report from the Rome CSAGI meeting, undoubtedly largely written by Markowitz, gave the most detailed description of the theory and objectives of the Moon camera program. "The aims of the International Geophysical Year program, which are of astronomical, geophysical and geodetic interest, include the determination of the following: a) changes in the speed of rotation of the Earth; b) the size and shape of the earth, independent of gravity; c) corrections to the orbital elements of the moon, the lunar parallax, and the distance to the moon; d) the three rectangular geocentric co-ordinates of each observatory, to about 40 meters probable error; e) deflections of the vertical to about 1" [one arcsecond] probable error." Markowitz left no doubt about the ultimate goal: "The above data will allow geodesists to form a World Geodetic System. The size and shape of the earth determined geometrically from this program may be combined with the determinations given by the usual methods to obtain the most probable size and shape."[31]

On April 29, 1955 the Technical Panel on Longitude and Latitude met at the USNO, with Gerald Clemence, Markowitz and Donald Rice (U.S. Coast and Geodetic Survey) present. Markowitz was elected Chairman of the Panel. After discussing the two main elements of the program, they affirmed that the general solution from the measurement of the Moon camera plates would yield values of delta T, rotation of the Earth, and geodetic results; there was no discussion of attainable accuracies. The overall budget from the IGY was only $16,300 for the operations of the Coast and Geodetic Survey operations in Hawaii; all other U.S. operations were funded by the Office of Naval Research.[32]

On November 30, 1956 the USNO received official notice from the U.S. National Committee of the National Research Council that the USNO would be designated the U.S. Data Center for latitude and longitude.[33] Markowitz was eventually appointed chair of the Latitude and Longitude section of IGY.

On May 1, 1957 Markowitz issued the first Circular on the Moon position program. He reported that the 20 cameras were being finished in the instrument shop of the USNO, and 4 measuring engines had been completed and tested. Markowitz's deputy, astronomer Glenn Hall, was to carry out the computer reductions with the IBM 650, and for this purpose the Yale Zone star catalogues were placed on punch cards. The cameras were to be distributed by naval attaches nearest the participating observatory, beginning May 3 with most of the others to follow by June, 1957. Plate specifications were given, with an estimate of 60 dozen fast plates and 60 dozen slow plates for the entire program, with addresses for two suppliers (Kodak and Ilford).[34] Detailed "Instructions for Operation of the Moon-Position Camera" were published in 1957, along with instructions for the other methods by astrolabe and PZT.[35]

In a progress report dated 4 January, 1958 Markowitz noted the program would commence about January 1958 and continue for 18 months, in accordance with decisions reached in Barcelona. He reported 20 cameras had been distributed around the world, and that both Moon cameras and Danjon astrolabes had been installed at the three US stations: USNO, San Diego State College, and US Coast Survey station in Hawaii,[36] 9 miles east of Honolulu.[37] The data indeed came rolling in, as evidenced by numerous monthly reports and files of data still in the USNO archives. The question is, what were the results?

## Moon Camera Results

In discussing Moon camera results, it should be noted that the USNO agreement with the IGY was to provide, for each plate, apparent right ascension and declination of the Moon and the universal time of observation in the *Astronomical Journal* or in the USNO circular series.[38] This goal was fulfilled. Beyond that, one searches in vain for any geodetic results from the Moon camera program in the published reports or the archives of the IGY at the National Academy of Sciences in Washington, D.C, the archives of the Naval Observatory, or professional publications such as the *Astronomical Journal* where one would expect to find them. My conclusion is that none were published. As late as 1958, Markowitz stated in an article that the USNO hoped to complete a general solution that would yield not only corrections to the orbit of the Moon and Ephemeris time but also the size and shape of the geoid.[39] In 1959 a summary of Moon camera observations was indeed published in the IGY General Report series, giving numbers of plates received, measured and reduced, but no results, nor were results given in updates in January and July 1961. IGY data collection officially terminated December 1, 1961, and the final Catalogue of Data in the World Data Centers, covering 1 July, 1957- 31 December, 1959, appeared in 1963.[40] Final results were published in the IGY Annals for 1964, but they were not results that would yield geodetic accuracies. In fact, they were no more accurate than the Danjon astrolabe determinations.[41] There was a statement that more would be published in 1965. But Markowitz retired in 1966, and, although the program for Ephemeris time continued successfully until 1974, no geodetic results ever appeared.

The interesting historical question is what happened to a program so highly touted in the IGY literature by its leaders, including Sydney Chapman and Joseph Kaplan, by its participants, including Elliott B. Roberts, and by its popularizers, represented by Walter Sullivan? First of all, scientists' queries as to exactly how geodetic results were to flow from the Moon camera observations demonstrate there was some skepticism about the method; this precipitated Markowitz to write an article in 1958 with more details, supplemented by articles from others he had solicited.[42] Secondly, USNO astronomer David K. Scott, who had a background and considerable work in the field of geodesy and worked with the Moon camera reductions at USNO, stated that he believed geodetic results from the Moon camera were not possible in principle because of uncertainties in the motion of the Moon that could not be disentangled from the uncertainties in the observers' positions on Earth: "...the corrections to the moon's position, which was essentially Delta T, were 100 percent correlated with the corrections to the observer's position, so the scheme would not work.[43] It was later found that artificial satellites were close enough that Doppler techniques could be used to solve the problem where the Moon camera could not. Markowitz himself had remarked on the difficulties of using the Moon compared with an artificial satellite. And lest we jump to the conclusion that Markowitz was pushing his Moon camera invention beyond its capabilities, Scott also stated that "the Air Force thought that we could get Earth-centered geodetic positions from Markowitz's moon camera. Well, Markowitz was not particularly interested. He wanted to get Delta T from it."

Finally, and most compellingly, a single handwritten letter by Fred Whipple in the USNO archives titled "For Markowitz by Whipple" and dated July 25, 1960, indicates that Markowitz's method was being superseded by Space Age results. Whipple (head of the SAO Moonwatch/Baker-Nunn program) wrote that "The SAO has been preparing for the precise determination of geoid shape and world-wide ties among geodetic systems by observations of artificial satellites. A major effort is under way and should begin to produce results during the year 1960."[44] Whipple went on to describe the 12 stations around the world where photographic observations were being made of artificial satellites since mid-1958, and that some 3000 "precision observations" had been reduced by mid-1960, with an expected accuracy of 10–12 meters. The Whipple archives at the Smithsonian Institution contain ten items of correspondence between him and Markowitz from 1956 to 1961. While there is no direct discussion of the Moon camera, Markowitz was keenly aware of Whipple's potentially competing program, and asked him in February 1956 to participate in a meeting to describe his proposed work on artificial satellites. Later correspondence between the two was limited to more perfunctory matters, such as Markowitz asking for timing accuracies required for Moonwatch, a natural question from the Director of Time Service at USNO, the nation's timekeeper. (Whipple replied in 1961 that he needed .001 seconds (one millisecond) immediately, and half that (.0005 seconds) within the next year.[45]) The bottom line is that while Moonwatch and Baker-Nunn results became part of the National Satellite Geodetic Program and were used to create the SAO Standard Earth Model,[46] the Markowitz Moon camera data was not accurate enough to be incorporated into any geodetic system.

Neither correspondence at USNO or at the National Academy of Sciences IGY archives contains any smoking gun document stating that geodetic results from the Moon camera were a failure. Pembroke Hart at the National Academy wrote to Markowitz in August 1959 simply that his obligations would be discharged by updating his summary of Moon camera observations in a fourth 6-monthly catalog.[47] This was perhaps a de facto recognition that not every program proposed for the IGY would or could be successful.

As for post-IGY observations with the Moon camera, in a letter from Markowitz to Odishaw dated Dec 4, 1958 Markowitz wrote that a recent Moscow meeting had decided that astronomical observatories would in general continue their latitude and longitude observations throughout the world.[48] In March, 1960 he reported that longitude and latitude observations with the PZT at USNO up to October 1959 had been published in Time Service Bulletins A 130 to A 170, and that astrolabe results would be sent to Paris as part of the broader latitude and longitude program. As to the Moon program, Markowitz wrote "The loan of moon cameras will be continued to those observatories which wish to keep on making observations. The measurement and reduction of the plates are proceeding at a moderate but satisfactory rate. It is expected that corrections to the limb of the moon will be utilized during 1960 in order to obtain the apparent positions of the moon as observed."[49] The program for Ephemeris time continued until 1974, employing Time Service personnel at the U.S. Naval Observatory and graduate students at participating observatories around the world.

The Moon camera program, as well as the broader IGY latitude and longitude program, also left another legacy. Because timing was essential for latitude and longitude observations, already at the first CSAGI meeting in Brussels in 1953, it was recommended that all observatories associated with the broader latitude and longitude program make concurrent observations of time and latitude; that "two-way intercommunication of time signals shall be arranged during the IGY, for the precise determination of the transmission time of the signals and of the variations of this transmission time;" and that the measurement of the transmission times between stations be made with utmost precision, including ionospheric measurements along the propagation path to determine conditions affecting propagation time. Furthermore, the CSAGI recommended that URSI send the following communication to all concerned organizations:

> CSAGI, considering (a) the necessity to ensure the utmost precision of international longitude measurements during the IGY; (b) the advantages afforded for the interchange of time between astronomical observatories, by the emission at frequent intervals of time signals capable of being received over the entire surface of the earth, including the southern hemisphere, draws the attention of the responsible organizations, and especially of CCIR [Comité Consultatif International des Radio Communications, International Radio Consultative Committee] and URSI [Union Scientifique Radio Internationale, International Union of Radio Science], to the importance of organizing a world system of emission of time signals during the IGY, and asks that this be established and in working order before the IGY begins.[50]

The Latitude and Longitude program, of which the Markowitz Moon camera was a part, thus emphasizes that behind the scenes of the IGY was an alphabet soup of international organizations—including the French Bureau Internationale de l'Heure (BIH), the International Latitude Service (ILS), the CCIR, URSI, and the CSAGI itself—involved in ensuring that time was properly measured and disseminated, not only for latitude and longitude, but for all IGY observations.[51]

## Summary and Conclusions

The conclusion, then, is that the Markowitz Moon camera was successful for time determination and calibration of the atomic second, but not for geodesy. So why study a program that failed? As Peter Galison has emphasized in his pioneering book *How Experiments End*, history is full of failures, and in the realm of science and technology historians are interested not only in successful programs, but also in programs that failed because they were too ambitious, or conceptually flawed, or overtaken by events.[52] In this case all three played their role. The Moon camera was too ambitious because its intended accuracies could not be reached, in part because only 12 of 20 observatories reported results due to technical difficulties, personnel shortages or funding problems. It was conceptually flawed because in the end it appears the uncertainties in the Moon's complicated motion could not be separated from uncertainties in the observer's position. And it was overtaken by events when the Space Age made possible satellite geodesy. None of this could have been totally foreseen before the effort was made. Sometimes an experiment has to be tried before success or failure becomes clear. Failed experiments are an important part of the history of science and technology, and it is important to realize that not all the ambitious plans for the IGY succeeded.[53]

The scope of the Markowitz Moon camera program was also considerable, at least in terms of the Navy's involvement in time and money, an indication of its importance to military needs. Some 20 Moon cameras were constructed in the Naval Observatory instrument shop, as well as 4 measuring engines. Some of these Moon cameras are still at the Naval Observatory and elsewhere, as are thousands of the photographic plates, and voluminous data records and correspondence—all testimony to a once-vibrant program.

The international scope of the program was also impressive. Securing the cooperation of 20 observatories around the world was no small task, and the overall latitude and longitude program of the IGY was even larger, involving some 39 observatories using astrolabes, transit circles or photographic zenith tubes. The continuing international nature of the program is evident from the results reported to the National Academy of Sciences in 1970. Results poured in from 12 observatories, together with the number of plates measured at the Cape, Greenwich and Paris. More generally, the BIH, ILS, CCIR and URSI were involved with the IGY latitude and longitude program.[54]

The technology and technique for the Moon camera program were also innovative. Markowitz developed a new technology that had not been built before, though he cited earlier attempts by H. N. Russell 50 years earlier using a similar technology at a single station.[55] In addition, the program involved long-standing astronomical problems analogous to the method of lunar distances three centuries earlier: accurate star catalogues, accurate theory and observation of the motion of the Moon, and robust computing techniques, though now at an order of magnitude or more better accuracy.[56] Though the accuracies were still not enough to address geodetic demands, the program nonetheless shows science marching toward ever greater accuracies required for its increasingly refined problems.

Finally, the Markowitz Moon camera was in many ways a success, certainly in terms of the determination of Ephemeris time, delta T and the calibration of the atomic second to astronomical reality. For six years from 1960 to 1967 Ephemeris time was basis for the official international definition of the second. The Moon camera continued to determine Delta T until 1974. And even when Atomic time superseded Ephemeris time in 1967, its definition of the second as 9,192,631,770 vibrations of the cesium atom had been calibrated to the real world of astronomy through the Moon camera. The Moon camera was not, however, accurate enough to produce the hoped-for geodetic results such as the determination of the geoid and detection of continental drift; for the latter no one at the time knew the magnitude of the movements, and it was 3 decades before the actual centimeter level shifts of the continents could be measured. Even the early satellite results, while revealing the slight pear-shape of the Earth, could not detect continental drift.[57]

Finally, it turns out the Markowitz Moon camera results were important for geodesy in other ways, again related to time. In his book on *Satellite Geodesy* published in 1964, Ivan Mueller wrote that "two kinds of time are required for satellite tracking. The geocentric motion of any satellite, the data in the satellite ephemerides, is expressed as a function of Ephemeris time. For this in practice Atomic time is substituted."[58] The Moon camera was important for both of these, determining Ephemeris time and calibrating Atomic time, the latter a community response to the challenge posed by the atomic clocks of the physicists. The other kind of time important for satellite tracking was Universal time in the form of UT1, which takes into account polar motion. Again, Markowitz's work was important, and indeed it was he who had set up the system of UT0, UT1 and UT2 at the IAU meeting in Dublin in 1955.[59]

In the end, the Markowitz Moon camera was an interesting ground-based attempt at geodesy implemented as part of the IGY. Its successes and failures highlight the importance of the new methods that superseded it, methods made possible only by the Space Age. The geodetic failure of the program highlights NASA's important early role in geodesy with the launch of geodetic satellites like Anna 1B (1962), LAGEOS (1976, 1992) and others. The extent of the subsequent multi-institutional satellite effort in geodesy is evident in the massive and detailed volumes of the *National Geodetic Satellite Program,* which in themselves could form the basis for a history of space-age geodesy.[60] NASA (and the U.S. Naval Observatory) remained very much involved in geodesy with programs in Satellite Laser Ranging (SLR) and Very Long Baseline Interferometry (VLBI), as well as its use of data from the DoD's Global Positioning System (GPS). In light of these programs, the once ambitious hopes for geodetic results using the ground-based Markowitz Moon camera faded into history.

## Acknowledgments

I wish to thank Dennis McCarthy (U.S. Naval Observatory), Brent Archinal (Astrogeology Research Program of the U.S. Geological Survey), Patrick McCray (University of California Santa Barbara) and David DeVorkin (National Air and Space Museum) for their very useful comments, as well as Sally Bosken and Gregory Shelton at the U.S. Naval Observatory Library, Janice Goldblum at the National Academy of Sciences archives, and Mary Markey at Smithsonian Institution archives in Washington, D.C.

## Notes

1. G. Waldo Dunnington, *Gauss: Titan of Science* (New York: Exposition Press, 1955, reprint Mathematical Association of America, 2004), especially 113–138. Geodesy is a huge area in need of a critical history. Some of the general history is given along with the scientific concepts in James R. Smith, *Introduction to Geodesy: The History and Concepts of Modern Geodesy* (New York: John Wiley and Sons, 1997).

2. On the first Washington–Paris radio determinations of longitude in 1913, and on the two World Longitude Operations of 1926 and 1933, see Steven J. Dick, *Sky and Ocean Joined: The U.S. Naval Observatory, 1830–2000* (Cambridge, UK: Cambridge University Press, 2003), 472–473.

3. John Cloud, "World Geodetic System," in Stephen B. Johnson, gen. ed., *Space Exploration and Humanity* (in press).

4. John Cloud, "World Geodetic System," *Space Exploration and Humanity* (in press), Stephen B. Johnson, General Editor. Cloud, one of the few historians of geodesy, describes the development and work of perhaps the most significant early geodetic sciences group during the first period, that at Ohio State University, in John Cloud, "Crossing the Olentangy River: The Figure of the Earth and the Military-Industrial-Academic Complex, 1947–1972," *Studies in the History and Philosophy of Modern Physics*, vol. 31, no. 3 (2000), pp. 371–404.

5. Deborah Jean Warner, "Political Geodesy: The Army, the Air Force and the World Geodetic System of 1960," *Annals of Science* 59 (2002): 363–389.

6. Warner, "Political Geodesy," *Annals of Science*, 59 (2002), 363–89; and Warner, "From Tallahassee to Timbuktu: Cold War Efforts to Measure Intercontinental Distances, *Historical Studies in the Physical Sciences*, vol. 30: 2 (2000), 393–415. See also John Cloud, "American Cartographic Transformations during the Cold War," *Cartography and Geographic Information Science*, vol. 29, no. 3 (2002), pp. 261–282. For a list of "Geodetic Reference Spheroids" preceding the WGS 84, and the current context in which WGS 84 is used, see U. S. Naval Observatory, *The Astronomical Almanac for the Year 2006* (U.S. Government Printing Office, 2004) "Reduction of Terrestrial Coordinates," p. K13.

7. These concepts are defined, with diagrams, in Smith, *Introduction to Geodesy*, especially 85–87.

8. William Markowitz, "Use in Geodesy of the Results of Lunar Observations and Eventual Observations of Artificial Satellites," *Bulletin Geodesique* 49 (1958): 33–40; and "Geocentric Co-Ordinates from Lunar and Satellite Observations," *Bulletin Geodesique*, 41–49.

9  W. Patrick McCray, "Amateur Scientists, the International Geophysical Year, and the Ambitions of Fred Whipple," *Isis*, 97, 4 (2006), 634–658. See also chapter 15.

10. W. Patrick McCray, "Amateur Scientists, the International Geophysical Year, and the Ambitions of Fred Whipple," *Isis* 97, no. 4 (2006): 634–658.

11  "Proposal for the consideration of the USA National Committee on the Third Geophysical Year, 1957–58." USNO archives on-site, Moon Camera files.

12. Joseph Kaplan, "The Scientific Program of the International Geophysical Year," in *Symposium on Scientific Aspects of the International Geophysical Year, 1957–1958, Proceedings NAS* 40 (1954), 926–931, 929.

13. Sydney Chapman, *IGY: Year of Discovery, The Story of the International Geophysical Year* (Ann Arbor: University of Michigan Press, 1959), 926–931, especially 929.

14. Elliott B. Roberts, "The IGY in Retrospect," *Annual Report of the Smithsonian Institution for 1959*, 282.

15. Walter Sullivan, *Assault on the Unknown: The International Geophysical Year* (New York: McGraw Hill, 1961), 395.

16. Steven J. Dick and Dennis D. McCarthy, "William Markowitz, 1907–1998," *Bulletin of the American Astronomical Society* 31, no. 4 (1999): 1605, online at http://adsabs.harvard.edu/cgi-bin/nph-bib_query?bibcode=1999BAAS...31.1605

D; Steven J. Dick and Dennis D. McCarthy, "William Markowitz, 1907–1998," in Dick, McCarthy, and B. Luzum, *Polar Motion: Historical and Scientific Problems* (San Francisco: Astronomical Society of the Pacific, 2000), 335. On Markowitz's work at the USNO, see Dick, *Sky and Ocean Joined*, 480–485.

17. W. Markowitz, "Redeterminations of Latitude and Longitude," *Transactions of the AGU* 26, no. 2 (October 1945): 197.

18. For details see Dick, *Sky and Ocean Joined*, ref. 2 above, pp. 482 ff. As Paul Forman has written, "The seasonal variations in the rate of the earth's rotation, which the pendulum was barely able to detect, were made fully apparent by quartz oscillators in the years before the Second World War. Still, the evidence tended rather to be overlooked by both astronomers and horologists unready for the conceptual revolution which it implied." It was only in the 1950s that they took action. Paul Forman, "Atomichron: the Atomic Clock from Concept to Commercial product," *Proc. IEEE*, 73 (1985), pp. 1181–1204, also at http://www.ieee-uffc.org/fcmain.asp?page=atomichron

19. See F.R. Stephenson, *Historical Eclipses and Earth's Rotation* (Cambridge, UK: Cambridge University Press, 1997); F.R. Stephenson, "Variations in Earth's Clock Error Delta T between AD 300 and 800 as Deduced from Observations of Solar and Lunar Eclipses," *Journal of Astronomical History and Heritage* 10 (2007): 211–220.

20. William Markowitz, "Reminiscences of the U.S. Naval Observatory, 21 October, 1986," based on oral history interview by Dennis McCarthy. Substantially the same story is told in the Markowitz interview by Steven J. Dick, 18 August, 1987. Both interviews are available at the U. S. Naval Observatory library. See also Dick *Sky and Ocean Joined*, p. 480, and Paul Forman, "Atomichron," note 15: "In the mid 1950s, astronomers remained disinclined to cede primacy in time determination to physicists and engineers. The steady improvement of quartz oscillators and the advent of the first atomic frequency standards pushed astronomers to try to reestablish uniformity in astronomical time by shifting from the Earth's daily rotation to its annual revolution to define the second; 'Ephemeris Time,' as it was called was formally adopted in 1960 by the General Conference of Weights and Measures."

21. William Markowitz, "Reminiscences of the U.S. Naval Observatory, 21 October, 1986," based on oral history interview by Dennis McCarthy. Substantially the same story is told in the Markowitz interview by Steven J. Dick, August 18, 1987. Both interviews are available at the U. S. Naval Observatory library.

22. William Markowitz, "Photographic Determination of the Moon's Position, and Applications to the Measure of Time, Rotation of Earth, and Geodesy," *Astronomical Journal* 59 (1954): 69–73: 70. Markowitz cites (69) a similar program by A.S. King and H.N. Russell from 1911–1917, but here a snapshot of the Moon was taken in the middle of a 10-minute exposure. On the history of the 12-inch, see Dick, *Earth and Sky Joined*, 374 and following, and T.J. Rafferty, "Refurbishing the US Naval Observatory's 1892 Saegmuller 12" Refractor," *Telescope Making* 15, 24–29.

23. W. Markowitz, R. Glenn Hall, L. Essen and J. V. L. Parry, "Frequency of Cesium in Terms of Ephemeris Time," *Physical Review Letters*, 1 (August 1, 1958), 204L, 1–2 and 105–106.

24. "Proposal for the Consideration of the USA National Committee on the Third Geophysical Year, 1957–58." USNO archives onsite, Moon Camera files. The document is undated, but internal evidence indicates late 1953 or early 1954; it preceded publication of the technique in the March 1954 *Astronomical Journal* (note 16 above). On the use of solar eclipses and occultations for geodesy, see Ivan Mueller, *Introduction to Satellite Geodesy* (New York: Frederick Ungar, 1964), Part I. As mentioned above, the Moon had been used for three centuries to determine geographical longitude through the famous "lunar distances" method. But the results were of low precision.

25. This stated accuracy, or at least the conversion from arcseconds to feet also quoted in the IGY Annals, vol. IIA, p. 142 (ref. 23 below), does not make sense. 0.15 arcseconds is about 15 feet on the surface of the Earth. 0.15 arcminutes (9 arcseconds) is 900 feet,

but accuracies should have been better than that. Other descriptions (e.g. Chapman, ref. 10, pp. 18–19) state distances on the Earth's surface will be measured "with errors of the order of no more than 100 to 200 feet).

26. "Plan for Participation of the U. S. Naval Observatory in the International Geophysical Year, 1957–1958," USNO onsite archives, Moon Camera files. The estimated date of this document is 1955. The IGY Moon camera proposal was one part of the plans for USNO participation in IGY. The first program was the Moon camera, which would produce geocentric latitude and longitude and changes in the rate of rotation of Earth. A second program was the determination of astronomical latitudes and longitudes using the prismatic astrolabe (invented by Danjon). The third program was observation of solar activity, already long under way at USNO. Total cost was estimated at $165,000, of which the Office of Naval Research had already funded $49,200 for fiscal years 1954 and 1955.

27. *Annals of the International Geophysical Year* IIA (London: Pergamon Press, 1959), 22. Markowitz (1954), note 16 above, 73, references this Brussels meeting.

28. William Markowitz, "Photographic Determination of Moon's Position."

29. On the PZT see Dick, *Sky and Ocean Joined*, ref 2 above, pp. 458–461, and on the Danjon astrolabe, p. 460. The best contemporary overview of the PZT is Markowitz's article in the *Telescopes* volume, cited in note 18.

30. *Annals* IIA (1959), 139–145; the citations are on 140–141.

31. *Annals* IIA (1959), 143–144. This volume was not published until 1959, but shortly after the Rome meeting Markowitz wrote Odishaw of these results, Markowitz to Odishaw, October 29, 1954, National Academy of Sciences (NAS) IGY archives, Washington, D.C., series 5.8. Odishaw was the executive secretary of the U.S. National Committee of the IGY; this was the first letter in a series of correspondence on this subject between them. Appendix B describing the Moon camera program was originally written by Markowitz and dated September 10, 1954, with copies both at the National Academy and USNO archives. Andre Danjon was the president of the Latitude–Longitude group in 1954 and circulated the adopted resolutions and Markowitz's Appendix B shortly after the meeting.

32. Minutes of the First Meeting, NAS IGY archives, series 4. Rice was the Chief of the Gravity and Astronomy branch of the Coast and Geodetic Survey. The Survey had participated in the two previous world longitude campaigns, operating the station in Hawaii in both 1926 and 1933. It would now operate the Hawaii station (known as NIU longitude, 9 miles east of Honolulu) for both IGY programs, astronomical and geodetic latitude and longitude. The Survey also operated two of the five permanent observatories for determining variation of latitude under the International Latitude Service. See NAS IGY archives, series 8.1, folder on latitude and longitude, project 7.1, 1956–57.

33. Hugh Odishaw to Captain Charles Freeman, superintendent USNO, November 20, 1956, USNO onsite archives, Moon camera files, and NAS archives, IGY records, series 8.1, World Data Centers, USNO, 1956–1959.

34. Markowitz, Letter Circular No. 1, Moon Position Program of the IGY, May 1, 1957, NAS IGY archives.

35. *Annals* IV, parts I, II, and III (London, 1957), 195–196.

36. Markowitz to Odishaw, January 4, 1958, NAS IGY archives, series 8.1.

37 Markowitz to Odishaw, 4 January, 1958, NAS IGY archives, series 8.1.

38. C.L. Freeman, superintendent USNO, to U.S. National Committee for IGY, February 13, 1957, NAS archives, series 8.1. E.O. Hulbert, senior scientist for USNC–IGY to chair and executive committee, USNC–IGY, February 14, 1957. See also IGY *Annals* VII, parts I, II, and III (London, 1959), 283, where the USNO also hoped to discuss results in its publications. A full list of latitude and longitude observatories is given on 286–288, broken down by astronomical versus geodetic observations.

39. Markowitz, "Geocentric Coordinates," 47.

40. *Interim Catalog of Data in IGY World Data Center A*, IGY General Report Series, no. 7, November 30, 1959, 78. Results are reported for July 1, 1957–September 30, 1959. Also *Seventh 6-monthly Catalog of Data in IGY World Data Center A*, IGY General Report, no. 13, January 1961, 98. An eighth catalog, July 1961, the final *Catalog of Data in the World Data Centers* was published (Pergamon Press, 1963).

41. IGY *Annals* XXXVI, 294.

42. Markowitz, "Use in Geodesy…" and "Geocentric Coordinates…", reference 6 above. Markowitz wrote that "Although much has been published which describes the various techniques it is not clear to geodesists just how the moon is to be used to determine the size and shape of the earth. Prof. P. Tardi requested that at Toronto specific details be given to show just how this was to be done. Accordingly, I invited several of my colleagues to prepare reports covering eclipses, occultations, and the earth satellite, while I undertook to describe the lunar photographic method," ref. 6, p. 34. The reports Markowitz refers to were "On the Geodetic Application of a Solar Eclipse," by A. Goldstein, O.F. Mattingly and F.J. Heyden, S.J.; "The Occultation Method of Line Measurements," by J.A. O'Keefe; "The IGY Optical Satellite Tracking Program as a Source of Geodetic Information," by F.L. Whipple and J.A. Hynek, and Markowitz's own article "Geocentric Co-Ordinates from Lunar and Satellite Observations," all published in ref. 6 above. Markowitz further noted that these reports would concentrate on their respective programs because "It would be assumed in each case that observations could be made to the degree of precision anticipated." p. 34.

43. Steven J. Dick, OHI, with David K. Scott, February 26, 1988, 19, 27–33, USNO library.

44. "For Markowitz by Whipple," July 25, 1960, USNO onsite archives, Moon camera files. The manuscript letter is erroneously dated "July 25, 190," but internal evidence makes it clear that the year is 1960.

45. Fred L. Whipple to Wm. Markowitz, January 19, 1961, Unit 7431, Fred Lawrence Whipple papers, series 1 (General Correspondence), Box 3, Accession 98–132, Smithsonian Institution Archives. There are ten pages of Whipple/Markowitz correspondence from 1956–1961 in Box 3.

46. SAO results are given in *National Geodetic Satellite Program* (NASA: Washington, DC, 1977), Pt. II, p. 885. NASA SP-365.

47. Pembroke Hart to Markowitz, August 7, 1959, NAS IGY archives, series 8.1.

48. Markowitz to Odishaw, December 4, 1958, NAS IGY archives, series 5, latitude and longitude, 1954–1960.

49. Markowitz to Hart, March 25, 1960, "Data Center A Longitude and Latitude," NAS IGY archives, series 8.1, folder WDC project 45.10, 1960."

50 IGY *Annals*, vol IIA (1959), (reference 22 above), p. 22.

51. On the history of the BIH and ILS see *Polar Motion: Historical and Scientific Problems*, Steven Dick, Dennis McCarthy and Brian Luzum, eds. (San Francisco: Astronomical Society of the Pacific, 2000), especially Part II.

52. "Historians of science, principally those interested in the development of ideas, have often insisted that any historical reconstruction that ignores what seems in retrospect to be erroneous will be an inadequate account." Peter Galison, *How Experiments End* (University of Chicago Press: Chicago and London, 1987), p. 5. Galison cites Alexandre Koyré and Thomas Kuhn in this regard, noting that when they refer to historically held beliefs as "mistaken," is not to judge past theories anachronistically, but rather to appreciate forgotten frames of mind that may reveal precepts were held to be of central importance in their time. With regard to experiments and observations being superseded, I was involved in one such situation at the U. S. Naval Observatory when a 10-year transit circle program to produce fundamental positions for Southern Hemisphere stars at the level of a few hundredths of an arcsecond was superseded even before publication by the Hipparcos satellite, which produced positions at the milliarcsecond level, ten to a hundred times better accuracy. The satellite almost failed to achieve orbit, however, and if it had not worked the transit circle results, using new

electronic "image dissector" technology, would have been considered cutting edge science. Dick, *Sky and Ocean Joined*, ref. 2 above, pp. 239–41.

53. While the Moon program was only a small part of the IGY, it would be interesting to know how many other IGY programs did not fully attain their goals.

54. "Report of the Moon Program, 1 January, 1957 to 1 January, 1970, R. Glenn Hall to Richard Y. Dow, 6 February, 1970, USNO on-site archives, Moon camera files.

55. See ref. 18 on Russell's method.

56. On the method of lunar distances, see Dick, *Sky and Ocean Joined*, ref. 2 above, pp. 16–18 and sources cited therein.

57. Deborah Jean Warner, "From Tallahassee to Timbuktu: Cold War Efforts to Measure Intercontinental Distances, *Historical Studies in the Physical Sciences* 30 (2000): 393–414, especially 404–405.

58. Ivan Mueller, *Introduction to Satellite Geodesy* (Frederick Ungar: New York, 1964), 290–93.

59. See also Paul M. Muller, "Time and Polar Motion in Early NASA Spacecraft Navigation," in Dick, McCarthy, and Luzum, *Polar Motion*, 215–219.

60. Fred L. Whipple to Wm. Markowitz, January 19, 1961, Unit 7431, Fred Lawrence Whipple papers, series 1 (General Correspondence), Box 3, Accession 98–132, Smithsonian Institution Archives.

# Part Five

# Legacies of Global Science: Space Science, Anthropology, and Earth Science

# Chapter 17

# The International Geophysical Year and Planetary Science

## Erik M. Conway

The International Geophysical Year (IGY) is famous for a number of important legacies. One is for the iconic graph of rising $CO_2$ emissions, the Keeling Chart.[1] Another is for its stimulus to polar exploration and the establishment of permanent research facilities in Antarctica.[2] The third is the "space race" between the United States and the Soviet Union, initially to launch the first satellite and then to put humans into orbit.[3]

Each of these three storylines could spawn a number of different histories and, in the case of the space race, already has. Several histories have explored the human spaceflight legacies of the IGY. McDougall has examined the politics of the early space race. Newell, a participant in the adventure, looked back at early space science.[4] Divine and Dickson trace the details of the IGY satellite competition between the United States and the Soviet Union.[5] Green and Lomask examined Project Vanguard, the original U.S. satellite program.[6]

Planetary exploration was one of the major scientific legacies of the IGY. It, too, has drawn some study. Clayton Koppes published a study of the Jet Propulsion Laboratory (JPL) planetary missions through 1976.[7] I wrote an essay summarizing U.S. and Soviet planetary exploration to 2004.[8] Siddiqi provides a reference to every planetary mission through 2000.[9] There are project-level studies of many, though not all, American planetary flights, but only two works look at planetary exploration in terms of its effects on scientific disciplines.[10]

The first of these is Ronald E. Doel's history of solar-system astronomy in America before 1960.[11] Doel traced interest in planetary subjects within four different research fields: astronomy, meteorology, geochemistry, and geophysics. Among practitioners of these fields during the 1950s, he found a great deal of fruitful collaboration on questions of cosmogony, meteoritics, geology, and planetary atmospheres. He saw resistance to planetary astronomy forming at the nation's observatories at the end of the 1950s, but he views that as a consequence of the large influx of new resources to geophysics

**Figure 17.1** Montage of the planets and four large moons of Jupiter in our solar system is set against a view of the Rosette Nebula. The light emitted from the Rosette Nebula results from the presence of hydrogen (red), oxygen (green), and sulfur (blue). Most of the planetary images in this montage were obtained by NASA's planetary missions, which have dramatically changed human understanding of the solar system in the past 50 years. (Courtesy of NASA Jet Propulsion Laboratory, Pasadena, California)

during the IGY. This, in his view, caused stellar and galactic astronomers to hold on much more tightly to their precious observing time against the influx of newcomers, driving a wedge between those fields and planetary astronomy that had not existed earlier. Frustration about this caused famed astronomer Gerard Kuiper to leave the Yerkes Observatory to found, in 1960, the second dedicated interdisciplinary planetary-science program in the United States, the Lunar and Planetary Laboratory of the University of Arizona. Doel's story ends that year, but it seems clear that he saw planetary science continuing to be a research field, not a discipline, for many years thereafter.

Joseph Tatarewicz's *Space Technology and Planetary Astronomy* explores the impact on planetary astronomy of the IGY inauguration of spacecraft-based exploration during the 1960s and early 1970s.[12] He perceived, within the community of American astronomy, a great deal of resistance to planetary exploration during the first decade after the IGY, and argues that NASA had set out to create a new discipline (and constituency) around a relative handful of interested scientists. He drew data from various National Academy studies of astronomy through 1980, with the bulk of the data from before 1975. By that time, the two major scientific societies involved in "space methods" (as geophysicist Homer Newell had called them), the American Astronomical Society (AAS) and the American Geophysical Union, had each established specialized planetary-science sections. In 1973 the AAS Division for Planetary Sciences adopted the journal *Icarus*, then edited by astronomer Carl Sagan, as the primary publication venue for planetary science. On the basis of these accomplishments, Tatarewicz argues that planetary science "had become an established disciplinary 'entity' by the early 1970s."[13]

## Context: Planetary Exploration since the IGY

Between 1960 and the early 1980s, only the United States and the Soviet Union carried out planetary missions. In 1985 western Europe and Japan joined these two space pioneers with missions to Halley's Comet and have remained active in exploring the inner planets since.[14] Only the United States has undertaken missions to the outer planets, unless one counts the Jovian gravity assist trajectory flybys of the European solar physics mission *Ulysses*. All of the eight currently accepted planets have received one or more robotic visitors, and newly-minted "dwarf planets" Pluto and Ceres are targets of currently in-flight spacecraft.[15]

Intellectually, planetary exploration has been revolutionary. In the past half century, the scientific community has been forced to revise, repeatedly, its understanding of the solar system as new data has been generated from spacecraft, and from new ground facilities.[16] Far from discovering more living worlds, as Mars and (by some) Venus had been thought to be, they have found no evidence that life exists anywhere else in the solar system. This has caused some scientists to argue that life may be extremely rare in the universe (though some continue to insist on the opposite) and to realize how unusual Earth is.[17] It has inspired the environmental movement and has been politically influential, too.[18]

NASA is by far the dominant supporter of planetary science in the United States. Private foundations, such as the Carnegie Institute of Washington, support a small number of planetary scientists, but do not carry out planetary missions. The Department of Defense develops technologies useful to planetary science and space astronomy but has carried out only a handful of missions itself.[19] So, planetary science is dependent on NASA's fortunes. But planetary exploration's finances do not move in lockstep with the agency's overall budget. In some presidential administrations, and in some Congresses, planetary exploration has been favored over NASA's other activities; in others, it has met with disfavor. The first Ronald Reagan administration nearly eliminated planetary exploration entirely, believing that sciences with no value to military affairs should be reduced dramatically or eliminated as inappropriate expenditures.[20] After lobbying by JPL, which had little other business at the time, Congress intervened to prevent this.[21] So, the intellectual success of the field at gaining new knowledge has little to do with its financial fortunes.

Planetary missions have involved four different technological styles that have had implications for the scientific techniques used and the related disciplines interested in them. The earliest American and Soviet planetary spacecraft carried out "flyby" missions, spending only a few hours close to each of their targets. Data was collected by remote sensing instruments, primarily derived from techniques employed in astronomy, astrophysics, and geophysics. The second of these technological styles involved planetary orbiters, beginning with the American *Mariner 9* spacecraft in 1971. These, too, relied primarily on remote sensing techniques but enabled planetary-scale mapping and geological effort. The third style has been planetary landers and atmosphere-entry probes. Both techniques permit the use of in situ measurements, using techniques most common to geochemistry and atmospheric chemistry.

The rarest technological style has been sample-return. Lunar samples have been brought to Earth by American astronauts and by Soviet robotic landers, and samples of the solar wind and of a comet's coma have been returned by American spacecraft. Sample-return from Mars has been a long-cherished goal of planetary scientists, but they have not as yet been able to obtain the several billion dollars required to accomplish this objective. Returned samples, like in situ samples, are subjected to the normal tools of geochemistry.

Planetary science, finally, is also distinct from the effort to find what are generally called "extra-solar planets," and sometimes "exoplanets," planet-sized bodies orbiting other stars. (Non solar system bodies are not subject to the International Astronomical Union definition.) The first of these was discovered in 1992 using a radio telescope; since then, most of the 300 or so known have been found by ground-based techniques.[22] Methodologically this "planet-hunting" effort is more akin to traditional astronomy and astrophysics than it is to planetary science, which increasingly draws its techniques from the geosciences, as suggested above. Extra-solar planet-hunting currently draws a great deal of interest and excitement, in part at least because it reopens the possibility of finding other living planets. Two proposed space-based astronomy missions, the American *Terrestrial Planet Finder* and the European *Darwin*, are explicitly directed at finding planets with Earthlike atmospheres. An atmosphere consisting of oxygen, nitrogen, and water vapor, the primary constituents of Earth's atmosphere, cannot exist without bioprocessing, so an Earthlike atmosphere is a clear "fingerprint" of life (at least of the current forms of life on Earth).[23] Yet their methodology is directly astrophysical, using infrared spectra to distinguish atmospheric chemistry across hundreds of light-years.

## Discussion

Robert S. Kramer refers to the period 1971–78 as a "golden age of planetary exploration."[24] Yet, that decade was marked by inconsistent funding and nearly terminated by serious NASA budget cuts. There were no planetary missions approved between 1978–82. Further, I was interested in exploring the evolution of the field quantitatively, using available publication, dissertation, and funding data. While this sort of quantitative exploration is no longer popular in history, it can help to illuminate trends. It does have its dangers, though, especially when the object of research is not a traditional, well-defined scientific discipline. Because planetary science is interdisciplinary, with practitioners identifying themselves as belonging to several other primary disciplines, the discipline-based statistics published by the National Science Foundation (NSF) and American Institute of Physics are not very helpful. Publication databases use the same discipline-based search terms as do these statistical works, so they likewise do not allow unambiguous identification of planetary papers. Manipulation of these terms, however, can allow the extraction of some useful information.

Immediately following the IGY, funding for planetary science grew rapidly. Between the middle 1960s and late 1970s, however, financially the field suffered a "rough patch" of rapid ups-and-downs in the NASA planetary-science budget line. It reached a low ebb in 1977 and approached the bottom again in 1982–83. Since then, it has seen sustained, if slow, growth. The financial turmoils of the 1970s and early 1980s are also clearly reflected in the low growth rate in publications during the same period and in the relative paucity of dissertations. Since the early 1980s, planetary-science publications have been on a relatively stable upward trend, as have finances. Space science has been growing at a fraction of the NASA budget since the late 1980s, and planetary science has grown with it. See table 17.1 on Ph.D. thesis abstracts.

In terms of publication, the 1990s dominate. There is substantial growth between each decade, with the lowest rate of growth between 1981 and 1991. This is the decade in which there were no new U.S. planetary missions and new data only from the *Voyager 2* spacecraft heading toward Uranus and Neptune, in addition to data from the European, Soviet, and Japanese missions to Halley's Comet. Yet, despite the lack of new data, there was still growth in publications. Rapid growth in publications resumed in the 1990s, though not reaching the growth rate shown during the 1960s. See Table 17.2.

The increase in publication in the 1990s cannot be explained by the emergence of large amounts of new data. Because of the failure of *Mars Observer*

**Table 17.1** Ph.D. thesis abstracts, by National Science Foundation quintile and by subject planet

| Quintile | Mercury | Venus | Mars | Jupiter | Saturn | Uranus | Neptune | Comets | Total |
|----------|---------|-------|------|---------|--------|--------|---------|--------|-------|
| 1975–1979 | 8 | 7 | 4 | 16 | 5 | 2 | 1 | 3 | 43 |
| 1980–1984 | 9 | 9 | 15 | 20 | 13 | 3 | 3 | 9 | 72 |
| 1985–1989 | 9 | 17 | 17 | 24 | 13 | 6 | 4 | 30 | 90 |
| 1990–1994 | 14 | 44 | 42 | 33 | 19 | 13 | 20 | 36 | 185 |
| 1995–1999 | 15 | 39 | 50 | 45 | 22 | 13 | 17 | 37 | 201 |
| 2000–2004 | 19 | 18 | 64 | 37 | 15 | 7 | 13 | 18 | 173 |

*Source*: Firstsearch Dissertation Abstracts

in 1993, the only new planetary data available prior to 1996 came from the *Magellan* radar mapping of Venus.[25] There is a corresponding publication "spike" for the veiled planet. Interestingly, though, despite the lack of new data there is also a substantial publication increase for Mars. So, while the availability of new data is clearly one influence on publication rate, it is not the only one. Charles Kennel, an astrophysicist by training and a former head of NASA's Earth science program, suggests that the high growth in publication in the decade might be the result of familiarity with computing technology. All data returned from space is digital, so the commoditization of computing power may well be responsible for a considerable portion of the increase in publication. NASA also implemented a new "open data" policy requiring investigators, beginning in the late 1980s, to make their data public, and it is now available via the Web-accessible Planetary Data System. The combination of data-accessibility and widespread computing power likely explains part of the publication boom.

Another potential, though likely only partial, explanation for the increase in publication post-1992 is the availability of another major journal, the *Journal of Geophysical Research—Planets*. An offshoot of the already-large *Journal of Geophysical Research*, the "Planets" volume separated papers on the non-Earth planets from several of the other thematic volumes (Atmospheres, Oceans, Solid Earth, and Space Physics) and expanded the total annual page-count.

Also, data obtained from the Lunar and Planetary Institute (LPI) in Houston, Texas, indicate that the number of practitioners of planetary science has increased significantly. Annual meeting attendance varied between 500 and 700 during the 1970s and 1980s, but since 1995 attendance has been on a steadily increasing trend. More than twice as many people attended in 2006 as had in 1989.[26] In addition to the availability of more venues for publication, there has also been substantial growth in the number of practitioners. See Table 17.3 on annual attendance.

Comparative study of the planets has also grown dramatically, especially in the 1990s. Before the late 1970s, NASA had no Earth-science program, as such, instead flying Earth orbiters as part of a now-defunct "Applications" program."[27] This applications model of science did not work for NASA, as it assumed other government agencies would fund research on datasets from Earth-orbiting missions. This did not happen, at least not to NASA's

**Table 17.2** Publication totals, as gleaned from the Web of Science Database, using search terms as given in the first column

| Search Term | Year Range | Nr. of Articles | Totals |
|---|---|---|---|
| Planetary | 1945–1957 | 93 | |
| | 1958–1969 | 324 | |
| | 1970–1980 | 816 | |
| | 1981–1991 | 1418 | |
| | 1992–2002 | 5547 | 8198 |
| Mercury | 1945–1957 | 3 | |
| | 1958–1969 | 51 | |
| | 1970–1980 | 125 | |
| | 1981–1991 | 66 | |
| | 1992–2002 | 317 | 562 |
| Venus | 1945–1957 | 25 | |
| | 1958–1969 | 434 | |
| | 1970–1980 | 799 | |
| | 1981–1991 | 934 | |
| | 1992–2002 | 1466 | 3658 |
| Mars | 1945–1957 | 96 | |
| | 1958–1969 | 454 | |
| | 1970–1980 | 1051 | |
| | 1981–1991 | 1042 | |
| | 1992–2002 | 3002 | 5645 |
| Jupiter | 1945–1957 | 33 | |
| | 1958–1969 | 307 | |
| | 1970–1980 | 683 | |
| | 1981–1991 | 554 | |
| | 1992–2002 | 2610 | 4187 |
| Saturn | 1945–1957 | 10 | |
| | 1958–1969 | 80 | |
| | 1970–1980 | 259 | |
| | 1981–1991 | 434 | |
| | 1992–2002 | 1159 | 1942 |
| Uranus | 1945–1957 | 12 | |
| | 1958–1969 | 35 | |
| | 1970–1980 | 198 | |
| | 1981–1991 | 361 | |
| | 1992–2002 | 477 | 1083 |
| Neptune | 1945–1957 | 12 | |
| | 1958–1969 | 25 | |
| | 1970–1980 | 100 | |
| | 1981–1991 | 319 | |
| | 1992–2002 | 715 | 1171 |
| Pluto | 1945–1957 | 11 | |
| | 1958–1969 | 12 | |
| | 1970–1980 | 48 | |
| | 1981–1991 | 114 | |
| | 1992–2002 | 239 | 424 |

satisfaction. When set up, the new Earth-science directorate (it has had several names) emphasized comparative studies. At least in quantitative terms, this seems to have been successful. But comparative studies have not been exclusive to Earth. See Table 17.4 on Comparative studies.

**Table 17.3** Annual attendance at the Lunar and Planetary Science conference. Years for which data was unavailable are left blank. Sources: data from 1975 to 1991, compiled from the annual conference proceedings. From 1995 to 2007, courtesy the Lunar and Planetary Institute

| Year | Attendees |
| --- | --- |
| 1975 | 600 |
| 1976 | 700 |
| 1977 | 685 |
| 1978 | 710 |
| 1979 | 745 |
| 1980 | 650 |
| 1981 | 600 |
| 1982 | 557 |
| 1983 | 525 |
| 1984 | 500 |
| 1985 | 550 |
| 1986 | |
| 1987 | |
| 1988 | |
| 1989 | 770 |
| 1990 | |
| 1991 | 777 |
| 1992 | |
| 1993 | |
| 1994 | |
| 1995 | 820 |
| 1996 | 742 |
| 1997 | 851 |
| 1998 | 1033 |
| 1999 | 1082 |
| 2000 | 1116 |
| 2001 | 1173 |
| 2002 | 1164 |
| 2003 | 1179 |
| 2004 | 1317 |
| 2005 | 1460 |
| 2006 | 1582 |
| 2007 | 1508 |

Dissertation data suggests that planetary science remains a field caught between two parent disciplines. Using Mars as an example, 175 Ph.D. theses were reported to Dissertation Abstracts between 1975 and 2006 as belonging to astronomy and astrophysics, while 112 were entered as geology. Only 37 of the 250 Mars dissertations were categorized as belonging to both. Further, I've omitted the subfield of atmospheric science, which has its own subject descriptor in the FirstSearch database. Including this subfield would make this disciplinary snapshot still more complex.

**Table 17.4** Comparative studies. Drawn from the Web of Science Database, using the given search terms

| Search Term | Date range | Total |
|---|---|---|
| Earth AND Mars | 1945–1957 | 3 |
| | 1958–1969 | 15 |
| | 1969–1979 | 41 |
| | 1980–1990 | 53 |
| | 1991–2001 | 1019 |
| Mars AND Venus | 1945–1957 | 1 |
| | 1958–1969 | 49 |
| | 1969–1979 | 75 |
| | 1980–1990 | 41 |
| | 1991–2001 | 533 |

Examination of NASA's science budgets over the past few decades makes clear that planetary science remains a small part of the larger space-science endeavor. In 2006, planetary exploration was about 30 percent of NASA's overall space-science budget. While it is intellectually fascinating, and often draws substantial public attention, in terms of PhD production, it is a small enterprise. NSF PhD data for the 1995–99 quintile shows 931 astronomy and astrophysics PhDs granted. I found 1,809 astrophysics abstracts in that same quintile, and my planetary abstracts data shows 201 planetary dissertations completed in that same period (not including 37 comet-related abstracts). For further comparison, in the United States during that time frame 206 PhDs were awarded in history and philosophy of science and technology.[28]

## Conclusion

Planetary science is still not a discipline in its own right, but in the decades since the IGY it has become a relatively stable enterprise with institutional support, a few specialized university programs, and specialty journals. As historian Stephen Pyne has argued, planetary voyages opened a new era of exploration, with machines acting as agents in places humans cannot (yet) go.[29] It is not clear, however, how many Americans are "aboard" for this new era of exploration. Currently, JPL media affairs and mission outreach staff are happy to reach audiences of a few hundred thousand with Web-based features, and they rarely get the kind of front-page newspaper coverage that was common in the 1960s—not even from its local newspaper, the *Los Angeles Times*. The hundreds of reporters that once appeared for planetary encounters now number in the dozens. So, while planetary science has grown tremendously as a profession and has a devoted base of "fans" among the larger public, it also has become routine and not newsworthy.

While Tatarewicz perceived the beginning of a new discipline of planetary science in the early 1970s, I think what has happened since then is much different—and more interesting. There has been a clear trend toward integration of planetary science into geoscience programs at the university level. The California Institute of Technology created the first integrated Geological and Planetary Science Division in 1951, but it is no longer unique. Similar programs exist at the Massachusetts Institute of Technology (Earth, Atmospheric, and Planetary Science); Harvard University (Department of Earth and Planetary Sciences); University of Washington (Department of Earth and Space Sciences); and several campuses in the University of California system. Yet, this trend is either not universal or, more likely, is incomplete. For example, Kuiper's final home, the University of Arizona, maintains astronomy, Earth science, and the Lunar and Planetary Laboratory as separate institutional entities, while its immediate rival, Arizona State University, has done precisely the opposite, merging its Earth, planetary, and astronomy efforts into a single entity. But Arizona State only embarked on unification in 2006.

In the 20 years since Tatarewicz wrote, there has been a transformation of the geosciences to include planetary science as a key part of their enterprise, not the creation of a new planetary-science discipline. Disciplinary lines are being erased, not formed. The IGY legacy was transformative, but it is the sciences of Earth that are being transformed in its wake.

## Acknowledgment

The research described in this paper was carried out at the Jet Propulsion Laboratory, California Institute of Technology, under a contract with the National Aeronautics and Space Administration.

## Notes

1. James Rodger Fleming, *Historical Perspectives on Climate Change* (New York: Oxford University Press, 1998), figure 9–5; Spencer Weart, *The Discovery of Global Warming* (Cambridge, Massachusetts: Harvard University Press, 2003). Also online at www.aip.org/history/climate/Kfunds.htm.

2. Dian Olson Belanger, *Deep Freeze: The United States, the International Geophysical Year, and the Origins of Antarctica's Age of Science* (Boulder: University Press of Colorado, 2006).

3. See, for example, Walter A. McDougall, *The Heavens and the Earth: A Political History of the Space Age* (New York: Basic Books, 1985), 118–125; Roger D. Launius et al., eds., *Reconsidering Sputnik: Forty Years since the Soviet Satellite* (New York: Harwood Academic Press, 2000); Clayton Koppes, *JPL and the American Space Program* (New Haven, Connecticut: Yale University Press, 1982), 78–85.

4. Homer M. Newell, *Beyond the Atmosphere: Early Years of Space Science* (Washington, D.C.: NASA SP-4211, 1980).

5. Robert A. Divine, *The Sputnik Challenge: Eisenhower's Response to the Soviet Satellite* (New York: Oxford University Press, 1993); Paul Dickson, *Sputnik: The Shock of the Century* (New York: Walker, 2001).

6. Constance McLaughlin Green and Milton Lomask, *Vanguard: A History* (Washington, D.C.: NASA SP-4202, 1970).

7. Koppes, *JPL and the American Space Program*, 185–215.

8. Michael S. Reidy, Gary R. Kroll, and Erik M. Conway, *Exploration and Science: Social Impact and Interaction* (Santa Barbara: ABC-CLIO, 2007).

9. Asif A. Siddiqi, *Deep Space Chronology: A Chronology of Deep Space and Planetary Probes* (Washington, D.C.: NASA SP-2002–4524, 2002).

10. For example, Edward Clinton Ezell and Linda Neuman Ezell, *On Mars: Exploration of the Red Planet, 1958–1978* (Washington, D.C.: NASA SP-4212, 1984); Michael Melzer, *Mission to Jupiter: A History of the Galileo Project* (Washington, D.C.: NASA SP-2007–4231, 2007); Henry C. Dethloff and Ronald Schorn, *Voyager's Grand Tour: To the Outer Planets and Beyond* (Washington, D.C.: Smithsonian Books, 2003).

11. Ronald E. Doel, *Solar System Astronomy in America: Communities, Patronage, and Interdisciplinary Science, 1920–1960* (Cambridge: Cambridge University Press, 1996).

12. Joseph N. Tatarewicz, *Space Technology and Planetary Astronomy* (Bloomington: Indiana University Press, 1990).

13. Tatarewicz, *Space Technology*, 123.

14. A complete inventory of planetary missions through 2000 is available in Siddiqi, *Deep Space Chronicle*; Reidy, Kroll, and Conway, *Exploration and Science*.

15. In 2006 the IAU adopted a definition for the word "planet" that excludes Pluto, and created a new "dwarf planet" category that encompasses Pluto and several large asteroids and Kuiper Belt Objects. However the IAU definition for planet has a number of flaws (strictly read, it excludes Earth, Mars, and Jupiter too) and is likely to be significantly revised in the future. See Govert Schilling, "Underworld Character Kicked Out of Planetary Family," *Science* 331, no. 5791 (September 1, 2006): 1214–1215; Mark V. Sykes, "The Planet Debate Continues," *Science* 319, no. 5871 (March 28, 2008): 1765. There is rich historical material in this struggle about what is and is not a planet.

16. For some examples of how scientific knowledge about the planets has changed, see Ronald A. Schorn, *Planetary Astronomy: From Ancient Times to the Third Millennium* (College Station: Texas A&M Press, 1998); William Sheehan, *The Planet Mars: A History of Observation and Discovery* (Tucson: University of Arizona Press, 1997); Stephen G. Brush, *Fruitful Encounters: The Origin of the Solar System and of the Moon from Chamberlin to Apollo* (New York: Cambridge University Press, 1996).

17. Historian Steven Dick calls this the "extraterrestrial life debate." See Dick, *The Biological Universe: The Twentieth-Century Extraterrestrial Life Debate and the Limits of Science* (New York: Cambridge University Press, 1996); also see Peter Ward and Donald Brownlee, *Rare Earth: Why Complex Life Is Uncommon in the Universe* (New York: Copernicus, 2000).

18. William Bryant, "The Re-Vision of Planet Earth: Space Flight and Environmentalism in Post-Modern America," *American Studies* (Fall 1995): 43–63; Sheila Jasanoff, "Image and Imagination: The Formation of Global Environmental Consciousness," Clark A. Miller and Paul N. Edwards, eds. *Changing the Atmosphere: Expert Knowledge and Environmental Governance* (Cambridge, Massachusetts: Massachusetts Institute of Technology Press, 2001), 309–337; Erik Conway, "Satellites and Security: Space in Service to Humanity," in Steven J. Dick and Roger D. Launius, eds., *Societal Impact of Spaceflight* (Washington, D.C.: NASA SP-2007–4801, 2007), 267–289.

19. The 1993–94 *Clementine* mission to the Moon carried out by the Ballistic Missile Defense Organization is the sole recent example, although in the late 1950s the Department of Defense carried out several missions before the establishment of NASA.

20. A good overview of the financial fortunes of planetary exploration is Amy Paige Snyder, "NASA and Planetary Exploration," in John M. Logsdon et al., eds., *Exploring the Unknown Vol. V.: Exploring the Cosmos* (Washington, D.C.: NASA SP-2001–4407, 2001), 263–300. Also see document II–31 in this volume.

21. Peter J. Westwick, *Into the Black: JPL and the American Space Program, 1976–2004* (New Haven, Connecticut: Yale University Press, 2007), 45–56.

22. For an overview of the subject, see Fabienne Casoli and Thérèse Encrenaz, *The New Worlds: Extrasolar Planets* (New York: Praxis Publishing Ltd., 2007).

23. The pre-Cambrian Earth had life for about two billion years but no significant atmospheric oxygen, so this life-detection method would not have worked for most of Earth's own existence. Lack of an oxygenated atmosphere does not prove a planet is lifeless. NASA has created a research field called "astrobiology" that is aimed at figuring out what other sorts of life might be possible and how they might be detected. See Steven J. Dick and James E. Strick, *The Living Universe: NASA and the Development of Astrobiology* (New Brunswick, New Jersey: Rutgers University Press, 2004).

24. Robert S. Kraemer, *Beyond the Moon: A Golden Age of Planetary Exploration, 1971–1978* (Washington, D.C.: Smithsonian Institution Press, 2000).

25. The year the U.S. probe *Galileo* reached its target, Jupiter, after nearly seven years in flight.

26. Data for the years 1975–91 compiled by the author from the annual LPI *Proceedings* volumes; missing data is years in which the *Proceedings* did not provide any information on attendance. Data from 1995–2007 provided by Mary Cloud at LPI, Houston, Texas.

27. There's little published to date on NASA's Earth-science research. One summary is John H. McElroy and Ray A. Williamson, "The Evolution of Earth Science Research from Space: NASA's Earth Observing System," in John M. Logsdon et al., eds., *Exploring the Unknown Volume IV: Space and Earth Science* (Washington, D.C.: NASA SP-2004–4407, 2004), 441–473; Erik M. Conway, *A History of Atmospheric Science in NASA* (Baltimore, Maryland: Johns Hopkins University Press, 2008); Erik M. Conway, "Drowning in Data: Satellite Oceanography and Information Overload in the Earth Sciences," *Historical Studies in the Physical and Biological Sciences* 37, no.1 (2007): 127–151.

28. NSF data from NSF, Division of Science Resources Statistics, Lori Thurgood, Mary J. Golladay, and Susan T. Hill, *U.S. Doctorates in the 20th Century* (Arlington, Virginia: NSF 06–319, 2006).

29. Stephen J. Pyne, "Seeking Newer Worlds: An Historical Context for Space Exploration," in Steven J. Dick and Roger D. Launius, eds., *Critical Issues in the History of Spaceflight* (Washington, D.C., NASA SP-2006–4702, 2006), 7–36; also see Roger D. Launius and Howard E. McCurdy, *Robots in Space: Technology, Evolution, and Interplanetary Travel* (Baltimore, Maryland: Johns Hopkins University Press, 2008).

# Chapter 18

# Polar Politics, Historical Narratives, and Saami Prehistory

## *Noel D. Broadbent*

This essay discusses the political context of historical narratives and Saami prehistory in Sweden. The Saami, natives of the North Nordic region, have lost challenges to their traditional winter reindeer grazing rights because of a lack of written history and convincing archaeological evidence. The essay discusses the background to this situation and presents new archaeological evidence of the diversity of Saami culture—evidence that can broaden the narrative of Saami land uses and better facilitate recognition of their place in Nordic history, including recognition of the Saami as indigenous people.[1]

The focus of the 2007–08 International Polar Year (IPY) was on large-scale environmental change, human-natural systems, and benefiting residents of northern regions. In this context the Arctic social sciences have an important role to play, as the period 1986–2006 was a time of unprecedented change for indigenous peoples. The fall of the Soviet Union in 1990 opened the Arctic to greater access, but also to new social, cultural, economic, political, and environmental challenges. The Arctic Council and the International Arctic Social Sciences Association were formed in 1990. Major land claims and advances in the process of research through the formulation of ethical principles have led to more applied and collaborative endeavors by the social-science community. The ultimate benefit has been the raising of research standards, and the increased applicability of regional studies to global science and policy development. According to one expert on recent Arctic politics: "The Circumpolar North is emerging as a distinctive international region endowed with political dynamics of its own that are not only interesting in their own right but that are also suggestive for those seeking to understand the political aftermath of colonialism or internal colonialism in other parts of the world."[2]

Polar exploration has been motivated by nationalism and by the ambitions of both individuals and nation states. Research has often served as a means for asserting these interests, politically and culturally, and polar science has also been used as a vehicle for promoting environmental awareness—the polar bear has become the international icon of global-warming politics. In a previous article I have discussed three expeditions that were motivated by politics.[3] The first was that of the Solomon August Andrée Ballooning Expedition to the North Pole in 1897. Sweden was in a political union with Norway at the time—Norway gained its independence in 1905—and Norway, whose prowess in polar exploration was almost a given, was watching this attempt with considerable disdain. The Andrée expedition failed and was also a political disaster.[4] The second example was the U.S. Antarctic Service Expedition in 1939–41. This expedition was a direct response by President Franklin D. Roosevelt to German Antarctic claims, most notably Germany's dropping of claim markers with swastikas in Norwegian sectors. Germany subsequently invaded Norway in 1940.[5] The building of a 10,000-foot runway by the Americans at Marble Point across from the American base at McMurdo on Ross Island, Antarctica, was never finished but the runway's existence led to a quick reclaiming of the place in the 1980s, when other countries started looking for a suitable site for their programs.[6] Marble Point is still likely to be developed, but is an environmental hot potato because planes the size of Boeing 747s could land there year-round.

This essay is also about polar politics, but within a nation-state. As an archaeologist, one hopes to not be involved in political or legal debates, but it is naive to pretend that what we do does not have political implications. In my case, my 30 years of academic interest in the prehistory of the Bothnian coastal zone has led me also to consider the role of Saami prehistory in this region. In the following, I will explore the politics of the Saami in Sweden and the curious state of affairs concerning their rights as indigenous people. Sweden is a world leader in human rights, but for historical reasons faces a difficult dilemma when dealing with indigenous rights within its borders. This discussion emanates from the response the "Search for a Past" research project has elicited from archaeologists and political scientists who were unaware of the impact archaeology has had in Saami court cases. The project also responds to two of the social science and humanities topics for the IPY: (1) How can historical studies and records of the polar regions enhance understanding of contemporary social and cultural problems? And: (2) What do the polar sciences contribute to global diversity and the political status of indigenous people worldwide?[7]

## Myths and Narratives

It is hypothesized that many of the most fundamental values that led to progressive social policies in Sweden, such as equity under the law, have at the same time rendered minority rights problematic: All citizens should have the same rights, and no group should have separate rights. This is, in concert with immigration conflicts, one of the major political dilemmas of many western European countries faced with cultural diversification.

Sweden, however, differs from other Nordic and Arctic nations with respect to indigenous rights. The Swedish Saami have not been recognized as indigenous people in accordance with United Nation policies (ILO no. 169, 1989). The International Labour Organization (ILO) is a specialized United Nations agency. Convention 169 comprises regulations and obligations for the protection of indigenous peoples, including their institutions, property, land, culture, and environment. The regulations also encompass education and social well-being. Indigenous people are defined as people with their origins in ethnic groups who lived in the country when national borders were formed, and who have wholly or partly retained their social, economic, cultural, and political organizations.

To make matters worse, archaeological testimony has recently contributed to traditional reindeer-herding rights being lost in the Swedish courts. Hunting and fishing rights have also been challenged and lost. A recent Swedish government report, *Samernas sedvanemarker* (*Saami Traditional Lands*, 2006), has reaffirmed most of these existing policies. These policies equate Saami reindeer-herding with other economic sectors, such as forestry and tourism. The linkage of identity with economy overrides indigenous claims based on any other criteria, including continuity of settlement, self-identity, or language. The reasons for this situation are complex and relate to Swedish policies developed during the past four centuries. It was believed the Saami were in danger of extinction and needed protection. As a people, they were believed to be suited only to a nomadic lifestyle.

**Figure 18.1**   A Saami family in Norway. (Courtesy of Library of Congress. From *Landscape and Marine Views of Norway and Sweden* [Detroit Publishing Company, 1905])

According to Swedish historian Dr. Lennart Lundmark,[8] this was an off-shoot of race biology in Sweden and of a "Lapp shall be Lapp" policy. While based on the good intentions of the Swedish state, this is a rare, perhaps unique, case of an indigenous identity being defined by an exclusive economic activity.

The Saami relationship to archaeology is especially interesting because of the ways archaeological evidence and argumentation have been applied in land-use conflicts, and by the ways new archaeological evidence is challenging both Saami and Nordic historical narratives. In these contexts archaeology has had a profound, not always positive, impact.[9]

## The Saami

The Saami (Lapps) number as many as 80,000 people in Norway, Sweden, Finland, and the Kola Peninsula in Russia. About 20,000 Saami live in Sweden, of whom approximately 2,000 are involved in reindeer-herding today. The Saami speak nine Finno-Ugric languages that are unrelated to the Indo-European language family. One of the most pervasive myths is that they migrated to Scandinavia less than 2,000 years ago, although this is no longer supported by archaeological evidence. Lapland, as an administrative region, dates only to the eighteenth century, created as a means of preserving the nomadic herding lifestyle and for separating the Saami from Swedish settlers.

Johan Turi, a Swedish Saami reindeer herder, wrote *Muittalus Samid Birra* (*A Book on Lappish Life*), published in 1911; it was one of the first indigenous ethnographies ever written. Turi wrote the following widely quoted text on Saami origins and settlement:

> It has not been said that the Lapp came from somewhere else. The Lapp was settled all over Lapland; and the Lapp lived on the seacoast and there were no other dwellers besides them, and that was a good time for the Lapps. And the Lapp also lived everywhere on the Swedish side and there were no settlers anywhere; the Lapps did not know there are other people besides themselves.[10]

These ideas can be tested using archaeology, although reindeer-herding as a way of life itself is often difficult to document. Nevertheless, Professor Evert Baudou, the most prominent Swedish archaeologist specializing in northern Swedish prehistory, argues that cultural identity cannot be ascribed to archaeological material. Professor Baudou's testimony in the Reindeer Grazing Case from 1995, in which the Saami lost their traditional rights to winter grazing in Härjedalen in southern Lapland, stated (in translation): "Who lived in Norrland [northern Sweden] 9,000 years ago? Our definition of an ethnicity today cannot apply, and the question must remain unanswered. The very diverse material from different time periods implies that many different folk groups lived in Norrland's prehistory, as in southern Sweden."[11]

According to Norwegian anthropologist Thomas Eriksen,[12] the major goal of indigenous minorities is to survive as a culture-bearing group. This

issue is therefore of a political nature and about minority and majority relations. The history of nationalism itself is a history of these kinds of conflicts
over competing narratives that seek to define a social community.[13] The
beginnings of Nordic archaeology in the seventeenth century were intimately
connected with the Swedish nation-building process. Unity and nationalism
were important goals. Myths were created to glorify the realm, and the field
of Nordic archaeology came into being to do just that. It was founded on
the desire to create a glorious national ethos. Myths about prehistory have
served to perpetuate attitudes about nationalism and racial stereotypes; cultural, economic, and social evolution; territoriality, and identity,[14]

The deep past has therefore been the prerogative of the Scandinavian
majority through the field of Nordic archaeology, which often emerges as
self-evident. Indeed Saami culture has been studied largely through the lens
of Nordic culture. Saami ethnicity (their ethnogenesis) is believed to have
only come into being 2,000 years ago through interactions with other Nordic
peoples.[15] Saami prehistory/ethnicity is, therefore, equally about the people
they have interacted with. Because the Saami do not have written histories
of their own, without prehistory they are, indeed, as Baudou stated, without
identity. The government, by default, ends up holding all the cards.

Post–World War II Swedish and German archaeologists have generally
avoided taking positions on ethnicity. This attitude is motivated by a need to
be dissociated from the fascist and Nazi archaeology of the 1940s, but also to
avoid the often-uncomfortable intellectual connections between Sweden and
Germany since the 1920s. In 1922, for instance, Sweden was the first country
in Europe to establish a State Institute for Racial Biology. While Sweden was
not pro-Nazi, the role of the Swedish welfare state, a product of the 1930s, is,
however, crucial for understanding Swedish domestic policies toward minorities.[16] Racial hygiene and education were core elements of the Swedish welfare system. The Sterilization Act, implemented from 1941, and affecting
some 63,000 citizens, mostly women, was only discontinued in Sweden in
1975, 30 years after World War II.[17] Special nomadic schools were established
for Saami children to help preserve the nomadic lifestyle and perpetuate the
interpretation of the Saami as primitive, vulnerable, and consequently falling
under the protection of the state. They were not sterilized. The Germans
occupying Norway during the war also seemed to have had a romantic view
of the Saami, allowing them to continue with trans-border herding.[18]

The goal of Swedish education, according to former Social Democratic
prime minister Ingvar Carlsson, is to make "good members of society that
respect consensus." The collective mentality of Nordic society, particularly
through institutionalization in a welfare-state system, has thereby created an
intrinsically problematic environment for minority groups, especially those
who claim a stake in history and prehistory.

Sweden is clearly undergoing a major reappraisal of its history and its
cultural policies. For example, Swedish historian-journalist Lundmark has
published landmark studies of Saami history (translated titles): *As Long as We
Have Territory: The Saami and State during 600 Years* (1998); "*The Lapp Is*

*Fickle, Unstable and Uncomfortable": Swedish State Saami Policies in the Age of Racism* (2002), and *Saami Tax Lands* (2006). *Eugenics and the Welfare State,*[19] published in English, had an immediate international impact. A number of popular books in Sweden have expanded on this theme, most notably a 2006 work by Maja Hagerman titled, *Det rena landet* (*The Pure Country*), a study of Swedish prehistory and racism.

## Knowledge as Power

Swedish politicians have long been aware of the challenges of diversity, particularly regarding immigrants, who now number 12 percent of the population. The Saami have, in fact, benefited from this concern, especially in education, with home-language (foreign languages spoken in the home) programs initiated in the 1970s. As one Swedish minister of culture put it: "It is completely clear that Swedish society is at a crossroads. How are we to handle the multicultural society we live in?"[20]

A government-mandated report released in 1996, *Knowledge as Power: Action Plan for How Museums in Their Work can Counteract Anti-Foreign Attitudes and Racism* (translated title), recognized the importance of history in contemporary Swedish society.[21] The following social dimensions of racism and xenophobia are listed in the report:

1. State uniformity under Lutheran orthodoxy has permeated Swedish social life since the 16th century and still steers both institutional and individual thinking and actions.
2. The concept of the Swedish folk (folkstam) was part of the creation of the "folk home," which led to, among other things, sterilization laws until 1975.
3. Sweden is continuing with an un-examined past from both before and after World War II.
4. Openly declared racism, which disappeared after the war, does not mean that it has disappeared from people's thoughts.
5. Norms and concepts about what it means to be Swedish have not changed along with social change.
6. Anti-Foreign attitudes and racism based on earlier generations' attitudes are more of an adult than a youth problem in Sweden.
7. Racism, which was originally focused on biological differences, has shifted today to mostly social, religious and cultural differences.
8. Anti-Foreign attitudes are a structural problem of the majority society.
9. In order to integrate indigenous and immigrant citizens, a common base of values and principles will be necessary.[22]

The report goes into considerable detail about the abuses of Swedish archaeology in the past, and describes a number of museums in which the Saami have been either marginalized or presented without reference to cultural conflicts. Another theme is that of the Vikings, who in addition to having been a pervasive icon of Swedish and Nordic identity, are also used as a symbol of Aryan, racist and neo-Nazi political movements in Europe and the United States.

One of the conclusions of the report is that museums must be more active in social debates and act as forums for cultural criticism. This report was met with strong reactions. For example, ten years later American anthropologist Janet E. Levy described some of the same paradoxes in her case-study regarding Saami representations in Swedish and Finnish museums:

1. In the Nordic countries, archaeological dialogues with indigenous peoples are underdeveloped, although the Saami are one of the rare populations in Europe that participates in the discourse of indigenism.
2. The majority Nordic populations, whose prehistories and continuities with the past are almost unproblematic, argue today that such connections cannot be defined. One consequence of this attitude is to deny the uniqueness and boundedness of Saami identity, which is, ironically, otherwise accepted by the public, the government and the Saami themselves.[23]
3. The nationalistic majority "master" historical narrative prevails by default and this is explicitly manifested in national museums. The call by Nordic archaeologists to depoliticize archaeology inevitably has different implications for majority [Scandinavian] and minority [Saami] populations.[24]

### Saami Prehistory

Against the background of the current situation of the Saami in Sweden, one must consider the enormous potential importance of new archaeological research. The interdisciplinary "Search for a Past" project has focused on Saami territory and its economy outside of Lapland and along the Bothnian coast to within 300 kilometers of Stockholm. The project has helped define this Saami past in terms of territory, economy, technology, and chronology.[25]

The project's hypothesis is that the ancestors of the Saami occupied the Swedish Bothnian coast during the Iron Age, and this was part of a long-term settlement system. The starting point for the project was to obtain representative material from different regions along the coast. Detailed mapping and archaeological excavations were, therefore, carried out at nine locales along a 460-kilometer north–south transect in three counties. The northernmost investigations were carried out on Stor-Rebben Island in Norrbotten County at latitude 65°11' N. Västerbotten County is located about 80 kilometers south of Stor-Rebben. This region is as much as 200 kilometers north of Germanic Iron Age settlement or settlement place names. Seven sites were investigated, two on islands. The southernmost excavations were carried out at Hornslandsudde in Gävleborg County at latitude 62°37' N, Hornslandet is 460 kilometers south of Stor-Rebben, and only 300 kilometers north of

Stockholm. Hans Westberg[26] relates the oral history of the Hornsland area as follows (in translation):

> According to tradition, which is still preserved among the older population who practiced fishing at Hölick's fishing village ca 2 km west of Hornslandsudde, the dwelling sites on the point derive from a fishing people of Lappish (Saami) origin. The Hornslandsudde site is, however, within an area of well known Germanic Iron Age settlement.[27]

This sampling strategy provided a basis for comparisons regarding chronology and economy and the full array of constructions well within German Iron Age settlement regions and well north of them. Thirty-one huts at 13 elevations between three and 20 meters above sea level were excavated and samples of bone and charcoal obtained from their hearths. More than 50 radiocarbon dates have been analyzed. The results can be summarized as follows:

1.  The chronology of these coastal sites extends from 100–1250 CE. There were two main periods of site use: 400–500 CE and 800–1100 CE. The sites were totally abandoned by 1279 CE.
2.  The sites are similar from north to south and consist of mostly rectangular and square hut foundations with central hearths. These cluster in groups of three to five dwellings and are found together with storage cairns and stone alignments that appear to be corrals.
3.  Animal bones from hearths show that seals were hunted in the spring and fall (both ringed seals and harp seals have been identified).
4.  Bones and teeth of sheep/goats, reindeer, and even cattle have been found in the huts at five different locales, three of which are north of Germanic Iron Age agrarian settlement.
5.  A ritual bear grave of Saami type was found embedded in a hut and was radiocarbon dated to the early eleventh century.
6.  Circular sacrificial features of Saami type have been found on or near six different hut sites.
7.  Of more than 1,100 place names referring to "Lapps" in Sweden, 87 percent are found in the Bothnian coastland. This includes the site of Hornslandsudde. There are 390 of these place names in Västerbotten.

The north-to-south transect extends from deep within known Germanic Iron Age settlement areas and more than 200 kilometers north of such areas. While there are distinctive traits of Saami or proto-Saami culture (the hut-clustering, the storage facilities, the ritual features, et cetera), there are also similarities to Germanic culture. This is actually not surprising because of the contacts, including marriage, that must have been long-standing between these communities. Nordic linguists, who noted the Saami had borrowed many Germanic words relating to agriculture, speculated the Saami must have lived lives similar to the north Scandinavians:

What we are confronted with is a class of Saami who, during ancient Nordic (primitive Scandinavian) times, presumably around the birth of

Christ, in addition to hunting and fishing (possibly reindeer-herding), supported themselves through a form of animal husbandry and farming, which was at about the same level and in the same region as the Norse.[28]

The settlement picture one gets from this sample is that there were Saami living along the length of the Bothnian coast, and that their territories overlapped those of both Germanic peoples to the south and Finnish (so-called Kvens) to the north. The Västerbotten region seems to be the most purely Saami territory and was the last to be abandoned. As hunter-gatherers and pastoralists, they did not compete, and instead probably existed in symbiotic relationships with other settlers. This system continued until around 1250–1300. Many Saami could have subsequently integrated into settled Swedish society, and there is oral history that speaks of this happening. The first settlers on Holmön Island in Västerbotten are, according to local history, three "fishing Saami." Their names are known, as are the locations of their farmsteads.

Seen as a whole, this evidence suggests that characterizing Swedish Saami as nomads and reindeer herders fails to recognize the diversity of Saami culture and its part in Nordic prehistory. For different reasons, 90 percent of the Saami in Sweden today are not reindeer owners. Oddly enough, this might be closer to the prehistoric situation.

Saami territory also extended far outside of Lapland. According to the Icelandic and Norwegian sagas from 1100–1200s, the Saami lived as far south as Hadeland, some 20 kilometers northwest of Oslo.[29] They formerly lived in all but southwestern Finland.[30] The ancestors of the Saami were even known as far south as the western Dvina (Daugava) River in Latvia.[31] In Sweden they are recorded living as far south as the Dal River and the Mälardalen region.

## Conclusions

The dominant symbol of Saami identity is that of nomadic reindeer-herding. Although a powerful symbol, this has also been a liability in Sweden, inasmuch as Saami indigenous identity and rights are totally restricted to reindeer ownership. But now even these traditional herding, hunting, and fishing rights, once guaranteed by the state, are being challenged. The Saami speak nine distinct languages, and one can assume there were at least as many regional cultural differences. The uniqueness of the Saami resides in their status as an Arctic indigenous people with thousands of years of prehistory in the Nordic region. Their languages, cultures, and relationships to the northern environment characterize them as one of the great indigenous peoples of the world.

The fundamental problem regarding Saami acknowledgment in Sweden is that they have been defined by a narrative that limits their history in time and their identity by activity. This model was originally created to protect the Saami within a prescribed economy and territory (reindeer-herding and Lapland). Based on the facts emerging from archaeological research, combined with the deconstruction of biased historical narratives, a more inclusive paradigm of Saami identity can be realized. It is hoped that archaeology will prove to be a useful tool for all northern peoples in search of their pasts.

## Notes

1. A full presentation of the archaeological evidence is published in Noel D. Broadbent, *Lapps and Labyrinths. Saami Prehistory, Colonization and Cultural Resilience* (Washington, DC: Smithsonian Scholarly Publications, 2010).
2. Oran Young, *Arctic Politics: Conflict and Cooperation in the Circumpolar North* (Hanover, New Hampshire: University Press of New England, 1992), 18.
3. Noel D. Broadbent, "From Ballooning in the Arctic to 10,000-Foot Runways in Antarctica: Lessons from Historic Archaeology," in I. Krupnik, M.A. Lang, and S.E. Miller, eds., *Smithsonian at the Poles: Contributions to International Polar Year Science* (Washington, D.C.: Smithsonian Institution Scholarly Press, 2009).
4. Urban Wråkberg, ed., *The Centennial of S.A. Andrée's North Pole Expedition* (Stockholm: Royal Swedish Academy of Sciences, 1999).
5. Noel D. Broadbent and Rose Lisle, "Historical Archaeology and the Byrd Legacy. The United States Antarctic Service Expedition, 1939–31," *The Virginia Magazine of History and Biography* 110, no. 2 (2002): 237–258.
6. Noel D. Broadbent, "An Archaeological Survey of Marble Point, Antarctica," *Antarctic Journal of the United States* 29 (1994): 3–6.
7. Igor Krupnik et al., "Social Sciences and Humanities in the International Polar Year 2007–2008: An Integrating Mission," *InfoNorth Arctic* 58, no. 1 (March 2005): 91–101.
8. Lennart Lundmark, "Lappen är ombytlig, ostadig och obekväm," Svenska statens samepolitik i racismens tidevarv (Umeå: Norrlands universitetsförlag, 2002).
9. Broadbent, Noel. D. and Patrik Lantto. 2008. Terms of Engagement: An Arctic Perspective on the Narratives and Politics of Global Climate Change. In *Anthropology and Climate Change: From Encounters to Actions,* edited by Susan A. Crate and Mark Nuttall, pp. 341–355. Left Coast Press, Walnut Creek, CA.
10. Johan Turi, *En bog om Lapperness liv. Muttalus Samid Birra* (Stockholm: A.B. Nordiska Bokhandeln, 1911).
11. Evert Baudou, *Norrlands Forntid* (Örnskoldsvik: Ett historiskt perspektiv, CEWE-Förlaget, 1995).
12. Thomas Hylland Eriksen, *Etnicitet och nationalism* (Nya Doxa AB. Nora: Bokbolaget, 1993).
13. Loyde S. Kramer, "Historical Narratives and the Meaning of Nationalism," *Journal of the History of Ideas* 58, no. 3 (1997): 525–545.
14. D.D. Carr, "Narrative and the Real World: An Argument for Continuity," *History and Theory* 15 (1968): 117–131; C. R.J. Richards, "The Structure of Narrative Explanation in History and Biology," in M. Nitecki and D. Nitecki, eds., *History and Evolution* (Albany: State University of New York Press, 1992); Hastrup, 1987; Hodder, 1991; Solli, 1996; Kramer, 1997; and Hagerman, 2006.
15. Fredrik Barth, ed., *Ethnic Groups and Boundaries* (Boston: Little, Brown, and Company, 1969); Lars-Ivar Hansen and Bjørnar Olsen, *Samenes historie fram til 1750* (Oslo: Cappelen Akademisk Forlag, 2004); Knut Odner, *Tradition and Transmission,* Bergen Studies in Social Anthropology (Bergen: Norse Publications, 2002); and Baudou, 1988.
16. Bozena Werbart, *De osynliga identiteterna* (Umeå: Studia Archaeologica Universitatis Umensis, 2002); Gunnar Broberg and Mattias Tydén, "Eugenics in Sweden: Efficient Care," *Eugenics and the Welfare State* (East Lansing: Michigan State University Press, 2005), 77–149; and Lundmark, 2002.
17. Broberg and Tydén, "Eugenics in Sweden," 2005.

18. Patrik Lantto, "Raising their Voices: The Sami Movement in Sweden and the Swedish Sami Policy, 1900–1960," in Art Leete, ed., *The Northern Peoples and States: Changing Relationships* (Tartu: Tartu University Press, 2005), 203–234.
19. Gunnar Broberg and Nils Roll-Hansen, eds., *Eugenics and the Welfare State* (East Lansing: Michigan State University Press, 2005).
20. Marita Ulvskog, Swedish Minister of Culture, 1996.
21. Swedish Department of Culture, Report Ds (1996), 74.
22. Swedish Department of Culture, Report Ds (1996), 15–16.
23. Janet J. Levy, "Prehistory, Identity, and Archaeological Representation in Nordic Museums," *American Anthropologist* 108, no. 1 (March 2006): 139–145.
24. Levy, "Prehistory, Identity, and Archaeological Representation," 136.
25. Noel D. Broadbent, "The Prehistory of the Indigenous Saami in Northern Coastal Sweden," *People, Material Culture, and Environment of the North* (Oulu: Studie humaniora ouluensia, 2006), 12–26; Broadbent, 2005; Wennstedt, Edvinger, and Broadbent, 2006.
26. Hans Westberg, "Lämningar efter gammal fångstkultur i Hornlandsområdet," *Fornvännen* 56 (1964), 24.
27. Lars Liedgren, *Hus och Gård i Hälsingland* (Umeå: Studia Archaeologica Universitatis Umensis 2, 1992).
28. K.B. Wiklund, *Lapparna. Nordisk Kultur X* (Stockholm: Albert Bonniers Förlag, 1947). 60.
29. Inger Zachrisson, "Arkeologi och etnicitet. Samisk kultur i mellersta Sverige ca 1–1500 e. Kr," *Bebyggelsehistorisk tidskrift* 14 (1987): 26.
30. I.T. Itkonen, "Lapparnas förekomst i Finland," *Ymer* 1 (1947): 42–57.
31. Kerstin Eidtlitz Kuoljok, *På jakt efter Norrbottens medeltid* (Umeå: Center for Arctic Cultural Research, Misc. Publications 10, 1991), 32.

# Chapter 19

# Stratospheric Ozone Depletion and Greenhouse Gases since the International Geophysical Year: F. Sherwood Rowland and the Evolution of Earth Science

*Dasan M. Thamattoor*

In preparation for the International Geophysical Year (IGY), the British Antarctic Survey (BAS) of the Royal Society set up a research station near Halley Bay (named in honor of noted astronomer Edmond Halley). During the IGY, the Halley Bay station played a critical role in discovering the polar vortex above the Antarctic and, nearly 30 years later, provided important evidence for massive loss of stratospheric ozone within the vortex that could be attributed to the release of anthropogenic chlorofluorocarbons (CFCs) into the atmosphere. These findings provided the experimental confirmation for the theoretical predictions made in the mid-1970s by F. Sherwood Rowland and his colleague, Mario Molina. Discovery of this "ozone hole" in 1985, made possible by the Dobson spectrometers set up during the IGY, led to the swift enactment of the important intergovernmental treaty brokered by the United Nations, known as the Montreal Protocol in 1987. The Montreal Protocol initially called for a 50 percent reduction in CFC production by the end of the twentieth century, but subsequent meetings set up an accelerated schedule for major industrialized nations to phase out these products completely by 1996.

Another major initiative of the IGY, which has since played a major role in informing views on global warming and climate change, was the effort led by Charles David Keeling to measure baseline atmospheric concentrations of carbon dioxide ($CO_2$) at the Mauna Loa Observatory in Hawaii and in the Antarctic. At the time this work was started, there were no reliable data for $CO_2$ measurements in the atmosphere, and many scientists were skeptical about such measurement being even possible. Initial data gathered by Keeling showed a slight seasonal variation in atmospheric $CO_2$ of about two percent but, during the course of the first decade of such measurements, it was clear that the average levels of atmospheric $CO_2$ were undoubtedly

**Figure 19.1** The broad, white expanse of East Antarctica has a sameness that can make it hard to distinguish the permanent coastline of the continent. This image from 2002, however, shows a subtle distinction, a line where the rippled texture of the actual land surface gives way to the smooth surface of the coastal ice shelf. The shelves and coastlines to their interiors have a variety of names. From east to west, the image shows Sabrina, Banzare, and Wilkes (Clarie) Coasts, all fronting the Dumont d'Urville Sea. (Courtesy of Jacques Descloitres, MODIS Rapid Response Team, NASA/GSFC)

on the rise. The measurements Keeling initiated are still being maintained at these two sites and have been augmented at multiple sites by numerous agencies from around the world. This global network, which also monitors a variety of other greenhouse gases, plays an important role in assessing global climate change.

## A Brief Biography[1]

F. Sherwood Rowland, an important figure in atmospheric science, was born on June 27, 1927, in the town of Delaware, Ohio. Rowland's parents had moved there the year before, when his father was appointed professor of mathematics and chair of the department at Ohio Wesleyan University. He received his elementary and high school education from the Delaware Public School system and graduated from high school before his 16th birthday. He was a member of the varsity tennis team in his junior and senior years, and also played varsity basketball in his senior year. During several summers, when his high school science teacher would go away on vacation, he entrusted Rowland with the operation of the local weather station. Rowland recalls that this was his "first exposure to systematic experimentation and data collection." In 1943, while most of his classmates joined the military after receiving their high school diplomas, Rowland, still two years under the mandatory enlisting age of 18, enrolled at Ohio Wesleyan. He attended the university year-round on an accelerated academic schedule, played varsity baseball and basketball, and contributed to the sports page of the campus newspaper. Though eligible for a student deferment for his final year at university in 1945, he decided instead to join the U.S. Navy, as fighting was still ongoing in the Pacific. The war ended, however, while he was in basic training, so he spent the next year working in several navy separation centers, helping sailors make the transition to civilian life. After being discharged, he returned to Ohio to complete his degree. Rowland's year away from academics convinced him to not seek a quick end to his undergraduate program, and instead he stayed two more years, graduating in 1948 with a triple major in mathematics, physics, and chemistry. In the fall of that year Rowland followed the footsteps and recommendations of his parents, enrolling at the University of Chicago, where he was randomly assigned, on an interim basis, to the noted nuclear chemist and future Nobel laureate, Willard F. Libby. Rowland was impressed by Libby and stayed on in the group to work in radio-chemistry. His regard for Libby is captured in this telling quote: "Almost everything I learned about how to be a research scientist came from listening to and observing Bill Libby."

The lanky and athletic Rowland played on the varsity basketball and baseball teams in his first year of graduate school. In what must have been a major time-commitment, he continued on the baseball team in the spring of his second and third years and played semiprofessional baseball in a Canadian summer league. In 1952, he married fellow graduate student Joan Lundberg and earned a PhD for his thesis on the chemical state of cyclotron-produced radioactive bromine atoms.

That year Rowland joined the chemistry department at Princeton University as an instructor. In the summer of 1954, while working at the Brookhaven National Laboratories in Long Island, he ground up some lithium carbonate and glucose with a mortar and pestle and placed the mixture in the neutron flux of the nuclear reactor. Remarkably the neutrons reacted with lithium-6 to produce tritium, which had sufficient energy to be incorporated into glucose molecules. About 15 percent of the tritium was found attached to the carbon in glucose. This work, which demonstrated the potential to incorporate tritium into molecules containing carbon-hydrogen bonds, was published in *Science* and helped Rowland win research support from the Atomic Energy Commission (AEC). At the time, nuclear atmospheric testing was a common practice, and the World Meteorological Organization was recommending using radioactive tracer technology, including tritium, as part of an IGY initiative to investigate circulation and mixing in the atmosphere and oceans. In 1956, Rowland moved to the University of Kansas as an assistant professor and continued his work on tritium chemistry. He received tenure after two years and quickly rose through the ranks to full professorship by 1963. The following year he accepted a position as chair of the chemistry department at the University of California, Irvine (UCI), an institution still in the planning stages. He founded the chemistry department at UCI, where he is now the Bren Research Professor of Chemistry and Earth System Science.

In 1974 Rowland and his postdoctoral colleague, Mario Molina, published a seminal paper in *Nature* describing the destruction of stratospheric ozone by CFCs.[2] This work led to the banning of CFCs in 1978 and the signing of the Montreal Protocol in 1987. The discovery resulted in his sharing the 1995 Noble Prize in chemistry with Molina and Paul Crutzen. Rowland, who in 1993 served as president of the American Association for the Advancement of Sciences, is a member of the National Academy of Sciences (NAS), for which he also has been a foreign secretary (1994–2002). During his term as foreign secretary of the NAS, he helped create the InterAcademy Panel on International Issues that he cochaired from 1995–2000.

## Stratospheric Ozone Depletion[3]

Throughout most of his early career, Rowland's research interests focused on the chemistry of "hot" radioactive atoms. In 1970, however, after stepping down as chair of the chemistry department at UCI, he was drawn to the ongoing discussion about the environment. In keeping with his desire "to instill some freshness into our research efforts by every few years extending our work into some new, challenging aspect of chemistry," Rowland traveled to Salzburg, Austria, to attend an International Atomic Energy Agency (IAEA) meeting on the environmental applications of radioactivity. After the conference, he shared a compartment on the train to Vienna with AEC program officer William Marlow. During their conversation Marlow became acquainted with Rowland's growing interest in atmospheric science. Rowland, in turn, learned that one of Marlow's responsibilities was the organization of a series of chemistry-meteorology workshops, with the intention of encouraging more cross-fertilization between the two fields. Shortly thereafter, Marlow invited Rowland to attend the second of these workshops, in Fort Lauderdale, Florida, in January 1972. There, Rowland heard Lester Machta, former director of the National Oceanic and Atmospheric Administration Air Resources Laboratory, describe the work of Jim Lovelock, a British scientist, who reported finding traces of trichlorofluoromethane (variously known as $CCl_3F$, CFC-11, or Freon) in Earth's atmosphere. CFC-11 was an artificial compound produced by the chemical industry, and, according to Lovelock's estimates, its total concentration in the atmosphere was approximately equal to the amount manufactured up to that date.[4] In his Nobel lecture, Rowland noted these results as the starting point of his own work in this area of chemistry. A picture of the handwritten notes Rowland took of Lovelock's data is shown in figure 19.2.

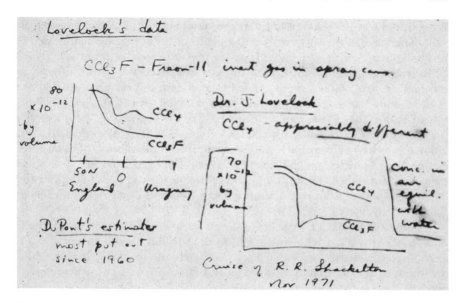

**Figure 19.2**  Notes made by Sherwood Rowland in January 1972 regarding James Lovelock's data. (Courtesy of Sherwood Rowland)

Lovelock's measurements were made with an extremely sensitive detector he had invented, based on electron capture (EC) by trace compounds. The EC detector was used to analyze individual gases exiting the gas chromatograph (GC), an instrument that separates a mixture of gases into its components. Using the ECGC instrument, which is especially sensitive for CFC-11 and similar molecules in its class, Lovelock detected CFC-11, not only in the atmosphere near his home in western Ireland, but also in all the air samples tested in both hemispheres on the 1971 voyage of the *R.V. Shackleton* from England to Antarctica.[5] Remarkably, Machta suggested the gas could cause "no conceivable harm in the atmosphere" and might be even advantageous as "an excellent tracer for air mass movements because its chemical inertness would prevent its early removal from the atmosphere."[6]

Rowland, however, held a different view: "As a chemical kineticist and photochemist, I knew that such a molecule could not remain inert in the atmosphere forever, if only because solar photochemistry at high altitudes would break it down." In early 1973 he included a page in his yearly proposal to the AEC, which had funded him since his days as a graduate student, proposing a predictive study of the atmospheric chemistry of CFCs. The AEC agreed to let him pursue his investigation.[7]

Later that year, Mario Molina, a native of Mexico City who had just obtained a PhD from the University of California, Berkeley, joined Rowland's group as a postdoctoral associate and began work on the CFC problem.

Interestingly, as Rowland noted, "At the time, neither of us had any significant experience in treating chemical problems of the atmosphere, and each of us was now operating well away from our previous areas of expertise."[8]

Rowland and Molina initially examined the physical- and chemical-removal mechanisms, or "sinks," for other chlorine containing compounds that are released into the troposphere, the lowest layer of the atmosphere, extending to about 10–15 kilometers above Earth's surface. These included: the photodissociation of colored species, such as molecular chlorine ($Cl_2$), into chlorine atoms by absorbing visible solar radiation; dissolution of polar molecules, such as hydrogen chloride (HCl) in water; their subsequent removal from the atmosphere by wet deposition processes, such as rainfall; and the oxidation of compounds containing C-H bonds by hydroxyl radicals in Earth's atmosphere.[9]

Commonly released CFCs, however, many of which only contain carbon, and halogens, such as chlorine and fluorine, do not absorb solar radiation encountered near the surface of Earth, are virtually insoluble in water, and are immune to oxidation in the lower atmosphere. So, with the usual removal mechanisms (noted above) unavailable to CFCs, Rowland and Molina wondered about the fate of these "survivor molecules."[10]

## The Chapman Cycle and the Ozone Layer

Most of the solar radiation reaching Earth arrives in a narrow band of wavelengths visible to humans in the range of violet (~400 nanometers, 1 nanometer = $10^{-9}$ meters) to red (~700 nanometers), in addition to the invisible near infrared (IR, >700 nanometers) and ultraviolet (UV, <400 nanometers). Further, the energy of radiation increases with decreasing wavelength.

In principle, any molecule is capable of decomposing if the radiation it absorbs is sufficiently energetic. CFCs, however, tend to absorb high-energy UV radiation of very short wavelength in the range that is typically not encountered near the surface of Earth. Protecting the surface of this planet from such damaging radiation is a complex set of reactions, involving oxygen ($O_2$) and ozone ($O_3$) molecules at higher altitudes of the atmosphere.

$O_2$ absorbs radiation below 242 nanometers and decomposes into two O atoms (equation 1). Each of these O atoms then combines with another $O_2$ molecule to produce $O_3$ (equation 2). Collision with a third molecule or equivalent entity, M, is necessary to absorb some of the excess energy of the reaction and stabilize the $O_3$ product. The $O_3$, in turn, absorbs high energy UV radiation below 290 nanometers and breaks down into $O_2$ and O (equation 3). $O_3$ can also combine with O to form two molecules of $O_2$ (equation 4). The noted geophysicist Sydney Chapman reported these reactions as early as 1930.[11]

Ozone formation

Equation 1    $O_2 + UV\ light \longrightarrow O + O$
Equation 2    $O + O_2 + M \longrightarrow O_3 + M$

Ozone destruction

Equation 3    $O_3 + UV$ light    $\rightarrow$    $O2 + O$

Equation 4    $O_3 + O$    $\rightarrow$    $O_2 + O_2$

Reactions, such as equations 1 through 4 above and others to be dis-cussed shortly, maintain a balance of ozone in the atmosphere so that about three parts in $10^7$ of the entire atmosphere are present as $O_3$, compared to almost 21 percent as $O_2$. Approximately 90 percent of these ozone molecules are formed in the stratosphere, the region extending from 10 to 50 kilome-ters above Earth. In some regions of the stratosphere, the mixing ratio of $O_3$ (the ratio of the ozone present to that of the entire atmosphere by volume) can be as high as one part in $10^5$.[12]

The presence of ozone in the atmosphere plays a critical role by absorb-ing high-energy solar energy, as in equation 3, and transferring some of that energy into heat, as exemplified by energy transfer to M in equation 2. The latter process contributes to the increase of temperature with altitude in the stratosphere. This process also contributes and to stratosphere's overall sta-bility, making it difficult for pollutants introduced there to be removed. The ozone layer, as Rowland notes, "performs two important physical processes: It removes short wavelength UV radiation, and changes this energy into heat, both creating and maintaining the stratosphere."[13]

It is known that $CCl_3F$ absorbs UV radiation <220 nanometers. Solar radiation of such high energy, however, is not available near the surface of Earth because the presence of $O_2$ and $O_3$ at higher altitudes prevent wave-lengths <290 nanometers from getting below the stratosphere. Thus, in order to undergo photochemical decomposition by solar radiation, the $CCl_3F$ molecule must first find its way to altitudes of 25–30 kilometers, which is well above where most of the $O_2$ and $O_3$ molecules are found. At such high altitudes, CFCs absorb short wavelength solar radiation and decompose to produce chlorine atoms (Cl) as shown in equation 5.[14]

Equation 5    $CCl_3F + UV$ radiation    $\rightarrow$    $CCl_2F + Cl$

Although the concentration of CFCs at such high altitudes is quite low, because of the stability of the stratosphere these molecules tend to persist for decades before being destroyed. Using three different eddy diffusion mod-els, Rowland and Molina calculated the vertical distribution of $CCl_3F$ in the stratosphere and noted the vertical profiles were similar, with the maximum destruction of these molecules occurring at an altitude of about 28 kilome-ters. According to their estimates, a $CCl_3F$ molecule in the atmosphere has an average lifetime in the range of 40–55 years, whereas a similar chemical, $CCl_2F_2$, had a lifetime of about 75–150 years.[15]

## The Fate of Chlorine Atoms in the Stratosphere

Examination of the known rate constants, measured in the laboratory, for the reaction of atomic chlorine with various species commonly found in the stratosphere, led Rowland to the conclusion that nearly all the Cl produced reacted with ozone, according to equation 6, to produce $O_2$ and another reactive intermediate called chlorine monoxide (ClO). The ClO formed in equation 6 then reacts with O, as shown in equation 7, or with NO, as shown in equation 8. In each case, atomic chlorine is regenerated. Importantly, Rowland and Molina noted that the sum of equations 6 and 7 is the equivalent of equation 4, which destroys $O_3$ and "constituted a free radical chain reaction in which the chlorine atom alternates between the Cl and ClO species." In this cycle, the first step (equation 6) removes one molecule of $O_3$, while the second, equation 7, captures an O, which could have otherwise reformed $O_3$ via equation 2.

Equation 6   $O_3 + Cl$    $\rightarrow$    $O_2 + ClO$
Equation 7   $O + ClO$    $\rightarrow$    $O_2 + Cl$
Equation 8   $NO + ClO$    $\rightarrow$    $NO + Cl$

Rowland and Molina estimated the catalytic efficiency of the Cl/ClO cycle removed about 100,000 $O_3$ molecules per chlorine atom. When this estimate is considered in the context of a million tons of CFCs released annually into the atmosphere at that time, a sobering thought emerged. In Rowland's own words, "the original question chiefly of scientific interest has now been converted into a very significant global environmental problem—the depletion of stratospheric ozone by the chlorine contained in chlorofluorocarbons."[16]

The chain reactions represented by equations 7 and 8 were discovered by Richard Stolarski and Ralph Cicerone in 1973[17] and are analogous to $NO_x$ equations 9 and 10, which also produce a pair of $O_2$ molecules from $O_3$ and O. The $NO_x$ species in this cycle are obtained by the decomposition of stratospheric nitrous oxide ($N_2O$) or by the release of NO and $NO_2$ directly into the atmosphere from the exhaust of high-flying aircraft.[18] The latter issue was discussed in the early 1970s, with respect to Concorde and Boeing aircraft, under the Climatic Impact Assessment Program.[19]

Equation  9   $O_3 + NO$    $\rightarrow$    $O_2 + NO_2$
Equation 10   $O + NO_2$    $\rightarrow$    $O_2 + NO$

Rowland notes "the distribution of chlorine is closely intertwined with the concentration of other stratospheric species, such as oxides of nitrogen (for example, equation 8), methane (equation 11), and oxides of hydrogen (equation 12)."[20]

Equation 11   $CH_4 + Cl$    $\rightarrow$    $CH_3Cl + HCl$
Equation 12   $HCl + HO$    $\rightarrow$    $Cl + H_2O$

Because there were no stratospheric measurements of CFCs available at the time, the predictions made by Rowland and Molina about their vertical distribution were based on model calculations. In 1975, however, two research groups sent up balloons with specially equipped cans to recover stratospheric air samples. It was found that the mixing ratios for CFC-11 measured from these samples were in close agreement with those calculated by Rowland and Molina.[21]

## Discovery of the Ozone Hole above Antarctica

The Halley Bay station (latitude 75.5° south) was equipped with Dobson spectrometers to measure total ozone. To make these measurements, the instruments determined the ratio of two preselected wavelengths of solar radiation, one that was absorbed by ozone (311 nanometers) and the other that was not (332 nanometers).

Total ozone concentrations are at a maximum near the polar regions in the Northern Hemisphere around March/April, reaching values well above 400 Dobson Units (1 Dobson Unit is about 1 part per billion by volume). In sharp contrast, ozone levels measured above the southern polar region were essentially constant through autumn and winter and into mid-spring. Then, during the next few weeks, the total ozone increased sharply. These observations, some of which were made during the IGY, established the presence of a strong Antarctic vortex.[22] As this vortex began to collapse around mid-spring, the ozone-laden stratospheric air arrived at the South Pole from the temperate regions.

This pattern of ozone levels above the South Pole (described above), was noted from the IGY through the early 1970s. Beginning in the late 1970s, however, a significant trend of decreasing average October ozone values was observed over Halley Bay. Indeed the total ozone concentration above the station had dropped from over 300 Dobson Units in the 1960s to under 200 Dobson Units in 1984. This data was published in 1985, with the suggestion that the ozone depletion correlated with the rise in atmospheric CFC concentrations.[23]

It soon became clear that the decreasing ozone levels observed above Halley Bay were characteristic of the entire South Pole. Data from the Total Ozone Mapping Spectrometer (TOMS) onboard the *Nimbus 7* satellite— providing about 100,000 daily measurements of ozone levels above the entire Southern Hemisphere—revealed a region of severely depleted ozone concentration that had developed above the Antarctic.[24] The *Nimbus* satellite eventually failed in 1993, but TOMS measurements have continued to be made on other satellites, notably the *Earth Probe* satellite. More recently, the Ozone Monitoring Instrument (OMI), onboard the *Aura* satellite, is continuing ozone measurements. It detected an ozone "hole" above the Antarctic in 2006 that was a double record-setter, both for area and concentration.[25] A completely different, chemical measurement of ozone was obtained using balloon sondes. Based on the oxidation of iodide to iodine by ozone, this method demonstrated that much of the ozone loss was occurring in the lower stratosphere.[26]

## Identifying the Cause of Ozone Depletion

Although it was becoming clear that substantial ozone loss was occuring in the stratosphere, there was much controversy at the beginning as to the cause of this depletion. Was the observed loss of ozone a part of some natural cycle, or was it primarily the result of human activity? Evidence soon began to emerge suggesting that the significantly higher concentrations of stratospheric chlorine in the mid-1980s, as compared to the period from 1950–70, was promoting reactions leading to ozone-loss above the Antarctic. Ground-based expeditions to McMurdo, Antarctica (latitude 78° south), led by Susan Solomon, revealed two maxima in the vertical profile of the critical intermediate ClO, the larger of which was in the lower stratosphere at altitudes coinciding with considerably reduced ozone.[27] In situ measurements of ClO and ozone inside the polar vortex by Jim Anderson's group, using high-flying E-2 aircraft launched from Punta Arenas, Chile (latitude 53° south), also showed that high levels of ClO were associated with dwindling ozone concentrations in the lower stratosphere.[28]

In the lower stratosphere ClO dimerizes into ClOOCl, as shown in equation 13. Subsequent decomposition of ClOOCl, according to equations 14 and 15, afford two Cl atoms,which partake in the reaction described by equation 6. The net consequence of combining equations 6 (times two) and 13 through 15 is the conversion of two ozone molecules into three molecules of oxygen, as shown in equation 16.[29]

Equation 13    $ClO + ClO + M \longrightarrow ClOOCl + M$
Equation 14    $ClOOCl + light \longrightarrow ClOO + Cl$
Equation 15    $ClOO + M \longrightarrow Cl + O_2 + M$
Equation 16    $O_3 + O_3 \longrightarrow O_2 + O_2 + O_2$

## Ozone Depletion in the Northern Hemisphere

Arosa, Switzerland, has kept a continuous record of atmospheric ozone measurements since August 1931. In 1986 Rowland and Neil Harris carefully reexamined this data. They noted that when the measurements from the period 1931–69 were compared with those from 1970–86, a startling conclusion emerged: The ozone concentrations for several autumn and winter months were substantially lower after the 1970s than in the nearly 40 years before.[30]

Data from the ozone stations in Caribou, Maine, and Bismarck, North Dakota, also indicated winter-time ozone losses during the period 1976–86, as compared to 1965–75.[31] The two consecutive 11-year intervals were evaluated to minimize the variations that could be introduced by the 11-year solar cycle. It soon was noted that a similar loss of ozone, in the period 1976–86 versus 1965–75, was also indicated by data from all 18 ozone stations located between latitudes 35° and 60° north.[32] Indeed, a report by the Ozone Trends Panel assessed that substantial depletion of ozone had already occurred over densely populated areas in North America, Japan, the former Soviet Union, and Europe.[33]

## Increase in Ultraviolet-B (UV-B) Radiation Reaching the South Pole

Instruments measuring the intensity of ultraviolet radiation reaching the surface of Earth showed a 150-percent increase in potentially damaging UV-B radiation (320 nanometers to 290 nanometers) coincident with 50-percent ozone loss above the South Pole in 1992–93.[34] Indeed, on one particular day, October 26, 1993, when there was an especially low ozone measurement above Palmer, Antarctica (latitude 64° south), the UV-B radiation exceeded by 25 percent the highest intensity recorded for any day that year. Furthermore, the highest weekly dose of UV-B radiation ever recorded by any of the measurement sites was also at the South Pole.[35]

## The Montreal Protocol

By the late 1970s regulations were in place in the United States, Canada, Norway, and Sweden to control the emissions of certain CFCs, especially those found in spray cans. Mounting evidence of the alarming loss of ozone above the South Pole and certain parts of the Northern Hemisphere, attributable to the rising concentration in stratospheric chlorine compounds produced by humans, prompted a 1985 United Nations meeting in Vienna to schedule a convention to address the protection of the ozone layer in the stratosphere. Two years later, in September 1987, an international agreement, commonly referred to as the Montreal Protocol, called for a 50-percent reduction in CFC production by the year 2000. The protocol also had a provision for revisiting the controls. Just three years later, given the gravity of the situation, the London meeting of 1990 put in place a plan for completely phasing out CFCs by the end of the twentieth century. A subsequent meeting in Copenhagen hastened the phase-out timeline to January 1, 1996, for the major developed nations, and also added a few other chemicals to the list of banned compounds.[36]

## Greenhouse Gases[37]

There is an energy balancing-act between solar energy coming in through Earth's atmosphere and radiant energy going back out of Earth. The wavelengths of the incoming and outgoing radiation are different, of course, since Earth and the Sun are different temperatures. If all the infrared radiation emanating from Earth could, in fact, escape, then calculations show the temperature of Earth would be 0°F (–18°C). However, the average temperature of Earth is considerably warmer and, according to a 1950s estimate, was at 57°F (14°C). The difference between the theoretical and actual estimates is the natural greenhouse-effect attributable to the radiative effects of trace

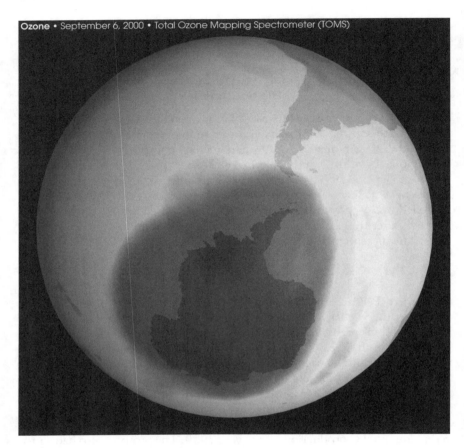

**Ozone** • September 6, 2000 • Total Ozone Mapping Spectrometer (TOMS)

**Figure 19.3** A NASA instrument that detected an Antarctic ozone "hole" three times larger than the entire land mass of the United States—the largest such area ever observed. The hole expanded to a record size of approximately 11 million square miles (28.3 million square km) on September 3, 2000. The previous record was approximately 10.5 million square miles (27.2 million square km) on Sept. 19, 1998. The ozone hole's size currently has stabilized, but the low levels in its interior continue to fall. The lowest readings in the ozone hole are typically observed in late September or early October each year. Ozone molecules, made up of three atoms of oxygen, comprise a thin layer of the atmosphere that absorbs harmful ultraviolet radiation from the Sun. Most atmospheric ozone is found between approximately six miles (9.5 km) and 18 miles (29 km) above the Earth's surface. Scientists continuing to investigate this enormous hole are somewhat surprised by its size. The reasons behind the dimensions involve both early-spring conditions and an extremely intense Antarctic vortex. The Antarctic vortex is an upper-altitude stratospheric air current that sweeps around the Antarctic continent, confining the Antarctic ozone hole. Since the discovery of the ozone hole in 1985, the Total Ozone Mapping Spectrometer (TOMS) has been a key instrument for monitoring ozone levels over the Earth. (Courtesy of TOMS science team and Scientific Visualization Studio, NASA GSFC, Greenbelt, Maryland)

gases, such as water vapor and carbon dioxide, the so-called primary greenhouse gases in Earth's atmosphere.[38]

The main components of Earth's atmosphere, molecular nitrogen and oxygen, contain two atoms and are transparent to infrared radiation. Greenhouse gases typically contain three or more atoms and are characterized by their ability to absorb infrared radiation. A limited list of the various gaseous components of Earth's atmosphere and their concentrations are shown with the greenhouse gases highlighted. Water vapor is by far the most abundant greenhouse gas, typically ranging between 1–4 percent concentration in the lower atmosphere, and especially abundant in the tropics. When the concentrations of greenhouse gases begin to rise in the atmosphere, they can play critical roles vis-à-vis global warming.[39]

## Carbon Dioxide ($CO_2$)

Beginning in 1938, the work of G.S. Callendar indicated that the combination of fossil fuel combustion and deforestation was causing the concentration of carbon dioxide in the atmosphere to rise and would likely result in anthropogenic global warming (AGW), yet an accurate and reliable measurement of this carbon dioxide was not available until the 1950s.[40] In the mid-1950s, however, a young American scientist by the name of Charles David Keeling, a post-doctoral researcher in the laboratory of Harrison Brown at Caltech, developed an instrument for measuring atmospheric carbon dioxide. In 1958, with the support of Harry Wexler, chief of Scientific Services of the U.S. Weather Bureau, and Roger Revelle, director of the Scripps Institute of Oceanography, Keeling set up stations to monitor atmospheric $CO_2$ in two remote areas far from sources of pollution, such as cities and other populated areas. One site was atop the volcanic mountain Mauna Loa in Hawaii, and the other at the South Pole. This work, an offical part of the IGY, soon would have a remarkable scientific, social, political, and economic impact on understanding human-induced climate change.[41]

At the time Keeling began his measurements, the atmospheric $CO_2$ level was around 315 parts per million by volume (ppmv). The Mauna Loa data also revealed that the land-rich, highly vegetated Northern Hemisphere showed a remarkable and reproducible seasonal variation each year in the $CO_2$ concentrations. This variation, induced by the growth and decay of land plants, and the attendant photosynthetic cycle, was manifested in a decrease in $CO_2$ concentration in spring and summer (plant growth) and an increase in concentration in the autumn and winter (plant decay).[42] Over time, however, it was noted that while the seasonal variations in the Northern Hemisphere followed the usual cyclical pattern, on average there was a steady rise in $CO_2$ levels as indicated by the iconic Keeling curve. This work provided the critical evidence for $CO_2$ accumulation in Earth's atmosphere and has motivated numerous efforts to understand and contain this phenomenon. Today, in addition to the original Keeling sites, there are nearly 100 stations worldwide that continue to monitor atmospheric $CO_2$ concentrations. As of 2010, the

$CO_2$ level had risen to 390 ppmv, up 23 percent since measurements began in 1958.[43] Its rise has been attributed to human activity, such as deforestation and the burning of fossil fuels.

## Methane ($CH_4$)

Water vapor and carbon dioxide, although major contributors to the greenhouse effect, are not the only gases linked to global warming. In 1978 Rowland and coworkers began sampling canisters of atmospheric gases, from the Northern and Southern Hemispheres, for other compounds including hydrocarbons. In January 1980 they observed that the methane concentration in the atmosphere was around 1.5–1.6 ppmv.[44] Rowland noted that the atmospheric concentration of methane has been going up since the 1750s, largely driven by population-growth and the industrial revolution. He attributed much of the recent increase in methane to agricultural sources, such as rice paddies and a substantial increase in the cattle population, expecially since the 1950s.[45]

Between January 1978 and December 1987 atmospheric methane concentration grew at a rate of nearly one percent per year.[46] There was a smaller, yet still significant, increase during the next seven years, as noted in December 1993. Since the year 2000, however, methane concentrations in the atmosphere has somewhat stabilized. In a recent interview, Rowland speculated that a more conscious effort by producers to control leaks from oil and gas fields, prompted by economics, may have played a role in containing methane concentrations in the atmosphere.[47]

The methane balance, notes Rowland, is easily perturbed and is not guaranteed to remain steady at the near-zero growth rate prevalent during the past eight years. For instance, two major spikes in tropospheric methane levels were observed in 1998 and 2002–03 and in both cases were accompanied by spikes in the hydrocarbon ethane ($C_2H_6$). The concomitant formation of ethane and methane is characteristic of burning biomass, according to Rowland.

## Synthetic Greenhouse Gases

Many synthetic compounds, including the CFCs discussed previously in connection with ozone depletion, also have the ability to absorb radiation and contribute to global warming. The increasing concentrations of CFCs in the 1970s–1980s, and their persistence in the atmosphere, could have played a significant role in global warming. Thanks to the Montreal Protocol and subsequent accords, however, the accumulation of such substances has been brought under control. The atmospheric concentration of CFC-12 has leveled off since about the year 2000, and the amount of CFC-11 has been actually declining since the mid-1990s.[48]

## Conclusion

Rowland's active career in environmental chemistry spans the five decades since the IGY. His work on CFCs led to the Montreal Protocol and won him a share of the 1995 Nobel Prize in chemistry. His perspective on environmental issues is both personal and extremely influential, and his continuing work in environmental chemistry has established links between two of the most pressing environmental issues of our time: stratospheric ozone depletion and anthropogenic global warming.

In addition to his scientific investigations of Earth's atmosphere, Rowland continues to be a vigorous advocate for changes in public policy to recognize and address environmental problems. He has testified before Congress on a number of occasions and urged public and private agencies to be proactive in protecting this planet. Activities set in motion during the IGY clearly played a major role in influencing Rowland's work and shaping his legacy of fundamental new discoveries and a more environmentally aware global citizenry.

## Acknowledgments

I am deeply indebted to Dr. Davida Gavioli for transcribing the audio recording of Professor Rowland's address at Colby. This essay would not have been possible without the catalysis provided by Professor James R. Fleming, who initiated this project and helped carry it through.

## Notes

1.  Much of this information is based on F. Sherwood Rowland "Autobiography" http://nobelprize.org/nobel_prizes/chemistry/laureates/1995/rowland-autobio. html (accessed January 19, 2009); and Bette Hileman, "A Giant among Chemists," *Chemical and Engineering News* 85 (December 24, 2007): 30–33.
2.  M.J. Molina and F.S. Rowland, "Stratospheric Sink for Chlorofluoromethanes: Chlorine Atom-Catalysed Destruction of Ozone," *Nature* 249 (1974): 810–812.
3.  F. Sherwood Rowland, "Nobel Lecture in Chemistry" (December 8, 1995); F. Sherwood Rowland, "Stratospheric Ozone Depletion" *Philosophical Transactions of the. Royal Society Bulletin* 361 (2006): 769–790; and F. Sherwood Rowland, plenary lecture, History of Science Society Annual Meeting, Crystal City, Virginia (November 1, 2007).
4.  Rowland, "Autobiography," 1995; and Hileman, "A Giant among Chemists," 2007.
5.  J.E. Lovelock, R.J. Maggs, and R.J. Wade, "Halogenated Hydrocarbons in and over the Atlantic," *Nature* 241 (1973): 194–196.
6.  Rowland, "Autobiography," 1995; and Hileman, "A Giant among Chemists," 2007.
7.  Rowland, "Autobiography," 1995; and Hileman, "A Giant among Chemists," 2007.
8.  Rowland, "Nobel Lecture," 1995.
9.  Rowland, "Nobel Lecture," 1995.
10. Rowland, "Nobel Lecture," 1995.
11. Sydney Chapman, "A Theory of Upper Atmospheric Ozone," *Memoirs of the Royal Meteorological Society* 3 (1930): 103–125. See also chapter 10.
12. Rowland, "Nobel Lecture," 1995.
13. Rowland, "Nobel Lecture," 1995.
14. Rowland, "Nobel Lecture," 1995; and Rowland, "Stratospheric Ozone Depletion," 2006.
15. F.S. Rowland and M.J. Molina, "Chlorofluoromethanes in the Environment," *Reviews of Geophysics and Space Physics.* 13 (1975): 1–35.
16. Rowland, "Nobel Lecture," 1995.
17. R.S. Stolarski and R.J. Cicerone, "Stratospheric Chlorine: A Possible Sink for Ozone," *Canadian Journal of Chemistry* 52 (1974): 1610–1615.

18. P.J. Crutzen, "The Influence of Nitrogen Oxides on the Atmospheric Ozone Content," *Royal Meteorological Society Quarterly Journal* 96 (1970): 320–325; H.S. Johnston, "Reduction of Stratospheric Ozone by Nitrogen Oxide Catalysts from Supersonic Transport Exhaust," *Science* 173 (1971): 517–522; and P.J. Crutzen, "Ozone Production Rates in an Oxygen-Hydrogen-Nitrogen Oxide Atmosphere," *Journal Geophysical Research* 76 (1971): 7311–7327.

19. "Environmental Impact of Stratospheric Flight, Biological and Climatic Effects of Aircraft Emissions in the Stratosphere" (Washington, D.C.: Climatic Impact Committee, National Academy of Sciences, 1975); A.J. Grohecker, ed., "The Natural Stratosphere of 1974," CIAP Monograph 1, Final Report (Washington, D.C.: U.S. Dept. of Transportation, DOT-TST-75-51, 1975).

20. Rowland, "Nobel Lecture," 1995.

21. A.L. Schmeltekopf et al., "Measurements of Stratospheric $CFCl_3$, $CF_2Cl_2$, and $N_2O$," *Geophysical Research Letters* 2 (1975): 393–396; L.E. Heidt et al., "Stratospheric Profiles of $CCl_3F$ and $CClF$," *Geophysical Research Letters* 2 (1975): 445–447.

22. G.M.B. Dobson, "Forty Years' Research on Atmospheric Ozone at Oxford: A History," *Applied Optics* 7 (1968): 401.

23. J.C. Farman, B.G. Gardiner, and J.D. Shanklin, "Large Losses of Ozone in Antarctica Reveal Seasonal ClOx/NOx Interaction," *Nature* 315 (1985): 207–210.

24. R.S. Stolarski et al., "Nimbus 7 SBUV/TOMS Measurements of the Springtime Antarctic Ozone Decrease," *Nature* 322 (1986): 808–811.

25. Rob Gutro, "NASA and NOAA Announce Ozone Hole Is a Double Record Breaker," (October 19, 2006), www. nasa.gov/vision/earth/lookingatearth/ozone_record. html (accessed January 22, 2009).

26. Rowland, "Nobel Lecture," 1995; and Rowland, "Stratospheric Ozone Depletion," 2006.

27. R.L. deZafra et al., "High Concentrations of Chlorine Monoxide at Low Altitudes in the Antarctic Spring Stratosphere, I, Diurnal Variation," *Nature* 328 (1987): 408–411.

28. J.G. Anderson, W.H. Brune, and M.H. Proffitt, "Ozone Destruction by Chlorine Radicals within the Antarctic Vortex: The Spatial and Temporal Evolution of $ClO$-$O_3$ Anti-correlation Based on In Situ ER-2 Data," *Journal Geophysical Research* 94 (1989): 11465–11479.

29. L.T. Molina and M.J. Molina, "Production of Cl2O2 from the Self-Reaction of the ClO Radical," *Journal of Physical Chemistry* 91 (1987): 433–436.

30. N.R.P. Harris and F.S. Rowland, "Trends in Total Ozone at Arosa," *Eos* 67 (1986): 875.

31. Harris and Rowland, "Trends," 1986.

32. F.S. Rowland, "Trends in Total Column Ozone Measurements," in "Report of the International Ozone Trends Panel," WMO Report No. 18, Volume I, 1990; F.S. Rowland et al., "Statistical Error Analyses of Ozone Trends—Winter Depletion in the Northern Hemisphere," in R.D. Bojkov and P. Fabian, eds., *Ozone in the Atmosphere* (Hampton, Virginia: Deepak Publishing, 1989), 71–75.

33. Rowland, "Trends," 1990; and Rowland et al., "Statistical Error Analyses," 1989.

34. C.R. Booth and S. Madronich, "Radiation Amplification Factors: Improved Formula Accounts for Large Increases in Ultraviolet Radiation Associated with Antarctic Ozone Depletion," in C.S. Weiler and P.A. Penhale, eds., *Antarctic Research Series* 62 (Washington, D.C.: American Geophysical Union, 1994), 39–42.

35. C.R. Booth et al., *Antarctic Journal of the United States* 39 (1994): 256–259.

36. Richard Elliot Benedick, *Ozone Diplomacy* (Cambridge, Massachusetts: Harvard University Press, 1991).

37. Rowland, "Plenary Lecture," 2007; and F. Sherwood Rowland, "Global Warming and Climate Change," Colby College Leadership Institute, March 14, 2008.

38. Rowland, "Global Warming," 2008.

39. Rowland, "Global Warming," 2008.

40. James Rodger Fleming, *The Callendar Effect: The Life and Work of Guy Stewart Callendar (1898–1964), The Scientist Who Established the Carbon Dioxide Theory of Climate Change* (Boston: American Meteorological Society, 2007).

41. James Rodger Fleming, *Historical Perspectives on Climate Change* (New York: Oxford University Press, 1998); Spencer W. Weart, *The Discovery of Global Warming* (Cambridge, Massachusetts: Harvard University Press, 2003); James Rodger Fleming, *Climate Change and Anthropogenic Greenhouse Warming: A Selection of Key Articles, 1824–1995, with Interpretive Essays* (NSF: National Science Digital Library, 2008), http://wiki.nsdl.org/index.php/PALE:ClassicArticles/GlobalWarming

42. C.D. Keeling, "The Concentration and Isotopic Abundances of Carbon Dioxide in the Atmosphere," *Tellus* 12 (1960): 200–203.

43. Scripps $CO_2$ Program, http://scrippsco2.ucsd.edu/home/index.php (accessed 22 January 2009).

44. Rowland, "Global Warming," 2008; I.J. Simpson et al., "Influence of Biomass Burning during Recent Fluctuations in the Slow Growth of Global Tropospheric Methane," *Geophysical Research Letters* 33 (2006): L22808.

45. I.J. Simpson et al., "Implications of the Recent Fluctuations in the Growth Rate of Tropospheric Methane," *Geophysical Research Letters* 29 (2002): 1479.

46. D.R. Blake and F.S. Rowland, "Continuing World-wide Increase in Tropospheric Methane 1978–1987," *Science* 239 (1988): 1129–1131.

47. Hileman, "A Giant among Chemists," 2007.

48. T.M. Thompson et al., "Halocarbons and Other Atmospheric Trace Species," in R.C. Schnell, A.-M. Buggle, and R.M. Rosson, eds., *Climate Monitoring and Diagnostics Laboratory*, Summary Report No. 27 (Boulder, Colorado: NOAA CMDL, 2004), 115–133.

# Index